Evidence and Interpretation in Studies on Early Science and Medicine

Evidence and Interpretation in Studies on Early Science and Medicine

Essays in Honor of John E. Murdoch

Edited by

Edith Dudley Sylla
William R. Newman

BRILL

LEIDEN • BOSTON
2009

On the cover: Fourteenth-century initial "C". MS British Library, Burney 275, fol. 166r.
Reproduced with kind permission of the British Library.

Originally published as Volume 14, Nos. 1–3 (2009) of Brill's journal *Early Science and Medicine*.

This book is printed on acid-free paper.

Library of Congress Cataloging-in-Publication Data

Evidence and interpretation in studies on early science and medicine : essays in honor of
John E. Murdoch / edited by Edith Dudley Sylla, William R. Newman.
p. cm.
Includes index.
ISBN 978-90-04-17878-6 (hardback : alk. paper)
1. Science, Medieval. 2. Medicine, Medieval. I. Murdoch, John Emery, 1927- II. Sylla, Edith Dudley. III. Newman, William Royall. IV. Title.

Q124.97.E95 2009
509'.02—dc22

2009034646

ISBN 978 90 04 17878 6

PRINTED IN THE NETHERLANDS

Contents

Introduction

William R. Newman and Edith Dudley Sylla

At a colloquium held in May 2007 at Boston College to honor John E. Murdoch on the occasion of his eightieth birthday, generously hosted by Stephen Brown and with help from the Society for Medieval and Renaissance Philosophy, some of the authors represented in this volume concluded that what had transpired was of sufficient novelty and interest to warrant a volume of essays reflecting on the present state of studies in early science. What appears here is not a planned project, such as the recent Cambridge Histories of Science or Philosophy, in which authors are assigned topics to provide coverage, but rather the product of the current scholarly interests of each of the individual authors involved. Many, but not all of those present at the colloquium and represented in this volume were former students of Murdoch's at Harvard and the papers in this volume only partly overlap with the papers presented at the colloquium. Nevertheless the very diversity of this volume is itself in part a testament to Murdoch's practice of encouraging his students to pursue their own interests, however distant from his own research they might be. The result is a collection of particularly lively essays, which cast light on each other in multiple ways. In this introduction we hope to delineate some patterns and larger themes that emerge from the range of contributions.

Despite being a loyal student of Marshall Clagett, Murdoch has always felt an affinity to philosophers as well as to historians of science. He wrote a dissertation that was an edition of Thomas Bradwardine's *De continuo* with a philosophical introduction.[1] But, as

[1] John Emory Murdoch, "Geometry and the Continuum in the Fourteenth Century: A Philosophical Analysis of Thomas Bradwardine's *Tractatus de continuo*." Ph.D. dissertation, University of Wisconsin, 1957.

he has often remarked, from the point of view of the philosophers William H. Hay and Julius Weinberg, who were on his committee, his was a philosophical dissertation, with a Latin text in an appendix. In the first part of his dissertation, Murdoch not only attempted "to describe, in precise, modern terminology, what Bradwardine's contentions were with respect to the function of geometry in analyzing the properties of continuous magnitude, but also endeavor[ed] to determine whether or not these contentions are substantially correct." It was perhaps his high standards of philosophical rigor that made Murdoch reluctant to publish his dissertation in its 1957 form despite the paleographical excellence represented by his Latin edition of Bradwardine's text. Interestingly, in his article "Mathematical vs. Experimental Traditions in the Development of Physical Science," Thomas Kuhn reported in a footnote that Murdoch had emphasized to him "the historiographic problems encountered if the classical sciences [like astronomy, optics, and statics] are conceived as continuing research traditions in the Latin middle ages."[2] If the essays in this volume show traces of Murdochean influence, one manifestation of such influence would be the insistent emphasis on careful and thoroughgoing analysis of texts with an exacting philosophical eye and a care not to exaggerate medieval accomplishments.

In this volume, the authors, like Murdoch, provide new evidence, largely from texts, concerning early science, but they also hazard new or revised interpretations of that evidence, and attempt to reach new truths about the history of science, if not about the content of science itself. Bolder than Murdoch himself in committing ourselves to print, we may be observed in these essays working in the vein of conjecture as well as demonstration. While we believe that these essays are 'substantially correct', we admit that there is more work yet to be done.

The second half of the twentieth century in the United States witnessed a broad current of work on the history of early and

[2] Thomas Kuhn, "Mathematical vs. Experimental Traditions in the Development of Physical Science," *Journal of Interdisciplinary History*, 7 (1976), 1-31, at 8 n. 7.

especially medieval science, for which Marshall Clagett and his students deserve much credit. Under Clagett's supervision at the University of Wisconsin, many Ph.D. dissertations were written containing editions of medieval texts. Several of Clagett's students—first of all Edward Grant and Murdoch—went on to supervise further Ph.D. work in history of medieval science, and those students had their own students. The University of Wisconsin Press published a classic series of editions and translations in the general area of medieval science. Clagett himself branched out to write a small textbook on Greek Science, and then magisterial multi-volume works on the influence of Archimedes in the Middle Ages and the Renaissance and on Egyptian Science. At the time of his death in 2005, Clagett intended to return to editing Richard Swineshead's *Book of Calculations*, a project which he had begun in the 1940s.[3]

Also in the second half of the twentieth century, the concept of The Scientific Revolution of the Seventeenth Century experienced first a dramatic rise with the immensely influential work of Alexandre Koyré and his followers, and then a subsequent decline, to the point at which Steven Shapin could start *The Scientific Revolution* (Chicago, 1996), "There was no such thing as the Scientific Revolution, and this is a book about it." While the concept of The Scientific Revolution of the Seventeenth Century was in full flower, large parts of the work on medieval science in effect attempted to fill in the background to the Scientific Revolution, constituting part of the debate over continuity vs. discontinuity in the history of late medieval and early modern science. Clagett's *The Science of Mechanics in the Middle Ages* (Madison, 1959) implicitly had Galileo's work as the goal to which it was heading. Now, however, to speak of medieval mechanics has been called anachronistic. Domenico Bertoloni Meli, in his *Thinking with Objects. The Transformation of Mechanics in the Seventeenth Century* (Baltimore, 2006), speaks of the "older historiography" represented by Clagett's book, and writes that "the use of these terms ['kinematics' and 'dynamics'] in the treatment of motion in the seventeenth century is fraught with

---

3) Marshall Clagett, "Richard Swineshead and Late Medieval Physics: I. The Intension and Remission of Qualities (1)," *Osiris* 9 (1950), 131-161.

danger. We are so used to thinking of kinematics and dynamics as part of mechanics that we may tend to think that the study of motion always belonged to mechanics, which is not the case. Moreover, we are so used to thinking of kinematics and dynamics as mathematical disciplines that we may assume the study of motion always to have been couched in mathematical language, which is also not the case." (p. 8)

In *Mechanics and Natural Philosophy Before the Scientific Revolution*, a book resulting from a 2004 workshop on "Mechanics and Natural Philosophy: Accomodation and Conflict," the editors, Walter Roy Laird and Sophie Roux, strike a balance between these two attitudes towards the idea of 'medieval mechanics', concluding:

> The history of mechanics—unlike the history of astronomy or optics—is the history of a variety of disparate topics and problems that were treated in different ways within very different disciplines and conceptual traditions. For this reason it is not enough for the historian of mechanics to trace these various problems and their solutions from one era to another until the right answer was hit upon. Rather, the history of mechanics is to a large extent the history of relations between different disciplines, and between theory and practice.[4]

There were several works on mechanics in antiquity, including the *Mechanica* of Pseudo-Aristotle and the text by the same name of Heron of Alexandria, not to mention Archimedes' works on statics, but the *Mechanica* of Pseudo-Aristotle was not known in the Latin Middle Ages, and it was not until the sixteenth century that it began to have an influence.[5] Meanwhile there was medieval work on statics and on motion (*de motu*) which had little to do with application to rigid bodies such as the lever, balance, or pulley. Thomas Bradwardine's *De proportionibus velocitatum in motibus* (1328), started a whole tradition of mathematical studies of motion, which

---

4) Walter Roy Laird and Sophie Roux, eds., *Mechanics and Natural Philosophy before the Scientific Revolution* (Boston Studies in the Philosophy of Science 254) (Dordrecht, 2008), 10-11.

5) Parts of the *Mechanica* were, however, known in Islam. See Mohammed Abattouy, "Nutaf min al-hiyal; A Partial Arabic Version of Pseudo-Aristotle's *Mechanica Problemata*," *Early Science and Medicine* 6 (2001), 96-122.

may be considered 'dynamical' studies of motion, and yet, to accept Bertoloni Meli's point, this was not part of 'mechanics' as understood in antiquity or in the early modern period.

In this volume the fundamental studies by Jean De Groot and Mark Schiefsky cast new light on the Pseudo-Aristotelian *Mechanica* itself, a necessary precondition for later study of its impact in the sixteenth century.[6] Concentrating on problem 1 of the *Mechanica,* Jean De Groot shows convincingly that the author's attempts to explain the labor-saving characteristics of the simple machines rest primarily on the geometry of the circle and not on any developed conceptualization of dynamics in general or of force in particular. De Groot's insight will make it necessary to reconsider the general framework of the history of mechanics, according to which the Pseudo-Aristotelian *Mechanica* takes a 'dynamic' approach to problems of statics, whereas the Euclidean-Archimedean tradition takes a 'mathematical-statical' approach.[7] In a rigid disk, such as a mill wheel, a single force applied to the disk will cause points nearer the center to move more slowly and points or parts near the periphery to move faster simply because such are the circles that they trace out in equal times.

Accepting De Groot's analysis of problem 1, Mark Schiefsky makes the important point that the author of the *Mechanica* was trying to use Aristotelian concepts to explain how the technology of the pre-existing simple machines had its surprising effects. As was typical in the early period, technology was in use well before theories were proposed to explain it—thus the steelyard was used to weigh goods in a public and verifiable way a century before the law of the lever was proposed to explain its operation. Schiefsky goes on to show how the thirty-four problems of the *Mechanica* are organized, with the circle the most fundamental explanatory principle, while the lever is reduced to the circle and the balance reduced to

6) In Laird and Roux, eds., *Mechanics and Natural Philosophy*, there are four papers, by Christiane Vilain, Walter Roy Laird, Mario Otto Helbing, and Egidio Festa and Sophie Roux, on the reappropriation and transformation of ancient mechanics largely in the sixteenth century through to the time of Galileo.

7) Clagett, *Science of Mechanics in the Middle Ages*, 4.

the lever. He shows that eighteen of the problems of the *Mechanica* explain apparently unnatural effects of rigid bodies by reducing them directly to the principles of the simpler machines, and ultimately to the circle, while other problems are indirectly so reduced.

Schiefsky's and De Groot's studies of the pseudo-Aristotelian *Mechanica* indirectly raise very interesting questions about the relationship of the science of mechanics to mechanical technology. In a word, it appears that the old truism that science and technology had little interpenetration in premodern times may be subject to revision. That issue gains further traction when we turn from the science of mechanics per se to the developments of mechanical contrivances, as John North did in his decades-long study of the late medieval abbot Richard of Wallingford. In the last years before his very recent death, North, another eminent scholar of the early mathematical sciences, trying to draw together in retirement the results of a lifetime of research, published a more accessible book on Wallingford, whose work he had previously studied in three very dense volumes.[8] In *God's Clockmaker*, North made a convincing case that Richard of Wallingford, Abbot of St. Albans and clockmaker extraordinaire, was the most original English scientist of the later Middle Ages.[9] But what are we to make of the fact that Wallingford had also studied theology at Oxford and that he was heavily involved in monastic administration? In an Epilogue to *God's Clockmaker*, North commented:

> There are those today who spend their lives writing about intellectual affairs, carefully filtering out all theological components, only to end by making the well-worn generalization that neither philosophy nor science was able to make progress before it was secularized, freed from theological bondage. Does the case of Richard of Wallingford support or run counter to this sweeping claim?[10]

---

[8] John North, *Richard of Wallingford: An Edition of his Writings with Introductions, English Translation and Commentary*, 3 vols. (Oxford, 1976).

[9] John North, *God's Clockmaker. Richard of Wallingford and the Invention of Time* (Hambledon and London, 2005), xv.

[10] Ibid., 381.

Pure astronomy and mathematics were very rarely anything but secular, North answers. But if Wallingford "regarded himself a student of God's creation," he goes on to ask, what was the interaction of his secular studies with his religious belief? North finds very little evidence relevant to answering such a question.

With these sorts of questions in mind, Abdelhamid Sabra's contribution to this volume takes on a definite sparkle. Almost as it were nailing theses upon the wall, Sabra's article presents the "simple ontology of Kalām atomism" as an alternative to Hellenizing *falsafa*. To those who are familiar with Sabra's lifetime of erudition, his remarkably clear and succinct depiction of medieval Kalām atomism must carry conviction—and yet it carries us in a direction quite opposed to the traditional emphasis on the exact sciences displayed by the older historiography. In Sabra's reading, kalām provided a view of the world that required God's ongoing creation at every instant. Since God's activity in the world was viewed primarily as that of eternal creation and re-creation, change was reduced to the creation of substances in different locations with accidents that were themselves products of a creative act lasting only for an instant. Such a theory of being could only be a form of atomism, in the strict sense, for both substances and their accidents were instantaneous manifestations of God's creative power. As Sabra shows, the matter-theory of the *mutakallimūn* was therefore driven by the logical consequences of its initial premise—that all change reduces to the creative activity of God. Hence, as Sabra points out, the system of kalām, whether practiced in its original Muᶜtazilite form or in its later Ashᶜarite version, represented a genuinely philosophical alternative to the *falsafa* of the followers of Greek rationalism, such as Avicenna and Averroes.

As compared to North in the case of Richard of Wallingford, Sabra has even less (more exactly, nothing) to say about the impact of kalām atomism on the development and fate of Islamic (or Arabic) science for better or worse. But those wishing to rewrite the history of Islamic science in the future will neglect Sabra's brilliant and coherent picture at their peril. Was it the case that the medieval Latin atomism against which Thomas Bradwardine argued in his *De continuo* had any sources in kalām atomism? Certainly, one textual source that John Murdoch has studied in this connection was the so-called *Metaphysics* of Algazel.[11] Unlike Richard of Wallingford, for whom we have very little in the way of theological writing, Thomas Bradwardine was famous for his theological work, particularly *De causa Dei contra Pelagium*. Here is a corner of the history of medieval science studied by John Murdoch with great erudition, about which more might still be said if one looked more thoroughly at the relations between medieval theology and medieval science.

But before turning to areas beyond the exact sciences, we must mention the second paper in this volume on Islamic science, namely Elaheh Kheirandish's paper tracing the 'footprints' of experiment in early Arabic optics, which seeks to deepen our knowledge of experiment in early Arabic optical literature beyond the previous work of Sabra and of Roshdi Rashed. Kheirandish's analysis focuses mainly on the early little known author Aḥmad Ibn ᶜĪsā, who may have flourished as early as the ninth century C.E. Ibn ᶜĪsā refers to metal sighting tubes in an attempt to test the belief that a putative visual ray emitted from the eye spreads out to form a sensitive cone. Apparently coming directly out of the tradition of pseudo-Euclid's *De speculis*, Ibn ᶜĪsā's experimental procedure modifies the view of previous scholars that the later and much more sophisticated experiments of Ibn al-Haytham derive from an astronomical rather than an optical tradition. At the same time, Ibn al-Haytham's celebrated technical term *iᶜtibar*, used to mean experiment, is also present in Ibn ᶜĪsā's *Book on Optics and Burning Mirrors* (in the verbal form, *iᶜtabara*), a fact that, as Kheirandish argues, further cements the continuity of later Islamic optical tradition with earlier models within the same discipline. Kheirandish concludes her paper by looking at some further examples of experiment in the Islamic astronomical and optical traditions, showing that similar terms were shared by both.

While cautious about bold claims, Kheirandish's paper adds detail and nuance to the history of experiment in Islamic optics.

---

11) [al-Ghazzali,] *Algazel's Metaphysics: A Medieval Translation*, ed. J.T. Muckle (Toronto, 1933). Of course, the text that went by this name actually represented mainly the metaphysics of Ibn Sina (or Avicenna).

Nevertheless, Ibn al-Haytham retains the prominence given to him by Sabra as an advocate of experiment as a method of proof. In contrast, medieval natural philosophy has been characterized as "empiricism without observation."[12] In an article with this phrase in its title, Edward Grant quoted Murdoch as stating a similar view:

> empiricist *epistemology* was dominant in the fourteenth century ... this did not mean that natural philosophy then proceeded by a dramatic increase in attention being paid to experience and observation (let alone anything like experiment) or was suddenly overwrought with concern about testing or matching its results with nature. On the contrary, its procedures were increasingly *secundum imaginationem* (to use an increasingly frequently occurring phrase) and when some "natural confirmation" of a result is brought forth, more often than not it too was an "imaginative construct."[13]

However accurate Murdoch and Grant may be about the "empiricism without observation" of medieval Aristotelian natural philosophy, Michael McVaugh gracefully shows in his paper in this volume that medieval physicians such as Arnau de Vilanova and Bernard Gordon can accurately be said to have tried to make use of the best empirical evidence available in recommending the use of drugs for their patients. Unlike most members of the faculties of arts of medieval universities whose theorizing occurred in the context of the education of future clerics, teachers of medicine were often practicing physicians themselves, teaching students who would also practice. Thirteenth-century medical theorists like Arnau weighed many different sorts of evidence: the testimony of the great physicians of antiquity such as Galen and Avicenna; the variable effects of drugs of different age and purity when given to patients of varying

12) Edward Grant, "Medieval Natural Philosophy: Empiricism without Observation," in *The Dynamics of Aristotelian Natural Philosophy from Antiquity to the Seventeenth Century*, eds. Cees Leijenhorst, Christoph Lüthy, and Johannes M.M.H. Thijssen (Leiden, 2002), 141-168.

13) John Murdoch, "The Analytic Character of Late Medieval Learning. Natural Philosophy without Nature," in *Approaches to Nature in the Middle Ages*, ed. L.D. Roberts (Binghamton, 1982), 171-213, quoted in Grant, "Empiricism without Observation," 167.

temperaments; the small margin between effectiveness and harm of strong drugs like opiates; and the observed interaction between mind and body when, for instance, the wearing of an amulet was followed by a patient's cure. If, as in the case of the optical authors discussed by Elaheh Kheirandish, one may wonder whether the experiments that are described were actually done in practice—or were only *secundum imaginationem*—we can be confident that physicians did treat their patients and, as in the case of Arnau, carefully observed the outcomes of their efforts, whether or not they could be certain what it was that had or had not caused an observed benefit. Here is more evidence to enrich the history of experiment, in particular investigation of the possible interactions between the traditions of methodical observation in medicine (going back to Galen and earlier), of comparative observation in astronomy, and of experiment in optics.

Attention to early traditions of experience or experiment leads naturally to Roger Bacon, the subject of Yael Raizman-Kedar's paper in this volume. Raizman-Kedar offers a revisionist interpretation of Bacon's view of intellectual species, arguing that for Bacon intellectual species are material. Bacon has been known to historians as a sharp critic of his contemporaries. McVaugh in his paper quotes Bacon as pointing out that the medical authorities of his time offer conflicting descriptions of the effects of drugs such as senna and rhubarb. In Bacon's terms, the effect of an Indian form of rhubarb is "against the law of the other medicines, and this is known by experience." Given the long history of writing about Roger Bacon as a forerunner of modern science and its methods, and given more recent arguments by historians such as David Lindberg that Bacon was actually rather typical of the natural philosophers of his time, even somewhat Augustinian in his epistemology, it is all the more difficult to discern Bacon's positions from his writings, without the overlay of previous interpretations, not to mention the factors that his views may have changed over time, and that the chronological order of his writings is not certain.[14]

---

14) Cf. David Lindberg, "Science as Handmaiden. Roger Bacon and the Patristic Tradition," *Isis* 78 (1987), 518-536, at 518, "... renowned for his attack on authority and his defense of mathematics and experimental science, Bacon has become the symbol,

With a boldness of youth comparable to Sabra's wisdom of age, Raizman-Kedar here argues that Bacon naturalized the intellect through his theory of species, which not only issue from objects and impress themselves on sense organs, but even end up, with modifications appropriate to the nature of the intellect, in the human intellect itself. Here we seem to have, within the person of Bacon, the struggle between Augustinian and empiricist views of the acquisition of knowledge, with a naturalistic conception coming out on top. As Raizman-Kedar describes them, the main lines of Bacon's theory of the acquisition of knowledge resemble the ideas of the *moderni* John Buridan and Albert of Saxony, described in Edith Sylla's paper. Raizman-Kedar's arguments for this view are based on careful and systematic interpretation of the evidence of Bacon's later texts, and should stimulate further detailed study. If Raizman-Kedar is right, then Bacon was in fact the empiricist he was long reputed to be, whether or not he actually performed anything that could be called an experiment.

While a number of the foregoing papers concern medieval attitudes toward experience and experiment, Hans Thijssen's paper on "The Debate over the Nature of Motion: John Buridan, Nicole Oresme and Albert of Saxony" focuses on an area that has concerned medievalists since Pierre Duhem argued at the beginning of the twentieth century that medieval theories of motion paved the way for Galileo.[15] Instead of looking to the medieval theory of motion, including the theory of impetus, as a step on the way between the Aristotelian idea that there must be a mover for every motion and the law of inertia, Thijssen looks at the related discussions of motion of Buridan, Oresme, and Albert of Saxony as a way of understanding how the relations of these three thinkers should be understood—if Oresme and Albert of Saxony were not members of a 'Buridan school', how can the similarities and differences between and among their views and Buridan's be understood? Again,

---

not exactly of medieval science, but of modern science struggling to emerge during the medieval period—prevented from doing so by ecclesiastical repression."

[15] See John Murdoch, "Pierre Duhem and the History of Late Medieval Science and Philosophy in the Latin West," in *Gli studi di filosofia medievale fra Otto e Novecento*, ed. Ruedi Imbach and Alfonso Maierù (Rome, 1991), 253-302.

by providing a text from the forthcoming edition of Buridan's *ultima lectura* on Aristotle's *Physics* and by arguing for a revised view of the relations of the three major figures he discusses, Thijssen provides both evidence and new interpretations, arguing that Buridan, Oresme, and Albert of Saxony should be considered not a 'Buridan school', but an intellectual community.

While many twentieth-century historians of medieval science were pursuing topics related in some way to the exact sciences, natural philosophy, or medicine, there have been other strains in the historiography of medieval science that have taken account of the institutional setting of scholastic science within universities. Anneliese Maier's marvelously detailed studies of the natural philosophy of the late scholastics have long served as a model for work that kept close to the manuscript sources. While Anneliese Maier always used commentaries on Peter Lombard's *Book of Sentences* as a key genre within which natural philosophical speculations might be found, scholars soon began to pay more attention to the theological contexts of these philosophical speculations. Following Duhem's assertion that the condemnations by Bishop Etienne Tempier at Paris in 1277 marked the birth date of modern science, scholars examined other condemnations and academic regulations in relation to their impacts on natural philosophy. Meanwhile, historians of philosophy more broadly pursued the many important developments within medieval logic. As early as the preparation of Grant's *Source Book in Medieval Science* (Cambridge, MA, 1974), Murdoch lobbied to have logic considered as a key discipline in relation to medieval science, much as mathematics became a key discipline or instrument in the early modern period. These concerns with institutional setting, theology, and logic broadened the focus of historians of medieval science and led to fruitful discussions over a wide range of topics. Behind the conference organized by Murdoch that led to the volume *The Cultural Context of Medieval Learning* there were intense discussions of how to reveal and portray the interplay between scientific texts and the social and cultural contexts in which they were produced.[16]

---

[16] John Emery Murdoch and Edith Dudley Sylla, eds., *The Cultural Context of Medieval Learning: Proceedings of the First International Colloquium on Philosophy, Science,*

In the most recent generation, the historiography of medieval and early modern medicine has burgeoned, animals, plants, geology, and natural history in general have received attention, once *outré* subjects such as alchemy and astrology have become mainstream for historical research, the history of medieval technology and scientific instruments has come into its own, and linkages between scholasticism, demonology, and popular religion have emerged.

In this volume, the papers of Craig Martin and William Newman are related in that both deal with different aspects of what the medievals categorized as 'meteorology', namely the study of the world beneath the sphere of the moon. Martin's contribution deals with macro-level phenomena such as weather and earthquakes, while Newman's concerns the microstructure of matter. Indeed, the thrust of Martin's paper lies in the irregularity of gross 'meteorological' occurrences such as tornadoes and volcanic eruptions. As he shows, a large number of medieval and early modern authors employed the *regressus*-theory most famously used by the early modern school of Padua, especially Giacomo Zabarella, in such cases. *Regressus* implied a preliminary discovery of causes from their effects, and then a stabilization of knowledge of the causes, followed by an attempt to demonstrate the effects from their newly revealed causes. In sublunar phenomena where the causes were not obvious, the gathering of preliminary factual material was difficult and subject to unreliability. Thus Renaissance natural philosopohers were content with explanations for meteorological phenomena that were free from contradictions and merely saved appearances. In short, medieval and early modern meteorology was a science of probable outcomes rather than certainty. Martin's findings seriously challenge the prevailing school of modern scholars such as Peter Dear, Simon Schaffer, Steven Shapin, Barbara Shapiro, and others who claim that early modern Aristotelianism was uniformly committed to an all-or-nothing picture where scientific knowledge had to attain to complete certainty in order to count as knowledge at all. The argument of these

---

and Theology in the Middle Ages (Boston Studies in the Philosophy of Science 26) (Dordrecht, 1975).

scholars that 'probabilism' was something introduced by *novatores* such as Robert Boyle is drastically undercut by Martin's analysis.

Newman's paper also challenges prevailing scholarly views by linking the little-studied matter theory of the alchemists with later chemists such as Antoine Laurent Lavoisier. As in Martin's paper, Newman's contribution brings in Aristotle's *Meteorology*, but focuses on a different part—namely the possibly pseudonymous Book IV. *Meteorology* IV has long been known to contain a corpuscular theory of matter wherein the *onkoi* and *poroi* composing material things are given considerable explanatory force. As Newman points out, this corpuscular theory was accompanied by a strikingly different sense of homoeomerity than the more famous locus classicus provided by *De generatione et corruptione* Book I, Chapter 10. Indeed, *Meteorology* IV, in conjunction with other Aristotelian *loci*, provided the inspiration for an analytical approach to the discovery of elements and principles that has come to be known as the 'negative-empirical' principle. This was simply the belief that things are made up of the final materials that technical analysis can yield by breaking them down. From such a perspective, homoeomerity itself becomes a warrant of elementarity, since division into more primitive components would indicate a lack of resistance to laboratory analysis. Newman goes on to show that this principle was transferred from medieval alchemy to early modern experimentalists such as Daniel Sennert and Robert Boyle, and through them (as well as other early modern chymists) to Lavoisier. This opens up a promising and hitherto unsuspected path from 'chymical atomism' to nineteenth-century chemical atomism.

As if to pass from the sublunar realm of Aristotelian meteorology to the superlunary world of the planets, three papers in the volume relate more or less closely to questions about the work of Copernicus. André Goddu's paper concerns Copernicus himself. Against the background of his long-standing research on Copernicus, Goddu here analyzes Copernicus's use of arguments based on part/whole relations, and, in particular, on the dialectical topic *from an integral whole.* He shows that Michael Mästlin clearly recognized Copernicus's use of this topic, as in the argument that what takes place in the heavens should have more weight in theorizing than what takes place on the surface of the earth, given that the heavenly regions comprise by far the largest part of the cosmos. And the same topos lies behind Copernicus's argument that previous astronomical theorists had failed because they put together disparate astronomical devices into what was more nearly a monster than the structured cosmos which God surely designed.

Edith Sylla's paper complements Goddu's in concluding that when there are conflicting candidates for the foundations of any science, it is dialectic rather than demonstrative science that must provide the aids to theory choice. In her paper, Sylla argues that the theory of science expressed in the *Questions* on the *Posterior Analytics* of John Buridan (and by extension the views of the *moderni* as a group) can be understood to embody 'critical realism', in the sense that Buridan believes that natural science, based on experience, is possible, despite the fact that its conclusions can never be absolutely certain. If at any given point of time, a demonstrative science based on experience is well confirmed and taken as true by its practitioners, nevertheless provision must be made for the occasions upon which the foundations are shaken.[17] Scholastic natural philosophers like Buridan were thoroughly accustomed to patterns of reasoning beginning from the assumption that God, by his absolute power, could violate the regular processes of nature, and if God did so, what then? This may have helped prepare them for situations in which the principles of a science were shaken for some other reason. As Sylla argues, many historians who examine the scholastic Aristotelian context of Copernicus's work use, perhaps simply for con-venience sake, versions of scholasticism that did not come into existence until after Copernicus's death. On the other hand, the ideas of the *moderni* were among the most prominent versions of Aristotelianism influential at Cracow in Copernicus's time. Thus

[17] Buridan's position in this regard turns out to imply something quite like Isaac Newton's fourth rule for the study of natural philosophy. Isaac Newton, *The Principia. Mathematical Principles of Natural Philosophy.* trans. I. Bernard Cohen and Anne Whitman (Berkeley, 1999), 796: "In experimental philosophy, propositions gathered from phenomena by induction should be considered either exactly or very nearly true notwithstanding any contrary hypotheses, until yet other phenomena make such propositions either more exact or liable to exceptions."

it is possible—though certainly not proved at this point—that the conception of science expressed by Buridan and the other *moderni* might have helped to fortify Copernicus's view that change and progress in astronomical theory are to be expected.

Coming at a similar inquiry from yet a third direction, Mike Shank, in his paper, seeks to connect Copernicus more tightly to the late-fifteenth-century university context in which he worked, specifically to the commentary of Giambattista Capuano da Manfredonia on the *Sphere* of Sacrobosco. Unlike the historians of science who see pre-Copernican astronomy as an exclusively mathematical science, Capuano explicitly argues that astronomy, although it has mathematical features, is a part of natural philosophy. Shank suggests that, when Copernicus addressed the physical objections to the Earth's motion in the initial sections of *De revolutionibus*, he may have used Capuano's commentary—since Capuano, who was quite likely known to Copernicus, had collected and reformulated in one convenient place many of the classic and late-medieval arguments pro and contra a moving Earth. Taken together, the papers of Goddu, Sylla, and Shank cast light on the long disputed question of the place of astronomy among the other disciplines in the early modern period. To what extent was there tension between and among the rival claims of natural philosophy and the mathematical sciences? Both Goddu's and Sylla's papers suggest that to resolve the rival claims of geocentric and heliocentric astronomical systems, dialectic rather than scientific demonstration might be called for. Goddu shows that one of Copernicus's most convincing arguments for the superiority of the heliocentric system relied upon the dialectical topic *from an integral whole*. It might still be asked, however, whether or to what extent the power of the application of this topos to the cosmos rested upon Aristotelian and/or Christian conceptions of the unity and design of the cosmos. All involved were apparently agreed that the cosmos reflects a far more glorious design than, say, the mere house that was a typical example used in explaining the topos *from an integral whole*.

Even farther afield from the traditional subject matter of history of medieval science in the Clagett vein are the exciting papers published here by Qiong Zhang on Jesuit attempts to accommodate scholastic science to the Chinese cultural context, in the contribution of Anne Davenport on angelology, and in Steven Marrone's study of evolving scholastic responses to magic. All three papers deal with the relationship between ordinary physical phenomena and a putative world of more spiritual beings—be they dragons, angels, or demons.

Qiong Zhang's contribution lays out the remarkable story of Chinese belief in size-altering dragons that persisted over several millennia, only to come under threat during the Jesuit mission to China that began in the late sixteenth century with Matteo Ricci. Traditional Chinese lore had seen the dragon as a weather-making being capable of sucking water up from terrestrial locations and sending it down again, often accompanied by thunder, lightning, hail, and other deleterious products of the upper regions. One of the interesting features of Zhang's article lies in her exposition of rationalizing attempts to explain or limit the dragon's power long before the Jesuits' arrival. The Han dynasty philosopher Wang Chong (27-97 CE), for example, denied that dragons were spiritual beings and claimed that they were merely powerful aquatic animals, but nonetheless retained the belief that they could produce rain. Like many subsequent Chinese philosophers, Wang explained the dragon's power in terms of a non-sentient vital energy or *qi*. The dragon's effect on the weather was due to a *qi*-based sympathy between this aquatic animal and the water, not to any draconian volitional activity. Wang's naturalizing explanation was mirrored in varying guise by a host of philosophers up until the arrival of the Jesuits, but the essential cluster of 'facts' centering on the dragon's ability to cause and alter weather remained intact until Ricci's Jesuit successors introduced their Chinese students to the works of Aristotle in the first half of the seventeenth century. This introduction proved to be the crucial 'modernizing' event that led to the dragon's overthrow and replacement by the entirely inanimate meteorology of the Stagirite, although its impact did not reach much beyond the tiny circle of Jesuit-inspired scholars. It is ironic to consider that the very Peripatetic physics that was being discarded in the West as hopelessly old-fashioned by Bacon and Descartes was at that moment serving to wean the Chinese from their world of self-moved, sentient agents of natural change.

A striking contrast may be found between Zhang's story and that of Anne Davenport. Her story samples some of the ways in which English Catholic recusants sought to defend and explain angelic appearances in a newly Baroque heaven. Davenport starts with Thomas White, Kenelm's Digby's co-theorist in the latter's attempt to reformulate scholastic natural philosophy in a post-Cartesian world. White, who postulated that angels produce elemental fire out of very pure air and, more generally, govern the transmutation of elements, argued that angels produce angelic appearances chemically, at God's command. White's friend, the Franciscan author Franciscus a Sancta Clara, focused on a more anxious issue. He resurrected medieval theories of angelic appearances in order to refute novel, and in his view, dangerous, claims concerning atmospheric beings and homunculi. Sancta Clara's younger confrere, in turn, the Cartesian Anthony Legrand, saw in angelic appearances a welcome occasion to confirm Descartes's mechanistic optics. What should we make of these fanciful ideas? Davenport's essay concludes with Robert Boyle's meditations on angels in order to argue that her Catholic authors nicely illustrate what Charles S. Peirce describes as the comforting, pleasing, and thus inherently conservative character of 'metaphysical' methods—which were soon to be replaced by deliberate efforts to rely exclusively on induction.

Steven Marrone's paper, though set in an earlier period than Zhang's or Davenport's, wrestles with some of the same issues. How much power and of what sort should we grant to magic? Is it due to activity of sentient beings, or is it a merely natural power? And what does it mean anyway for a power to be natural? Marrone follows these questions through their treatment by three key figures—William of Auvergne in the early thirteenth century, Thomas Aquinas up to his death in 1274, and Giles of Rome, often viewed as a follower of Thomas, in the generation after the Angelic Doctor. Marrone charts a development in these three authors' views, arguing that one can see the beginnings of a 'protomechanical' view of nature in this progression. He begins with William's claim that most natural change between bodies requires corporeal contact, but that some marvelous phenomena are actually the result of *magia naturalis*, a term created by the Parisian theologian to signify a seemingly wonderful power of nature not involving direct contact between bodies. Passing to Thomas, Marrone notes that the Angelic Doctor did not differ fundamentally from William in these views, although he placed a lesser emphasis on the hidden powers of nature. Finally, Marrone treats the case of Giles, who dealt at some length with the question whether a the corpse of a murder victim would begin to bleed in the presence of its murderer. Giles concluded that, as a matter of the regular course of action in nature, it would not, in large part because he could not identify a causal mechanism to explain it that did not call for direct bodily contact between the agent and patient. The need for bodily contact introduced by William had hardened by Giles's time into a maxim that could be used to deny the very evidence of a marvelous phenomenon itself.

The final contribution to our volume, Christoph Lüthy and Alexis Smets' "Words, Lines, Diagrams, Images: Towards a History of Scientific Imagery," builds on John Murdoch's *Album of Science: Antiquity and the Middle Ages* in order to problematize the issue of scientific images and their relationship to text. Arguing that previous scholars have assumed an unwarranted transparency in this area, Lüthy and Smets pass through a succession of problems raised by medieval and early modern texts in the attempt to demonstrate the radically complicated typology of images implied by them. Their first case-study shows how the graphic representations of ideas in early texts have a way of becoming integrated into frankly pictorial elements. Porphyry's Tree, for example, an abstract representation of Aristotelian genera and species, has a way of becoming a real tree in medieval manuscripts. Lüthy and Smets then progress to the case of similar-looking images with radically different meanings, such as the hexagonal groupings of six circles that represent such varying subjects as Giordano Bruno's atoms and Pietro Bongo's attempt to depict the six days of creation. They find an inverse to this problem in the case of alchemy, where multiple images are frequently used to describe the same chemical referent. Lüthy and Smets find the same complexity in the modern language used to described early scientific images. Given the dearth of systematic studies in this area, the impression one gets is one of massive confusion, where different scholars employ the same term to describe various sorts of images and diverse terms to categorize images whose function is in fact similar. Finally, Lüthy and Smets introduce what could be the most perplexing problem of all—the fact that images reflect the epistemological orientation of their author in fundamental ways. A Platonist like Bruno may see talismanic powers in an image where a Cartesian might view it simply as a representation of the microstructure of matter. In summary, Lüthy and Smets manage to bring the hugely complex typology of early scientific images to light in a way that has hardly been done before. In doing so, their work reveals the surprisingly diverse influence of Murdoch in an array of unsuspected areas. One can only hope that Lüthy and Smets will develop the promising start that they have made here in future venues.

In sum, even in areas that have been well-studied in the past, students of ancient and medieval science—and early modern scholasticism as well—are finding that there is much greater diversity, complexity, and accomplishment to early science than dreamt of by those who still think of Galileo's character Simplicio as a typical scholastic. In recent decades, as many more books and papers have been published about early modern science than about medieval science—publications that still assume that there was an origin to modern science in the seventeenth century if not a revolution however defined—there has been a tendency for specialists in early modern science to glance back briefly at what came before. These scholars nearly always assume that to be modern, science had to reject or overcome that earlier science, which must have had serious failings: otherwise the fact that it was superseded in favor of modern science would not have been necessary. Among many other misconceptions that result from treating the post-Tridentine Catholic Church as if it can be used to represent the Catholic Church of the Middle Ages as a whole is a general neglect of Islamic (and Jewish) scientific work, highly important throughout much of the Latin Middle Ages, but which tended to experience hostility after the Christian reconquest of Granada and, even more so, after Turkish conquest of Constantinople. Another misconception is to discount the role of the *moderni* (or the *Doctores Parisienses* so praised by Duhem), who may have been more influential in the fifteenth and early sixteenth centuries than students of early modern science now admit.

The present volume shows that the current oversimplified picture of medieval science is in some measure a self-fulfilling prophecy, resulting from a historiographical tradition that has culled out factors that were not obvious precursors to the grand themes of the Scientific Revolution as envisaged by mid-twentieth-century scholars. In reality, the canonical picture of the Scientific Revolution itself needs drastic rewriting, but before this can be done properly, the Middle Ages must be intensively revisited both for its own sake and for what it had to contribute to later science. How else can we learn what constituted significant and important change in early modern science, if we do not have an accurate picture of the science that came before in all its diversity? And how can we lay claim to an understanding of science over the *longue durée* without a solid understanding of the subject in its late antique and medieval incarnations? The present volume, by including the less abstract, more empirical side of Aristotelian and scholastic science and its relation to philosophy more generally, attempts to make the picture more complete. As students of John Murdoch, many of us have internalized his strictures that the history of science must be based on study of the science itself, and first of all on scientific texts, and that analysis of these texts must be thorough, painstaking and philosophically rigorous. The result of this analysis continues to challenge the commonplaces that populate the historiography of science, just as Murdoch himself has done over his long career. Anyone who still wants to maintain that the medievals and their early modern followers were unconcerned with experience and experiment, who asserts that they were locked into a demonstrative certainty that excluded the scientific discussion of nature in all its changeability, who sees them as slaves to rigid disciplinary boundaries, or who fails to account for the wide divergence in their views about the natural and supernatural, will have to come to terms with these collected articles. Although there is much current work on the history of early science extending into fields well beyond those ventured into here, we think that this book nevertheless reveals a surprisingly vigorous community of scholars working on early science and at the same time demonstrates the fruitfulness of John Murdoch's scholarly ideals.

Modes of Explanation in the Aristotelian *Mechanical Problems*

Jean De Groot*
The Catholic University of America

Abstract
Scholars have been puzzled by the central argument of *MP* 1 where the author addresses the basic principle behind the balance and lever. It is not clear what is intended to provide the explanation—the dynamic concepts of force and constraint or the geometrical demonstration. Nor is it clear whether the geometrical part of the argument carries any logical force or has value as a proof. This paper makes a case for the cogency of the argument as a kinematic, not dynamic, account. *MP* 1 proceeds systematically as it extends the explanatory power of the parallelogram of movements from rectilinear motion to circular motion. Euclid's *Elements* I.43 provides insight on the author's procedure. His general method is demonstrative, as described in *Posterior Analytics* I.1.

Keywords
Mechanical Problems, dynamics, kinematics, Euclid, force, constraint, demonstration, *ischus*, *ekkrousis*, *apodeixis*, François De Gandt, lever, balance, Aristotle's *Physics*, mechanics, ancient science, circle, concentric circles, moving radius

I.

Similarities have been drawn between Aristotle's ratios of speeds, weights, and times in *Physics* VII.5 and the ratios involved in the principle of the lever in *Mechanical Problems* 1.[1] The connection

---

* School of Philosophy, The Catholic University of America, 620 Michigan Ave., N.E., Washington, D.C. 20064, U.S.A. (degroot@cua.edu).

[1] The text used is M.E. Bottechia Dehò, *Aristotele / Problemi meccanici: Introduzione, testo greco, traduzione italiana, note* (Soveria Mannelli, 2000). For chapter designations, I follow the divisions in the Teubner text edited by Otto Apelt, *De plantis,*

between the two texts has been made on the basis of a common definition of speed and a law of motion derived from the proportional method of reasoning used in both treatises.[2] François De Gandt expressed doubt that the connection between *Physics* VII and *Mechanical Problems* (henceforth *MP*) is really very strong. He thinks the most fruitful explanatory feature of *MP* 1 lies outside the proportional method, which is itself unexceptional, being the universal ancient method for expressing laws of variation.[3] De Gandt's separation of the two texts as to theme and content is important, because it clears the ground for further evaluation of *MP* as one source for the origin of mechanics.

In interpreting *MP* 1, De Gandt places emphasis on the notion of constraint (ἔκκρουσις) in the author's decomposition of circular motion into rectilinear components. This analysis of circular motion into component motions is the greatest achievement of *MP*, he thinks, not surpassed until Huygens or Newton. Accordingly, he finds the central claim of the text—the assertion of a characteristic proportion for arcs of different-sized circles covered in the same

de mirabilibus auscultationibus, mechanica, de lineis insecabilibus, ventorum situs et nomina, de Melisso Xenophane Gorgia (Leipzig, 1888), since this is still the most frequently referenced text. Scholars disagree as to whether the author of the treatise was Aristotle himself.

2) Pierre Duhem understood this law to be that the force of a mover is measurable as the product of the weight of the body moved and the velocity impressed upon it. See Pierre Duhem, *The Origin of Statics: The Sources of Physical Theory*, trans. G. Leneaux, V. Vagliente, and G. Wagener (Boston, 1991), 11. This is not the law of motion derived more broadly from Aristotle's proportional reasoning and stated by Marshall Clagett in the following way, "… with time and force constant, the distance traversed is inversely proportional to the resistance, or with distance and force constant, the time is directly proportional to the resistance—with the proviso that the force is sufficient to cause movement" (*Science of Mechanics in the Middle Ages* [Madison, 1959], 428). Although the two statements of Aristotelian dynamics can be connected in many ways, Duhem's formulation highlights the possible connection of the statics of *MP* to the later principle of virtual velocities. De Gandt disputes that Duhem's law is an accurate account of the reasoning of *MP* 1 at all. See François De Gandt, "Force et science des machines," in *Science and Speculation: Studies in Hellenistic Theory and Practice*, ed. J. Barnes, J. Brunschwig, M. Burnyeat, and M. Schofield (Cambridge, 1982), 96-127, especially 124.

3) De Gandt, "Force," 125, 118.

time—to be less convincing than the claim that a combination of forces produces a resultant curved motion.[4] De Gandt's analysis is measured and perspicuous, providing a new and insightful perspective on the scholarly tradition that has sought a quantifiable Aristotelian dynamics to compare to modern dynamics.

This paper will take a somewhat different tack from De Gandt, however, concerning a theme treated in the course of his discussion of the decomposition of circular motion, namely the interplay of the geometric and the dynamic features of the explanation of the lever principle. The central principle of *MP* 1 is that points on a moving radius move at different speeds depending on their distance from the center of the circle. Accordingly, of revolving concentric circles, the circumference of the larger circle is longer and also moves faster in the same time. Ancient authors took this feature of the circle to explain why a small force exerted at the longer end of a lever can move a great weight at the shorter end.[5] De Gandt says that, in *MP*, the difference in velocities of larger versus smaller concentric circles is not taken as "un fait géométrique," since it is treated as requiring an explanation involving forces. The concept of force in the treatise is thus fundamental to the argument. Accordingly, the reference to the circle as the source of all mechanical marvels (847b16, 848a12) is a rhetorical flourish and not the enunciation of a fundamental principle.[6]

De Gandt's position on the significance of the circle in *MP* 1 is debatable. The argument to be made here is that the kinematic aspect of *MP* 1, which is based on the geometry of the circle, is the coherent and systematic account in the text and that the dynamic

---

4) De Gandt, "Force," 123-24. The passage at issue is 849a31-849b15.

5) The principle was known to Plato as a remarkable feature of the circle responsible for a variety of wonderful phenomena (θαύματα)(*Laws* X, 893c-d). Aristotle, in *Movement of Animals* (henceforth *MA*) 1 and 7, used the moving radius principle in explaining the motion of limbs and all manner of animal responses that exceed in degree their initial cause. See in particular *MA* 701b24-27, where he makes a comparison to the rudder. On this topic, see Jean De Groot, "*Dunamis* and the Science of Mechanics: Aristotle on Animal Motion," *Journal of the History of Philosophy* 46 (2008), 43-68.

6) De Gandt, "Force," 119-20.

aspects of the text are weaker and under-conceptualized. Indeed, the chapter presents two different accounts overlaid. The central account is kinematic. It treats of movements as distances covered in a time, with these distances related geometrically. Overlaid on the kinematic account is an explanation in terms of impulse or force (ἰσχύς) and constraint (ἔκκρουσις). This is the dynamic account to which De Gandt draws attention. The problem is that the fundamental notions of the dynamic account, force and constraint, do not contribute to a coherent theory that incorporates the kinematic account.[7] In the development of scientific ideas, the vagueness of a concept is hardly a fatal flaw. If we wish to understand the text of *MP* in its historical context, however, it is important to separate the two layers of the argument of chapter 1 and to notice that the mathematical account is what sustains the argument.

Returning to the original question of the relation of *Physics* VII.5 and *MP* 1, certainly the most striking difference between them is that in *MP* there is a central explanatory principle that provides a unified account of disparate phenomena. The moving radius principle explains the lever in *MP* 1, and then, in the rest of the treatise, the lever is shown to be the root phenomenon in a host of other devices. For good or ill in comparison to the later history of mechanics, movement of the lever is explained in *MP* by properties of the circle. The author's argument by proportionality depends on the geometry of the circle and is cogent understood in those terms.

De Gandt expressed doubt concerning the probative force of the central claim of *MP* 1, which is based on proportionality.[8] The

---

[7] For a definition of kinematics as contrasted with dynamics, see Clagett, *Science of Mechanics*, 163. Kinematics "studies movements taken in themselves, i.e., the spatial and temporal aspects or dimensions of movement without any regard for the forces which engender changes of movement" (163). Kinematics relies on geometry but treats lines and arcs as trajectories of movement. As a point of clarification, note that, although *MP* treats of the balance as well as the lever, it is not strictly speaking a work of statics, the science of weights in equilibrium.

[8] The problem of the adequacy of the proof of the lever principle in *MP* 1 cannot be completely resolved in this paper, the purpose of which must be limited to evaluating how the geometry of the circle sustains the argument developed in *MP* 1.

assumption of the present paper is that the kinematic account in *MP* is a demonstration (ἀπόδειξις), as demonstration is described by Aristotle in *Posterior Analytics* I.1. In this introductory chapter to his theory of science, Aristotle says that new knowledge is gained by recognizing an instance we do not yet understand as being a case of something already known.[9] This account is amplified later in terms of the structure of demonstration. In *Po An* II.8-10, it is clear that what is to be explained (e.g., the moon's being deprived of light) is an empirical observation deriving without intermediary from a cause (the earth's interposition between sun and moon). In this picture of demonstration, one phenomenon (interposition) is understood to be the reason why another more complex phenomenon (eclipse of the moon) takes place at all.[10] One phenomenon is present within another, so to speak. This is the mode of explanation the author of *MP* follows throughout the series of problems in *MP*. For example, a rudder, though small, has the power (δύναμις) to move a large ship, because it is a lever (850b31). The sea is a weight, and the helmsman is located at the fulcrum. The ship, inclining against the weight of the sea, moves in the opposite direction. Similarly, a ship moves faster the higher its yardarm, because the mast is a lever with its base as the fulcrum (851a40).

It will become clear in what follows that this pattern for demonstration—placing something better known within a more complex situation—is also the method of *MP* 1, where the Aristotelian author fills out the geometrical context surrounding the moving radius principle. In this case, the root 'phenomenon' is the parallelogram of movements for rectilinear motion (848b11-25). As the author presents it, a motion is rectilinear, if the proportion of sides of a parallelogram constructed from points along the locus of the movement remains the same. The movement itself is understood as

[9] See in particular Aristotle's subsuming of the angle inscribed in a semi-circle under the case of triangle, so as to take advantage of the properties of the triangle in proving every such inscribed angle is right (71a19-21) and his remarks on how one knows of a particular pair that it is even (71a31-34).

[10] Aristotle says that if we grasp, without the mediation of additional terms, a cause that accounts for a present phenomenon, then we know 'that it is' (ὅτι) and 'why it is' (διότι) at the same time (*Po An* II.8, 93a35).

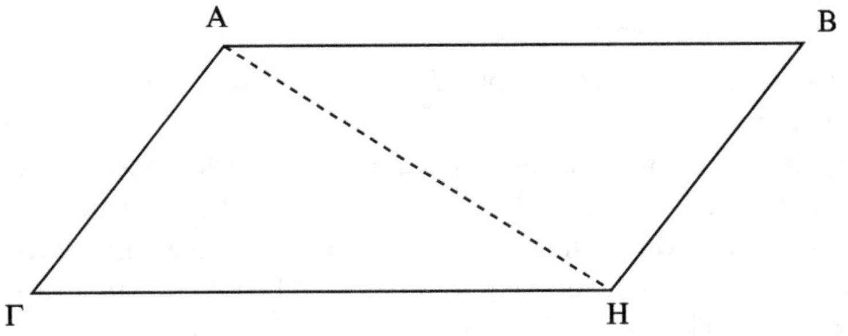

Figure 1

the diagonal (διάμετρος) of the parallelogram (Figure 1). As we shall see in an important transitional passage, the author contrasts motion along a diagonal to curved motion, as he seeks a sameness of ratio that will define circular motion in some way. He thus brings circular motion under a universal contrast of same ratio versus changing ratio within a scheme of rectilinear motion-elements that frame either a diagonal or a curve. The properties of the circle, revealed by the construction of perpendiculars and other straight lines within and around it, makes this contrast fruitful. Let me begin with the parallelogram of movements and then show how the author of *MP* explains his moving radius principle as a modification of the parallelogram of movements.[11]

[11] Although a case can be made that the argument of *MP* 1 carries a certain amount of conviction in its own terms as a demonstration in a mixed science, it cannot be said that the argument has rigor. There are a number of points where vagueness makes the modern reader doubtful about what the author intends. As a case in point, I use the term 'parallelogram of movement,' although the author uses only the term τετράπλευρον, quadrilateral, seemingly in reference to a rectangle. I use the term 'parallelogram,' because the geometry he describes is the geometry of any quadrilateral with parallel sides. In the important transitional passage treated below, however, he makes his argument in terms of the narrower case of a square whose diagonal is the motion under consideration. Fritz Krafft addresses terms used for the quadrilateral before Euclid in *Dynamische und statische Betrachtungsweise in der antiken Mechanik* (Wiesbaden, 1970), 91.

II.

Briefly put, the argument of *MP* 1 is this. Both rectilinear movement and movement in a circle, i.e., the distances covered in each case, can be characterized as a ratio of two other movements that are identified by straight lines. When these two movements are in the same ratio throughout a given time, the actual movement that takes place, M_a, is rectilinear (848b11). When their ratio varies, M_a is not a straight line.[12] The author is especially interested in arcs of a circle, for which (he claims) the ratio of component motions is different, whatever segments of the arc one compares. Whether M_a is straight or circular, however, the author is able to specify rectilinear distances, constructible in a geometrical picture surrounding M_a, which provide the ratio for the two contributing movements.[13] Accordingly, the argument, though compressed, is systematic in that it establishes a common mathematical treatment for rectilinear and circular motion. In both cases, straight lines index or represent two motions, which are unrealized separately but which nevertheless contribute to an actual motion.

There are three cases through which the author progresses in a short space:

a) motion in a straight line, whose ratio of surrounding equilateral sides remains the same (848b11-24),

---

[12] I use the symbol, M_a, for motion achieved, taking place, or projected to take place. M_a is the actual path of an object, as distinguished from the motion-elements the author supposes to contribute to M_a. This distinction is important, because the motion-elements are not realized apart from their combination in M_a. The author does not discuss how motions combine nor does he use the language of mixture. For the most part, he is careful to distinguish something moved or a line described by a moving object from the motion itself (φορά). Use of the cognate accusative, φέρεται δύο φοράς, is typical.

[13] In the case of circular motion, the continually changing ratio is meaningful because of its relation to the arc of a different but concentric circle. Because the context of explanation is the lever, the aim is to make a comparison of arc-distances on different circles. Arcs of circles covered in the same time will be those that have the same ratio of corresponding rectilinear elements for a given time taken. Accordingly, the fact that the ratio is continually changing in circular motion does not detract from the explanatory value of the account. On this point, see section III below.

b) curvilinear (περιφερές) motion, where the ratio does not remain the same (b24-34), and
c) circular motion, a more specific case of curvilinear motion (848b34-849a6).[14]

The transition from rectilinear to circular motion takes place in the passage cited in c), where the author applies the parallelogram of motions model to the chord of a quadrant of a circle.

In his initial treatment of rectilinear motion, the author says that when the ratio of the two movements is the same, an object moves in a straight line (b12). He takes a distance traveled and makes it the diagonal of the quadrilateral (σχῆμα, τετράπλευρον) whose sides are straight lines (848b12-14). Then, as in Figure 2, if A travels as

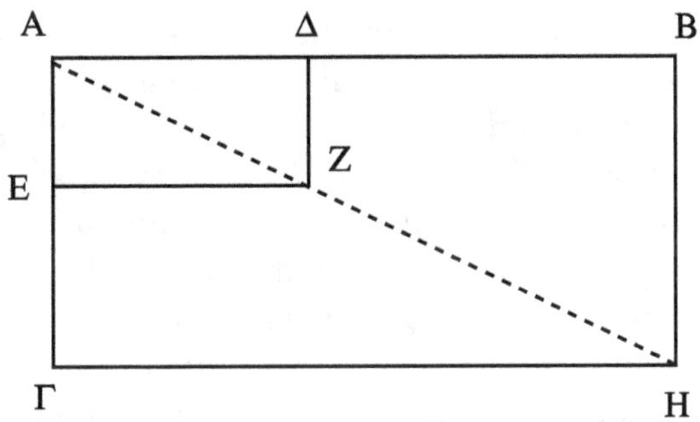

Figure 2

far as Δ in its motion along AH, it has also gone as far as E, and the ratio AΔ:AE is the same as AB:AΓ (b16-19). That the ratio of the sides meeting the diagonal remains the same all the way along the diagonal is a property of any parallelogram. The author expresses this by saying that the smaller parallelogram is similar to the larger (b19-20), and that the movement along the diagonal of the smaller is represented by a segment of the diagonal of the larger parallelogram (b21). By making these points, the author shows that it is possible to contextualize *any* rectilinear motion as the diagonal of

14) Commentators are divided as to whether the term περιφερές must refer to circular motion or the more general case of curvilinear motion. For references, see the commentary of Bottecchia Dehò on 848b34 (*Problemi*, 148).

a parallelogram. The parallelogram of movements is a kinematic generalization. The author's approach thus opens an avenue for treating any motion in terms of a ratio of element-motions contributing to it. He says it is impossible (ἀδύνατον) for motion to be in a straight line if there is no constant ratio for the sides of the constructed parallelogram (848b25-27). The first contrast the author makes, then, is between rectilinear and curved motion. Motion is curved when the two contributing motions do not form a consistent ratio for any time taken (b34).[15]

The generality of the argument he has given for sameness of ratio for any segment of a rectilinear movement makes the author sufficiently confident to offer a *reductio* argument against the proposition that any straight line could be composed from movements without the same ratio. He says that, if a line without a common ratio were a straight line, when this line is taken as a diameter and the sides of a quadrilateral are expanded or filled out around it (παραπληρωθεισῶν τῶν πλευρῶν), the moving object would be moved in accordance with the ratio of those sides (848b29-30). Then, the rectilinear distance covered both would have and would not have the same ratio for its different segments.

The terminology, παραπληρωθεισῶν, used by the author to describe construction of the sides of the parallelogram is interesting, because it is similar to terminology used by Euclid. In *Elements* I.43, Euclid speaks of parallelogram complements (παραπληρώματα) that fill out a larger parallelogram (παραλληλόγραμμον) whose diameter has been cut at a point K, as in Figure 3.[16] By drawing lines to two sides of the parallelogram from this point, two smaller parallelograms, AEKH and KGCF, are constructed within the larger

---

15) The author's language, δύο φερόμενον φορὰς ἐν μηθενὶ λόγ   μηθένα χρόνον, seems to rule out that a curve is composed of a series of very short straight lines. Any curvilinear line, whether irregular, elliptical, circular, etc., would be characterized in the same way in terms of its ratio of components *always* varying from one segment to another.

16) *Elements* I.43: "In any parallelogram the complements of the parallelograms about the diameter are equal to one another." If ABCD is a parallelogram, as in Figure 3, with diameter AC, and AC is cut at K, then EH and FG are parallelograms, and complements BK and KD are equal to one another.

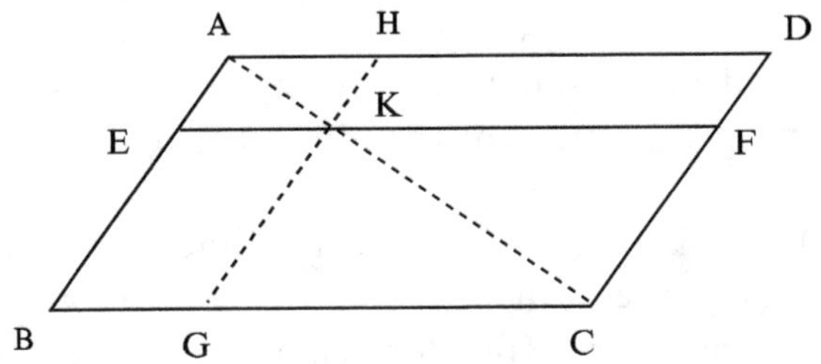

Figure 3

one. At the same time, there are created additional parallelograms, the complements EBGK and HKFD, which have areas equal to one another. The complements are, as Heath puts it, "figures put in to fill up (interstices)." Heath comments that the terminology—παραπληρώματα, complements—"implies that the complements as well as the other parallelograms are 'about' the diagonal."[17] That is, the diagonal is the focal element around which figures internal to the parallelogram are created and other lines within the parallelogram are brought into relation. Something like this—filling out a geometrical context around a diagonal—is the procedure of the author of *MP* 1 in constructing the parallelogram of movements. M_a is first placed as a diagonal, and then the sides of the parallelogram corresponding to that diagonal are instrumental to the quantification of the movement as a ratio.

The author of *MP* uses the terminology of complementarity loosely in comparison to Euclid.[18] The sides of the parallelogram are filled out, not areas, and the only smaller parallelograms considered are similar parallelograms constructed along shorter segments of the diagonal. Nevertheless, the term yields insight into the geometric method of the author and his kinematic conception of motion. Movements are only known by their distances completed. We place

---

[17] Thomas L. Heath, *The Thirteen Books of Euclid's* Elements, 2nd ed. (New York, 1956), vol.1, 340-41.

[18] The relation between *MP* and Euclid's *Elements* is a matter of conjecture. The similarities noted here are consistent with Heiberg' view (reported by Krafft in *Betrachtungsweise*, 91) that *MP* was written before Euclid had made mathematical terminology technical and consistent. The author of *MP* may have been familiar with theorems and methods that went into the *Elements*, however.

a movement, known in this way, within a geometric figure and then may take cuts of that distance, now taken as a diameter. Additional geometrical properties are brought into relief by lines related to the cut.

Having shown in this way that a rectilinear motion always has a consistent ratio, the author turns to circular motion. He says it is clear that from these things (φανερὸν ἐκ τε τούτων), i.e., from the contrast of a consistent ratio for distance covered versus a ratio that changes, *that a line describing a circle is moved with two motions*. Why is this clear? It seems that the author is thinking universally, as in *Po An* I.1, and is attempting to subsume circular motion under the contrast he has just established. If every motion either has or lacks the sameness of ratio characteristic of the sides of the parallelogram constructed around the movement, then a true circular arc must be characterizable in terms of this model of two motions and sameness or lack of sameness of ratio. So we must be able in some way to test the curve by the parallelogram of movements.

Accordingly, the author offers further support (καὶ ὅτι) for the composition of curved motion from the fact that a point borne in a straight line "arrives at a perpendicular (κάθετον), so that it is again perpendicular to the center" (849a1). This sentence has presented considerable problems of interpretation throughout the history of *MP*.[19] To understand it, the sentence must be taken together with what follows:

> That the line describing a circle moves with two motions simultaneously (φέρεται δύο φορὰς ἅμα) is obvious from these considerations, and because what moves along a straight line (κατ'εὐθεῖαν) arrives at a perpendicular, so that it is again (πάλιν) perpendicular [to a line drawn] from the center (ἀπὸ τοῦ κέντρου). Let ΑΒΓ be a circle, and let the endpoint Β [of a radius] (τὸ δ'ἄκρον τὸ ἐφ'οὗ Β) be moved to Δ; it will arrive (ἀφικνεῖται δέ ποτε) at the point Γ; if it moved in the ratio of ΒΔ to ΔΓ, it would move along the diagonal ΒΓ. But, as it is, seeing

[19] My reading follows the text of both Apelt and Bottecchia Dehò. The latter claims that the problem is not that the text is corrupt but that the meaning intended for κάθετον is not entirely clear. Κάθετον is ambiguous between 'perpendicular to' and 'tangent.' See her commentary in *Problemi*, 148-49. For more on the diversity of meanings taken for the passage, see below.

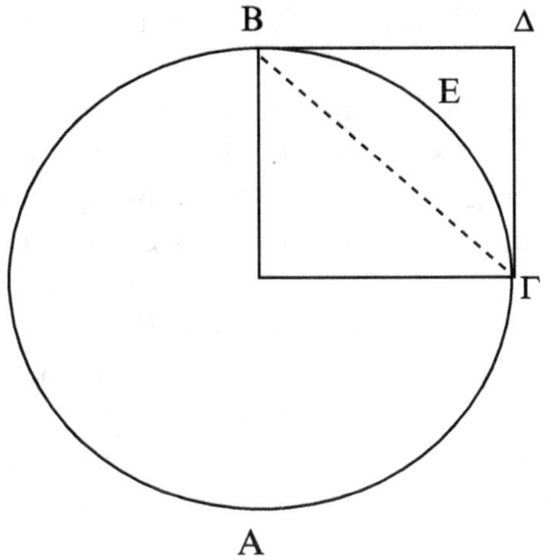

Figure 4

> that it moves in no ratio, it travels along the arc BEΓ (τὴν ἐφ' ᾗ BEΓ) (848b35-849a6) (see Figure 4).

Forster translates τὸ δ'ἄκρον τὸ ἐφ'οὗ B as "a point above the center," so that the πάλιν signifies a return to that position on the circumference.[20] This translation provides a coherent picture of motion constrained throughout a cycle of radius movement. To achieve this meaning for the passage, however, Forster omits κατ'εὐθεῖαν, when there is not a good textual basis for excising it. Furthermore, there is not a clear-cut reference in the text to motion returning to a particular point. The πάλιν may mean something else in the passage. One question has been—to which straight line does the author refer when he says, "what moves along a straight line arrives at a per-

[20] His translation reads, "... and because the point from being vertically above the center comes back to the perpendicular, so as to be again perpendicularly above the center," E.S. Forster in *The Complete Works of Aristotle*, ed. J. Barnes (Princeton, 1984), vol. 2, 1301. The movement of B is thus along the circumference of the circle, completing a full circuit. Hett keeps κατ'εὐθεῖαν and understands the passage as introducing two straight line movements that represent "a balance of centripetal and centrifugal forces," W.S. Hett, *Aristotle: Minor Works* (Cambridge: Harvard University Press, 1980), 341. Bottecchia Dehò translates the passage so as to refer to two tangents to the circle, at B and at Γ (*Problemi*, 65). Krafft takes the rectilinear movement to be from B to Δ (*Betrachtungsweise*, 25).

pendicular ..." (848b36)? Candidate straight lines are BΔ, ΔΓ, or the diagonal BΓ.[21]

It seems most likely that the straight line movement is along BΔ, since the author refers directly to this line ("...let the endpoint B be moved to Δ"). He constructs this line as a tangent with his description of the line concerned as κάθετον. From Δ, a second tangent to the circle can be constructed that touches the circle at Γ. It is possible that the author refers to this tangent, when he says, "... it [the point moving from B] arrives sometime at Γ (ἀφικνεῖται δέ ποτε ἐπὶ τὸ Γ) (849a4). On this interpretation, when he says something is again (πάλιν) perpendicular to a line from the center, he has in mind the path of a point moving along ΔΓ to regain a position on the circumference.

It is important, however, that the author uses these perpendicular lines in the way he used segments of sides of the parallelogram earlier. Sides of the parallelogram provide the measure of the motion-elements of the actually occurring motion.[22] Similarly, in the case of circular motion, the author does not suppose the point B is traveling the route of perpendicular lines. Rather, the lines BΔ and ΔΓ provide the ratio descriptive of the motion of B, if B moved along the straight line, BΓ, a chord of the quadrant. The author describes the path of B to Γ as a diameter (διάμετρον), if its ratio is BΔ: ΔΓ (849a4). In this case, the path of the moving object is straight.

It is clear, then, that the reason for drawing a line to Δ outside the circle is to create, in the style of the parallelogram of movements, a complement to this diameter, the chord of the circle. The rectilinear trajectory of a moving point to which he refers at 848b36 may be BΔ—or it may be BΓ, the actual movement accomplished along a chord which can be described by the ratio of its constructed sides, BΔ and ΔΓ. The important point for interpreting the passage

[21] It seems unlikely to be the rotating radius itself, since the phrase κατ'εὐθεῖαν refers not to displacement of a line from one location to another but to a type of line or to the path generated by movement of an endpoint. Reading ἐπ'εὐθεῖαν would not significantly change the meaning. Contrast 848b22, in the parallelogram of movements passage, where the author says, "Let the line AΓ move toward B."

[22] Compare 849a3 to 848b15-17.

is that an object moving from B—whether along the chord or the arc BEΓ—does indeed return to a tangent to the circle, i.e. a perpendicular to the radius at Γ (849a1).[23] This would be the meaning of πάλιν in the passage—returning to a tangent for the quadrant considered, not returning to the same point on the circle.

This passage, with its application of the parallelogram of movements to the chord of a quadrant, provides both continuity with the foregoing account and a sharp contrast to motion along the curve of the same quadrant. The author has thus narrowed the terms in which curved motion may be considered. The task, then, is to consider what are the geometrical complements in the case of the arc that will generate a ratio descriptive of curved motion, however much the ratio may be changing.

It is here that the author departs briefly from a kinematic account and introduces physical reasons for selection of the rectilinear elements of the circle that yield the appropriate ratio. He considers the case of two things moved by the same force but deflected in their movement by a constraint. When one is deflected (ἐκκρούοιτο) more and the other less, it is reasonable that the one deflected more moves more slowly (849a7-9). This is the case, he says, with larger and smaller concentric circles. The end of the radius is drawn more toward the center in the case of the smaller circle, because it is nearer the center (a11-12, 18-20). The author calls this 'being drawn in the contrary direction' (ἀντισπώμενον εἰς τοὐναντίον) though if we take a rectilinear tangent to the circle as the contrary, the deflection in fact operates along the normal to the projected motion. To say the nearer point is drawn more toward the center introduces a physical reason why a smaller circle moves more slowly. Indeed, the deflection toward the center is called contrary to nature (παρὰ φύσιν) (849a17), while the motion at a tangent to the circle is κατὰ φύσιν. Furthermore, the end of the shorter radius is dominated or influenced to a greater extent (κρατεῖται μᾶλλον) by the constraint toward the center (849a19).

[23] The author is concerned with an object that descends on a balance or a point on the circumference that commences moving. In the counterfactual case of that point moving in a straight line, we need not consider that the point remains the end of a radius, since a radius describes a curved path.

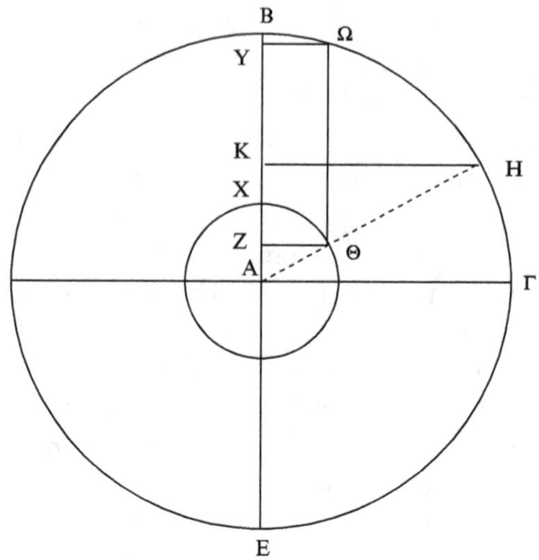

Figure 5

I will have more to say later about the idea of constraint toward the center. For the moment, note that he explains what constraint means by a return to a kinematic account (849a36-b1). The argument that follows is the heart of the demonstration of the principle of the lever and it has two parts, both of which turn on discerning *which* ratios of rectilinear elements constructible within concentric circles form a proportion of contributing motions for circular motion. In the first part (849a29-b7), it is shown which rectilinear elements of circles being compared in fact cannot enter into a proportion. In the second part (849b7-18), it is shown which elements do have the same ratio.

In Figure 5, X and B are endpoints of radii for smaller and larger concentric circles respectively. For arcs on concentric circles swept out by a single moving radius, the component of natural motion is represented by a perpendicular (κάθετος) dropped from the completion point of the arc movement to the radius taken at the starting position of the moving object—so, for the smaller circle, for instance, from Θ on the circumference to Z on the vertical radius. As representing a rectilinear component of arc-movement, this distance could also be a tangent to the smaller circle at X. As a tangent, it would more readily model the force or push exerted at a point on a wheel or at the end of a balance.[24] This perpendicular ZΘ divides the radius of the smaller circle into two segments, AZ and ZX. The segment ZX will turn out to be the measure of constraint exerted, but the author does not point this out immediately. Instead, he shows that there can be a natural component of motion for the larger circle, ΩY, that corresponds to ΘZ.

Being equal, the lines ΘZ and ΩY are the same component of natural motion contributing to some arc-movement on their respective circles. YB and its correspondent on the smaller circle, ZX, are taken to represent the amount of constraint undergone by B and X in moving to Ω and Θ respectively when moved with the same natural motion. The author says B is drawn less toward the center. The reason he gives, however, is simply geometrical: "For (γάρ) equal lines extended from unequal circles perpendicularly to the diameter, cut a shorter segment of the diameter for the larger circles" (849a36-b1). The construction shows that the effect of constraint is less for the larger circle, because YB is shorter than ZX (849a36). The author takes this to be the reason that BΩ and XΘ will not be distances covered in the same time. When the constraint is less, a greater distance will be covered in the same time. If this is a dynamic argument, it assumes what is to be proven. The idea of constraint seems here to contribute to ruling out a length as suitable for the ratio sought.

The comparison of YB and ZX is itself only mediate to the author's aim, however, and here the second part of the argument begins. The problem is which rectilinear components of the circle provide a ratio descriptive of actual motion and of arc-movements covered in the same time. If BΩ and XΘ are not arcs covered in the same time, this is because the motion κατὰ φύσιν is equal but the component παρὰ φύσιν is not the same. So there is no proportion between the ratios of these motion-elements for each circle. In fact, in the time a point moves along the arc XΘ on the smaller circle,

---

24) The kinematic account in *MP* does not rule out the mention of force (ἰσχύς or δύναμις), understood as a push, impulse, or initial impetus. In the demonstrative picture of argumentation outlined in section I, *explanans* and *explanandum* are drawn from the same universe of objects of ordinary experience. Accordingly, forces are bodies or attributes of bodies. This topic cannot be fully developed here.

a point moving along the circumference of the larger circle will have reached H. So, the appropriate motion-element entering into the ratio for the larger circle will be KH. And the motion-element of constraint is appropriately larger too. It is KB. What is interesting about this account is that constraint, as a cause of motion, has now effectively ceased to be an issue of interest, since even a faster movement has a larger component of constraint. The gist of the argument remains the taking of ratios of lengths. What matters is choosing motion-elements, representable as straight lines in some geometrical framing scheme, that will yield a sameness that can be tracked across simultaneous times.

The claim for a proportion across arcs for ratios that vary continually is the heart of the author's argument and is not unconvincing taken in its own terms. The segment KB of the diameter AB, cut off by dropping a perpendicular to AB from the end of an arc, approximates the difference between a rectilinear motion originating at B and the curved motion BH that in fact is accomplished. It accounts for the element of curvature toward the center. The perpendicular KH represents another contributor to arc-movement, the product of a force exerted at the circumference, the so-called natural motion from B. KH has been constructed by a method similar to that of *Elements* I.43, drawing lines from a point on the actual path of movement to the figure constructed around that path. The measures of constraint—ZX, YB, and KB—are like complements too, lines filled out around the actual path of a point along the circumference. They are not arbitrarily chosen as the measure of constraint but neither are they actually proven to be the appropriate measures. They are constructible as indirect measures of arc movement in each case. By the means described, the author has extended the proportionality characteristic of distances along a rectilinear motion to the case of the circle.

III.

There are two oddities about the central argument of *MP* 1 that make the connection between the parallelogram of movements and the circle less than obvious. In the first place, the sameness of ratio put forward by the author for a rectilinear motion applies to a single motion. The fact that proportionality in the case of an arc applies to a comparison of arcs instead seems to make it quite different from the original parallelogram. Furthermore, in the case of the circle, proportionality is brought into play to account for the action of a lever. Accordingly, the central demonstration by proportionality appears only loosely connected to the analysis of rectilinear motion. It seems *ad hoc*, and over-specialized or lacking generality. It seems to me true that the argument is targeted to the lever and, furthermore, that the beginning of mechanics in *MP*, viewed retrospectively, is limited in its being tied to the geometry of the circle. Section II has shown, however, that the move from analysis of a single motion to the comparison of arcs is based on a geometrical procedure that links the two accounts and increases the explanatory power of the parallelogram of movements.

The second apparent oddity runs deeper but along the same line as the first. In the analysis of motion in the text, it is clear that the neuter, τὸ φερόμενον, moving thing or what is moved, refers to τὸ ἄκρον, the extremity of a line.[25] In both the parallelogram of movements and the moving radius, the path of a moving object is conceived as the 'so far completed' path of an endpoint of a line drawn from a point of origin. In rectilinear motion, however, the moving endpoint extends the line of movement; in circular motion, the radius is constant and the endpoint describes a curved path around the point of origin. This seems like a big difference between the two cases. In *Physics* VIII.9, Aristotle had already noticed this difference and tended to minimize it for the purposes of analysis. He describes rectilinear motion in terms of the distance of a moving object from an origin (ἀρχή)(265a30 and b14). In this passage, he treats the circle as a special case in which the origin, the center, is not itself on the line of movement (265b3-8). Nevertheless, he regarded the center as part of the circle (b2-4).[26] This disanalogy in the relation of the moving endpoint to its origin did not

---

25) See for instance 849a3, a12.

26) See also *Physics* I.1, 184b3, where a circle is said to be known clearly when resolved into the particular elements that are parts of its definition.

prevent Aristotle from pointing out the many traits that rectilinear and curved motion share, and the author of *MP* seems to proceed on the same basis.

I have argued that the demonstration of the principle of the lever in *MP* 1 is best understood as a systematic kinematic argument and that it is questionable how much the addition of dynamic notions contributes to the force of the argument. Indeed, from a logical standpoint, the account involving physical notions like constraint is more than the author needs to construct an explanation. Aristotle had already enunciated the principle that a point further from its origin moves more quickly. He says this about an object moving in a straight line and makes a similar point about circular motion in the heavens.[27] A larger circle moves faster than smaller ones bound to the same center. The objection could be made that, to use this principle of Aristotle's, instead of a notion of constraint toward the center, would be a *petitio principii*, since the author of *MP* is attempting an explanation of just this principle. I have argued, however, that he is not so much resolving the phenomenon of the lever to other principles as he is making it a proper accident of the geometry of the circle taken as a whole.

To place emphasis on force or constraint as the explanatory element in the account strains the credibility of the text itself. How would constraint located at the center of the circle aid a physical explanation of the lever? First, how does a point, the center of the circle, constrain motion, since points have no power?[28] Surely, it is something about the rigid beam of the lever that does the constraining. If we geometrize the beam as a radius with points, why not consider any point along the beam as an origin of constraint? The answer, of course, is that the author's account continues to be grounded in a geometrical treatment of a circle with one center. But in that case, again, the natural direction of explanation would proceed *from* a correlation—centripetal force as a function of dis-

---

[27] See *Physics* VIII.9, 265b11-15 and *DC* II.8, 289b34-90a5.

[28] Aristotle says that the poles of the earth do not have power because they are points. See *MA* 3, 699a20-21. Additionally, in *DC* III.1, 299a27, points are said to have no weight.

tance from the center--*to* the fact and degree of constraint, rather than the other way around.[29]

My aim is not to contend that the dynamic notions of *MP* are without significance but rather to make a case for analyzing them in a way that would be compatible with the kinematic argument that remains firmly grounded in the circle. A revealing comparison text is found in *Physika Problemata* XVI, where the Aristotelian author considers why an unbalanced body, when thrown, rotates along its forward path (913a34-b4). The author conceives of the object as heavier and lighter parts set opposite one another, as if at the ends of a rod. Each part is inclined to move at its own pace, he says. But since each is constrained by its physical continuity with the other, the unbalanced object must travel in a circle. This is because "it is by this shape alone that these points always opposite one another (ταῦτα ἀεὶ κατάλληλα ὄντα σημεῖα) can cover unequal lines in the same time" (913b4-5). In this passage, it seems that it is the rigidity of the continuous body overall that constrains movement (913b1). The constraint is not located in a point. The moving radius principle is nevertheless integrated into the account as providing the only path for motion that will be compatible with the physical constraints of the rigid unbalanced body. So, some version of the principle taken as a single meaningful proposition is the explanation of the phenomenon. Even though this explanation in *Problemata* XVI remains well within the range of the circle's remarkable features, it is a promising beginning for combining geometrical properties with the physical reality of constraint. It would be worthwhile bringing this passage and others like it to bear on the relation of kinematics and dynamics in *MP*.

Grasping the prevalence of kinematics in *MP* 1 makes clearer, however, that the movement of thought in antiquity from wonder to science did not mean a shedding of the moving radius principle as an explanation. This principle was scientifically fruitful, taken as whole and without decomposition into dynamic components of a theoretical nature. Force in the passage examined is not a theoret-

---

29) Aristotle takes this approach, speed a function of distance, in the passages from *DC* II.8 and 10 already cited.

ical entity but connotes the pushes, pulls, and exerted pressure of ordinary experience. The moving radius principle helped to advance the science of motion from accounts dependent on weight to explanations of the behavior of rigid bodies as such.[30]

30) I am grateful to Edith Sylla for comments on an earlier version of this paper.

Structures of Argument and Concepts of Force in the Aristotelian *Mechanical Problems*

Mark Schiefsky*
Harvard University

Abstract
I argue that the main goal of the *Mechanical Problems,* a short treatise transmitted in the *Corpus Aristotelicum,* is to explain the working of technology in terms of the concepts of Aristotelian natural philosophy. The author's explanatory strategy is to reduce the thirty-five "problems" or questions that he discusses to one or more of three simple models: the circle, balance, and lever. The conceptual foundation of this reduction program is a principle concerning circular motion, viz. that a point on the circumference of a larger circle moves more quickly than one on a smaller circle, assuming that the circles turn about the same center at the same angular speed. I analyze the author's argument for this principle and his application of it throughout the text, especially to the analysis of the lever. The main conclusions are (1) that the author's justification of the circular motion principle is based on an innovative geometrical analysis of motion, not on a highly theoretical conceptualization of force; and (2) while the author is aware of a reciprocal relationship between weights and distances from the fulcrum in the case of the lever, his explanation of this fact makes no reference to the conditions for static equilibrium.

* Department of the Classics, 204 Boylston Hall, Harvard University, Cambridge, MA 02138, U.S.A. (mjschief@fas.harvard.edu). This paper owes its existence to a kind invitation from Peter McLaughlin to participate in a panel at the History of Science Society annual meeting in November 2007; I am grateful to all who participated in that event for their comments and questions. My approach to the *Mechanica* has been shaped by many discussions with colleagues in Department I of the Max Planck Institute for the History of Science in Berlin, especially Peter Damerow, Malcolm Hyman, and Jürgen Renn; I hope that they may find something of interest in the final product. I am also grateful to Bill Newman and Edith Sylla for their assistance and patience, and to Jean de Groot for sharing a draft of her paper with me. Finally it is a pleasure to dedicate the paper to John Murdoch, as a small token of appreciation for everything I have learned from him over the years and for his unstinting support.

Keywords
mechanics, technology, nature, natural philosophy, force, power, weight, reduction, models, lever, balance, circle, circular motion, steelyard

1. Introduction

The *Mechanical Problems* or *Mechanica*, a short treatise transmitted in the *Corpus Aristotelicum*, is a work of the first importance for the history of science. It consists of an introduction emphasizing the wondrous behavior of mechanical devices and arguing that that behavior can be traced back to the properties of circular motion; this is followed by a discussion of thirty-five "problems" or questions that are principally concerned with the operation of various technological devices. Attributed to Aristotle in antiquity and the early modern period, the *Mechanica*'s rediscovery in the Renaissance spurred the composition of an extensive commentary literature.[1] There is a direct historical link between the *Mechanica* and the emergence of classical mechanics in the work of Galileo and his contemporaries.

While historians of mechanics have generally recognized the importance of the *Mechanica*, they have not always interpreted the text on its own terms. The author's approach has typically been viewed as representative of a "dynamical" tradition, since it makes use of conceptions of motion and force, in contrast with the "statical" tradition that eschews considerations of motion and traces its origin to Archimedes' *On the equilibrium of plane figures*. Such a division is fundamental to the most thorough and incisive recent treatment of the *Mechanica*, F. Krafft's *Dynamische und statische Betrachtungsweise in der antiken Mechanik* (Wiesbaden, 1970). But the assumption of a division between statics and dynamics is a highly problematic starting point for analyzing the history of ancient mechanics, if for no other reason than that no ancient writer distinguishes explicitly between the two approaches. Even in the Renaissance, when such

1) The fundamental bibliographical survey of this literature remains P.L. Rose and S. Drake, "The Pseudo-Aristotelian *Questions of Mechanics* in Renaissance Culture," *Studies in the Renaissance*, 18 (1971), 65-104.

a distinction does emerge, writers tend to emphasize the points of agreement between the Aristotelian treatise and Archimedes' approach.[2] Stimulated in part by the question of Aristotelian authorship, scholars have tended to view the argument of the *Mechanica* against the background of Aristotle's scattered remarks on mechanics in works such as the *Physics* and *De caelo.* Thus Duhem, in his *Les origines de la statique,* explained the author's argument for the "law of the lever" (the inverse proportionality of weights and distances on the lever arm) as a direct application of the statements of proportionality between mover, moved, distance, and time found in *Physics* VII 5.[3] Contrary voices have been raised, but generally not by historians of mechanics.[4]

In this paper I will attempt to sketch out a more satisfactory interpretation of the *Mechanica*'s methodology and conceptual foundations. My starting point is the claim that the author's principal concern is to provide a theoretical explanation of the working of technology. The *Mechanica* thus reflects a distinction between theoretical mechanics—a body of knowledge consisting of relations between concepts such as weight, force, and motion—and practitioners' knowledge, i.e. the knowledge acquired in the design and use of technology. That the first text concerned with mechanics in the Greek tradition aims at the explanation of practitioners' knowledge is an important reminder that technology in Greco-Roman antiquity often developed in advance of theory. A striking example of this is provided by the steelyard or balance with unequal arms, the topic of problem 20 in the *Mechanica.* The existence of the

---

2) See e.g. the opening of Guidobaldo del Monte, *In duos Archimedis aequeponderantium libros paraphrasis* (Pesaro, 1588), esp. 4: *Quare Archimedes Aristotelem sequi videtur ...*

3) P. Duhem, *Les origines de la statique,* 2 vols. (Paris, 1905 and 1906), 1:5-6; 2:292-293.

4) H. Carteron offered a trenchant critique of Duhem in the introduction to his *La notion de force dans le système d'Aristote* (Paris, 1923; Eng. trans. as "Does Aristotle Have a Mechanics?" in J. Barnes et al., eds., *Articles on Aristotle,* vol. 1 (London, 1975), 161-174), though his main interest was not mechanics as such. The best recent critique of the Duhemian line is F. de Gandt, "Force et science des machines," in J. Barnes et al., eds., *Science and Speculation: Studies in Hellenistic Theory and Practice* (Cambridge, 1982), 96-127.

steelyard is reliably attested for the late fifth century BC by a passage of Aristophanes' *Peace* (performed in 421 BC, some 100 years before the *Mechanica* was written if not more).[5] It has often been assumed that the construction of the steelyard implies knowledge of the "law of the lever," i.e. the claim that static equilibrium requires the inverse proportionality of weights and distances from the fulcrum or balance point. But the fact that the use of the steelyard precedes by at least a century the first written attempt to explain its operation makes it seem quite likely that it was developed without knowledge of such general theoretical principles.[6]

2. Problems and Explanation

The *Mechanica* begins with some general reflections on the relationship of art or *technê* (the Greek term can refer to any kind of result-oriented expertise) and nature:

> Among things that occur according to nature (*kata phusin*), we wonder at those whose cause is unknown; among things that occur unnaturally (*para phusin*), we wonder at those that come about by means of art (*technê*) for the benefit of human beings. For in many cases nature acts in a way opposed to what is useful for us. For nature always acts in the same way and simply, while what is useful changes in many ways. Whenever, then, it is necessary to do something unnatural (*para phusin*), because of the difficulty we are at a loss (*aporia*) and have need of art (*technê*). For this reason, we also call that part of art that assists in such difficulties (*aporiai*) a device (*mêchanê*). For as the poet Antiphon said, so it is: "By means of art we gain mastery (*kratoumen*) over things in which we are conquered by nature." (847a11-21)[7]

---

5) Aristophanes, *Peace* 1240-1249, with the comments of S.D. Olson, *Aristophanes' Peace: Edited with Introduction and Commentary* (Oxford, 1998), 304. For the *Mechanica* a date in the late fourth or early third centuries BC seems the most likely possibility, though certainty is impossible.

6) For a modern parallel see J. Renn and M. Schemmel, *Waagen und Wissen in China: Bericht einer Forschungsreise* (Preprints of the Max Planck Institute for the History of Science 136) (Berlin, 2000).

7) Translations from the *Mechanica* are my own, based on the text of M. Bottechia, *Aristotele MHXANIKA* (Studia Aristotelica 10) (Padua, 1982). Unfortunately her edition includes no figures and does not number the individual problems; I follow the

The notion that *technê* can rescue human beings from a situation of need or helplessness (*aporia*) by means of a device or strategem (*mêchanê*) has a long history in Greek thought, and is brilliantly summarized in the quotation from the poet Antiphon that crowns the passage: *technê* enables human beings to overcome their natural disadvantages. The emphasis on the "unnatural" (*para phusin*) character of what is produced by *technê* has often been taken to suggest that the author conceives of the art-nature relationship as a fundamentally antagonistic one. In fact, however, the primary contrast between art and nature here is between the complexity and utility of the former as against the simplicity and regularity of the latter. The phrase *para phusin* has a number of senses in Aristotle, including (1) "beyond nature" in the sense of outside the order of nature, i.e. supernatural; (2) "contrary to nature," i.e. opposed to a thing's natural tendencies; (3) "forced" as opposed to unconstrained; and (4) "unusual" as opposed to what is natural or normal. I have argued elsewhere that the author's conception of the art-nature relationship is best understood as the idea that art goes "beyond nature" in the sense that it brings about results that nature left to its own devices does not. Insofar as art "masters" nature, it is by the creative combination of natural principles to go beyond what nature does on its own.[8]

The introduction continues by specifying the nature of mechanical problems more precisely. They are cases in which "the lesser master the greater, and things with small inclination (*rhopê*) move great weights (*barê*)" (847a21-23). Mechanical problems occupy an intermediate position between physics and mathematics: their subject matter (*to peri ho*) is revealed by physics, while their explanation (the 'how', *to hôs*) is made clear through mathematics (847a24-28).[9]

---

traditional designation as found in the editions of O. Apelt (Leipzig, 1888) and W.S. Hett (*Aristotle: Minor Works* [London and Cambridge, MA, 1936]).

[8] For the idea that art completes what nature leaves unfinished see Aristotle, *Physics* II 8, 199a8-20. I develop an interpretation of ancient mechanics along these lines in M. Schiefsky, "Art and Nature in Ancient Mechanics," in B. Bensaude-Vincent and W. Newman, eds., *The Artificial and the Natural: An Evolving Polarity* (Cambridge, MA, 2007), 67-108.

[9] This remark recalls Aristotle's notion of subordinate sciences, according to which mathematics supplies the explanatory element in disciplines concerned with the

Next, the author mentions the lever as the paradigm example of a mechanical problem:

> Among the puzzles in this class are those that concern the lever (*mochlos*). For it seems strange that a great weight (*baros*) is moved by a small force (*ischus*), and that, too, where a greater weight is involved. For the very same weight, which a man cannot move without a lever, he quickly moves by taking in addition the weight of the lever. (847a28-b15)

The key feature of mechanical problems is thus that they seem to violate the natural or normal relationship between force (*ischus*) and weight (*baros*). Normally, a large force is required to move a large weight, but the lever makes it possible to do so with a small force.

That the author's aim is to explain why this is so is made clear in the remainder of the introduction, which argues that the "original cause" (*tês aitias tên archên,* 847b16) of all mechanical phenomena lies in the remarkable nature of the circle. The circle unites opposites such as concave and convex, as well as opposite motions, forward and backward (since a point on a revolving circle that sets out in one direction ends up back at its starting point); but its most remarkable feature is that

> although the line from the center is one, none of the points on it moves at the same speed (*isotachôs*) as any other, but that which is farther from the fixed extremity always moves more quickly (*thatton*). (848a15-17)

That is, as a radius (the "line from the center") sweeps out a circle, a point on the radius that is farther from the center moves quicker than one that is closer in. The final part of the introductory section (848a19-37) describes a mechanical device whose operation depends on the fact that the circle combines opposite movements. It consists of a number of circles in contact with one another like a set of interlocking gears, so that when the first one rotates it

physical world (*Metaphysics* 1078a14ff.; *Posterior Analytics* 75b14 ff., 76a23-25). I discuss these passages in Schiefsky, "Art and Nature," 90-94.

causes the second to turn in the opposite direction, and so on. The section concludes with the following important remark:

> And so craftsmen, seizing on this natural property (*phusis*) of the circle, construct an instrument by concealing the principle (*archê*), so that only the wondrous character (*to thaumaston*) of the device (*mêchanêma*) is apparent, while its cause (*aition*) is unclear. (848a34-37)

Here several fundamental concepts of Aristotelian natural philosophy—nature, principle, and cause—are associated with one another and opposed to wonder: the operation of the device is wondrous only if the causal principle (*archê*) that explains it is concealed. Moreover the device is constructed not by working against the nature (*phusis*) of the circle but by making creative use of it. Mechanics, no less than any other *technê*, demands knowledge of natural principles.

The details of the author's explanatory program will concern us in the next section. But it should now be clear that his overall aim, as announced in the introduction, is to give an explanation of the working of technology in terms of the concepts of Aristotelian natural philosophy. The existence of a text like the *Mechanica* makes good sense against the background of the development of technology in the Greek world of the fifth and fourth centuries BC. I have already mentioned the steelyard, whose operation vividly demonstrates the ability of a small weight to 'overpower' a large one. But this is only one example among many. The surgical works of the Hippocratic Corpus refer to the use of machines for setting fractures and dislocations and reflect on their ability to achieve great effects.[10] The development of torsion artillery at the end of the fifth century provided a dramatic example of the possibilities of enhancing human strength by mechanical contrivance, and of the relevance of technology to military and political affairs.[11] No less important

10) [Hippocrates] *De articulis* (*On joints*) 72; *De fracturis* (*On fractures*) 31: "For of all the instruments devised by human beings, the strongest are these three: the turning of the winch, leverage, and the use of the wedge."

11) For this dating and for a general account of the development of torsion artillery see E.W. Marsden, *Greek and Roman Artillery: Historical Development* (Oxford, 1969).

than the rise of technology, however, is that of natural philosophy: for the "problems" generated by technology are only problems when viewed in light of a conception of what is natural or normal. The explicit formulation of relationships between such concepts as force and weight is thus also an important part of the background to the *Mechanica*. To invert Aristotle's famous dictum, the *Mechanica* shows that in an important sense, philosophy is the starting point of wonder.

3. The Reduction Program

The basic explanatory strategy of the treatise, announced in the introduction, is to "reduce" (*anagein*) each problem to the lever, the balance, or the circle:

> Now the things that come about with the balance are reduced (*anagetai*) to the circle, those that come about with the lever to the balance, and practically everything else that comes about with mechanical movements to the lever. (848a11-14)

The notion of reduction as the analysis of a complex entity into its simpler components is familiar from a number of Aristotelian contexts. In the *Prior Analytics* Aristotle claims that all syllogisms can be reduced to the first figure (29b1-2, 40b17-22, 41b3-5). In *De generatione et corruptione* II 2 (329b15-330a29) he claims that qualities such as hard and soft can be reduced to the four primary qualities of hot, cold, wet, and dry; the idea seems to be that the derivative qualities can be understood as kinds or forms of the primary ones. Compared to these somewhat obscure examples, the reduction program in the *Mechanica* stands out for its clarity and simplicity. The reduction is carried out by identifying correspondences between the device or situation in question and the lever, balance, or circle. Each of these three conceptual schemata or models has a fixed number of components or 'slots' designated by a standard terminology, as specified in the following table:

Model	**Slots**
circle	center (*kentron*), radii (*hai ek tou kentrou*)
balance	beam (*zugon*), cord (*spartion*) (the beam is suspended from the cord)
lever	lever (*mochlos*), fulcrum (*hupomochlion*), weight (*baros*), mover (*ho kinôn*)

While the terminology for the slots is quite consistent, each slot can be filled by different objects depending on the particular situation. Thus the 'mover' slot in the lever model can be filled by a weight (*baros*: problem 3), a power (*dunamis*: problems 5, 6), or a person (e.g. a sailor or workman: problems 5, 29).

As an example we may consider problem 4, which asks why rowers in the middle of a boat contribute most to its movement. The situation is analyzed using the lever model (references to the slots of the model are italicized):

> (a) Why do those in the middle of the ship contribute most to its movement? (b) Is it because the oar is a *lever*? For the thole-pin becomes a *fulcrum* (for this is stationary), and the *weight* is the sea, which the oar pushes away; the *mover* of the *lever* is the sailor. (c) And the *mover* of the *weight* always moves more *weight*, the more distant it is from the *fulcrum*; for in this way the radius is greater, and the thole-pin, being a *fulcrum*, is a center (850b10-16).

Slot	**Identification**
lever (*mochlos*)	oar
weight (*baros*)	sea
fulcrum (*hupomochlion*)	thole-pin
mover (*ho kinôn*)	sailor

The picture is that of a ship that is wider in the middle than at the bow or stern. In such a vessel, there is a greater length of oar inside the ship for a rower amidships than for one in the bow or stern; hence the greater mechanical effect achieved by the former. The order of exposition here is typical: the question (a) is followed by an identification of the slots of the relevant model (b), followed by an explanation (c) that gives the reason why the model in question accounts for the phenomenon. The questions (a) are normally

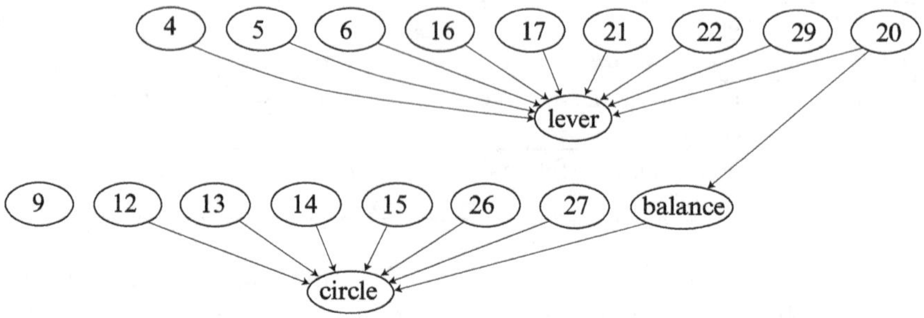

Figure 1: Reduction map

introduced by the phrase *dia ti* ('Why'), and the identifications (b) by the phrase *ê dioti* ('Is it because'). The model identification is not always so straightforward, and there is substantial variation in the structure of particular problems. But the general pattern of question, identification, and explanation is characteristic of the author's approach.

By analyzing the model identifications for each problem it is possible to build up a picture of the author's overall procedure, represented in graphical form in figure 1. The graph includes all problems for which there is an explicit model identification in the text, with the exception of problem 1 (where the balance is reduced to the circle) and problem 3 (where the lever is reduced to the balance). Of the 35 problems, 18 (including 1 and 3) meet this criterion. In the first few problems the author makes good on his promise (848a11-14) to reduce the balance to the circle (1), the lever to the balance (3), and all other mechanical phenomena to the lever (4-6). But this does not describe his procedure throughout: a significant number of problems are reduced directly to the circle. Moreover, the graph reveals that the balance plays a relatively unimportant role in the explicit model identifications. Aside from problem 3 (the lever itself) the balance is explicitly invoked only in the case of the steelyard (problem 20), which is identified with both the balance and the lever.

Several of the problems that lack explicit model identifications are nonetheless closely linked to the circle or lever models. First there are two cases of indirect reduction: problem 7, which refers back to the analysis of the rudder as a lever in problem 5, and problem 19, which refers to the analysis of the wedge as a lever in 17. Problem 8, "why are round and circular figures easier to move," draws heavily on the analysis of the circle in problem 1. In problem 18, the author claims that the compound pulley operates in a way analogous to the lever (853a37-b2); but this seems to mean only that it produces a large effect with little input, since the slots of the lever model are not identified. Problems 2 and 10 deal with aspects of the balance but without reducing it to the circle; problems 11 and 35 concern circular movement, but again there is no explicit identification of model slots. Problems 23 and 24 explore conceptual difficulties raised by the analysis of problem 1 (the composition of motions in the former case and the differential rotation of larger and smaller circles in the latter, the famous "wheel of Aristotle"). That leaves seven problems (25, 28, and 30-34) which have no connection with the circle, balance, or lever. Most of these, however, concern either technology (25, 28) or phenomena that are somehow problematic for Aristotelian natural philosophy, such as projectile motion (32-34) or the tendency of objects in motion to continue moving (31).

Thus there is a fair degree of coherence in the selection of problems. Approximately half contribute directly to the author's project of explaining technology by reference to circle, balance, and lever. Most of the rest have some connection with the basic models; only a few stand entirely outside the explanatory program announced in the introduction. Even so, they arguably have a place in a collection of problems concerned with phenomena involving forced motion that are *prima facie* difficult to explain in the context of Aristotelian philosophy.

To sum up, the reduction program of the *Mechanica* is a remarkable specimen of analytical procedure—a movement from the complex to the simple—as contrasted with the synthetic procedure, or movement from first principles to conclusions, that is typical of many Greek mathematical texts. The distinctive feature of the author's method is the identification of the circle, balance, and lever models; the structure of these models, expressed in a consistent terminology, guides his analysis throughout. I turn now to a closer examination of the conceptual foundations of the two most important models, the circle and lever.

4. The Circular Motion Principle

In keeping with the importance of the circle in the reduction program, the fundamental explanatory principle in the text concerns circular motion. It is introduced at the beginning of problem 1, which aims to explain the alleged fact that larger balances are more accurate than smaller ones:

> First of all a puzzle arises concerning the things that happen with the balance: for what reason (*aitia*) are larger balances more accurate than smaller ones? The principle (*archê*) of this is: why, in the circle, does the line that extends more from the center travel more quickly (*thatton*) than the smaller line which is close to the center, and moved by the same force (*têi autêi ischui*). (848b1-5)

At first this formulation might seem to differ from the author's earlier statement at 848a15-17: "although the line from the center is one, none of the points on it moves at the same speed (*isotachôs*) as any other, but that which is farther from the fixed extremity always moves more quickly (*thatton*)." But in fact the idea is the same. By "the line that extends more from the center" and "the smaller line which is close to the center" the author refers to two radii that sweep out concentric circles. He goes on to explain that "more quickly" means covering either the same distance in less time, or a greater distance in the same time, and that in the present case it is the latter sense that is relevant: the greater line traces out a greater circle in the same time (848b5-9). Thus, it is a basic assumption of the argument that the longer and shorter radii complete their motion in the same time. The upshot is that the principle should be understood as referring to two points on a single radius that traces out a circle, as at 848a15-17. What is new here, of course, is the reference to the fact that both points are moved "by the same force (*ischus*)"; the puzzle is that the same force yields different effects, depending on where it is applied along the radius.

Most of problem 1 is devoted to justifying this principle; the application to the balance follows only at the end (849b23-850a2). The goal of the argument can be understood by reference to figure 2 (based on the edition of Hett (above, n. 7)). We have two concentric circles ΜΞΧΝ and ΕΔΒΓ, and a radius AB which rotates over the arc BH in a given time so as to coincide with AH. Point

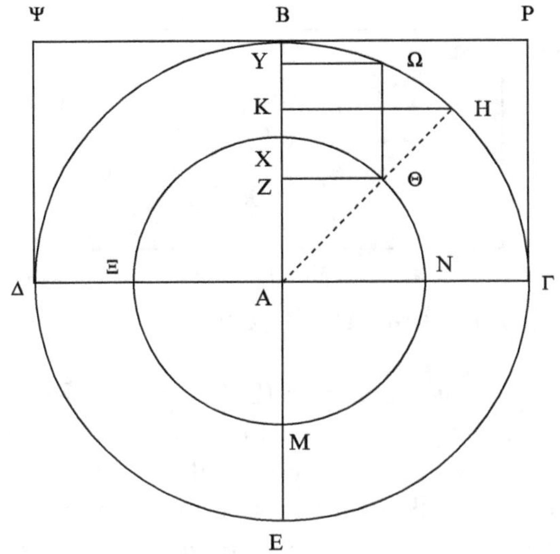

Figure 2: Circular motion in the *Mechanica,* Problem 1

B travels faster than point X, since it covers a greater distance in the same time (arc BH > arc XΘ). Points B and X are moved "by the same force (*ischus*)." The goal of the argument is to explain why, despite this, the points travel at different speeds.

The argument has three main parts. In the first, which runs from 848b9 to 849a6, the author presents a method of analyzing motion into components and argues that it can be applied to circular motion. In the graphical representation of this analysis, the lengths of lines represent distances covered in a given time. Figure 3 (also based on Hett's) illustrates the case of straight-line motion in which the two components are represented by AB and AΓ, and the object moves along the diagonal of the parallelogram formed by them (848b10-23). In such a case the ratio (*logos*) between the two components is constant for the time period in question. In the time it takes A to cover the distance AB, it also covers distance AΓ (and so ends up at H); similarly, in the time that A covers AΔ it also covers AE, where AΔ:AE is equal to AB:AΓ (thus A ends up at Z).[12] In the case of motion along a curve, the ratio between the

[12] From a modern point of view it is natural to take the lengths as representing the average speed of motion over a given time interval, but it should be recognized that

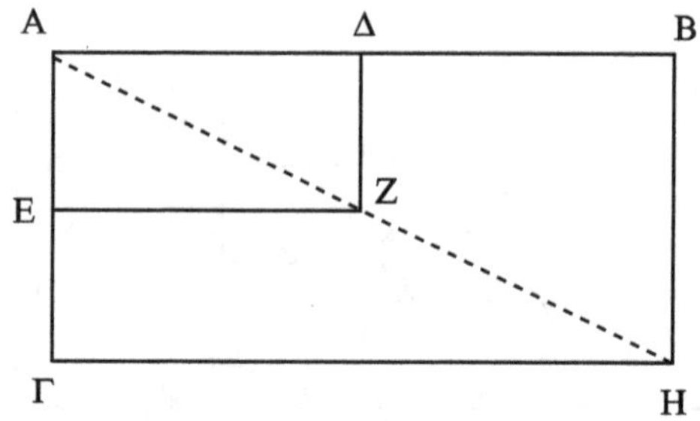

Figure 3: Rectilinear motion in the *Mechanica,* Problem 1

components is constantly changing (848b23-35). Using this method, circular motion is analyzed as consisting of one component directed to the center and a second at right angles to it. Thus in figure 2, in the time that point B moves to H along the circumference, the two components of its motion are represented by lines BK and KH: the distances covered in the vertical and horizontal directions, respectively.[13]

The second stage of the argument runs from 849a6-19. Having established that the line that traces out the circle "is carried with two motions" (*pheretai duo phoras*, 848b9-10, b36), the author states the following principle relating force and speed, which I will refer to as the "deflection principle":

> If, of two things moved from the same force (*apo tês autês ischuos*), one is deflected (*ekkrouoito*) more and the other less, it is reasonable that the one that is deflected more will move more slowly (*braduteron*) than the one that is deflected less. And this seems to happen with the greater and the lesser of the lines from the center that trace out the circles. For because the extremity of the lesser is closer to the fixed (center) than that of the greater, the extremity of the lesser travels more slowly

nothing like this is stated in the text. The assumption of a constant ratio between the components of motion does not imply that the object's speed is constant; it implies only that both components change in the same way. It is also worth remarking that the diagram (fig. 3) is quite underdetermined by the text. Nothing in the text establishes that the figure is a rectangle, nor is anything said about the exact ratio of AΔ to AB.

[13] On this section of the argument and on the problematic but crucial passage at 848b35-849a6, where the method of analyzing motion into its components is first applied to the circle, see the contribution of J. de Groot to this volume.

> (*braduteron*), as though drawn back in the opposite direction, toward the middle. (849a6-14)

Note that there is no claim here of a quantitatively precise relationship between force and speed. The idea is simply that, if two things are impelled with the same force and one is deflected or turned aside more than the other, the one that is deflected more will move more slowly (i.e. take more time to cover the same distance).[14] Evidently force or *ischus* is associated with the speed of movement: being moved "by the same force" (*ischus*) should result in the same speed, but it does not because of the greater deflection of the point closer to the center. At this point (849a14-19) the author identifies the component of circular motion directed towards the center as "unnatural" (*para phusin*) and that which is at right angles to it as "natural" (*kata phusin*); the point of this identification seems to be that the component of motion towards the center is the result of constraint.[15]

In the third and final part of the argument, which runs from 849a19 to 849b19, the author applies the deflection principle to the case of circular motion by arguing the point closer to the center is deflected more than the one that is farther out. He does this by arguing that the amount of "unnatural" or *para phusin* motion is smaller in the larger circle, *for a given amount of "natural" motion.* In figure 2, BY and XZ represent the "unnatural" components of motion of the larger and smaller circles for the same "natural" motion

---

14) Cf. 851a16-17: "the same magnitude moved by the same force (*têi autêi ischui*) advances farther in air than in water."

15) Many commentators (e.g. Krafft, *Betrachtungsweise,* 33; De Gandt, "Force," 122) understand *kata phusin* here to refer to a heavy body's motion towards its natural place, i.e. downwards. On this interpretation the diagram of figure 2 should be rotated 90 degrees clockwise. This is supported by the fact that problem 1 concerns the motion of the arms of a balance. However, it is clear from the sequel that the author understands this analysis of circular motion to be a very general one that is applicable to circles in any orientation, such as the potter's wheel which turns in a plane parallel to the ground (pr. 8, 851b19-21; cf. 852a1-13). It therefore seems more likely that the *kata phusin/para phusin* contrast here should be understood as a contrast between free and constrained motion: an object moving in a circle would fly off in a straight line, but for the constraint imposed by the deflection (*ekkrousis*) towards the center.

(ΩY = ΘZ). But BY < XZ, since ΩY and ΘZ are equal, and "equal straight lines inscribed in unequal circles at right angles to the diameter cut off a smaller segment of the diameter in the greater circles" (849a35-38).[16] Thus, point B is deflected less than point X; by the deflection principle, it travels more quickly. It is only at this point (849b1) that time enters the argument: in the time that X moves to Θ on the smaller circle, B moves to a point *beyond* Ω on the larger circle (since its motion is quicker).[17] But how far does it go? Here the assumption that the two points lie along a single radius is crucial. Because we have two points (X and B) that are fixed on the same radius, at the end of the motion, the ratio between the "unnatural" and "natural" components must be the same for both points. And this condition is satisfied only when point B has moved to H (for BK:KH = XZ:ZΘ).[18]

To recap: in part 1 the author argues that circular motion, like rectilinear motion, can be analyzed into two components; in part 2 he states the principle that if two things are moved by the same force, the one that is more deflected will move more slowly; and in part 3 he shows that this principle applies to the case of circular motion by demonstrating that the deflection is smaller in the larger circle. It is crucial to see, first, that this last claim depends on a comparison of motions completed in *different* times: when X has moved to Θ, B will be at H; but when B has completed as much "natural" movement as X, it will have completed less "unnatural" movement. Second, it is the assumption of proportionality, grounded in the geometry of the circle, which determines where the points end up when the motion is complete. The deflection

---

[16] This is stated without proof. For a proof based on Euclid 3.31 and 6.8, see T.L. Heath, *Mathematics in Aristotle* (Oxford, 1949), 233.

[17] The introduction of time at this point is strongly marked (848a38-849b4): "So in the amount of time (*en hosôi dê chronôi*) that AΘ traverses XΘ, in so much time (*en tosoutôi chronôi*) the end of BA has traversed a greater (distance) in the greater circle than BΩ. For the natural motion is equal, while the unnatural motion is less: BY is less than ZX."

[18] The importance of the proportionality requirement is strongly emphasized in this part of the argument; cf. 849b4-6: "it must be proportional (*analogon*): as the natural [motion] is to the natural, so the unnatural is to the unnatural."

principle says only that greater deflection yields a slower speed; it does not establish a precise relationship between forces and distances, and so cannot explain why there is proportionality between the natural and unnatural components when the motion is complete. What the deflection principle is supposed to explain is the surprising fact that although the same force is applied to the points closer and farther from the center, the closer point moves more slowly. But the application of the principle to circular motion is based purely on geometrical considerations. The analysis of motion presented in part 1 gives the author a way of comparing the motion of the larger and smaller circles; using this method, in part 3 he shows that the point on the smaller circle is more deflected in the way I have indicated. The upshot of the whole argument is that it is the geometry of circular motion that explains the different effects of the same force exerted at different distances from the center. It is therefore a mistake to suppose that the author is trying to explain the geometrical features of circular motion by reference to forces and their effects. In this argument the effect of a force is the *explanandum*, not the *explanans*.[19]

One consequence of this is that the concept of force or *ischus* that is relevant here is not a highly theoretical one, as commentators such as Duhem have claimed.[20] *Ischus* should be thought of

---

19) Another way of putting the point is that the kinematics of the situation explains the dynamics, rather than vice versa; here I am in fundamental agreement with the contribution of J. de Groot in this volume. Note that there is no reference in the *Mechanica* to a "force" from the center, except in a metaphorical sense (*hôsper antispômenon* 849a13).

20) Following Duhem, Krafft has proposed a quite different interpretation of the claim that the points closer and farther from the center are moved "by the same force." According to Krafft (*Betrachtungsweise,* 35-36), this corresponds to a situation on a balance beam where a smaller weight is attached at a greater distance from the center and a larger weight closer in, in such a way that the ratio of the weights is the inverse of that of their distances from the center. This is based on an application of the relationship stated in *Physics* VII 5 (249b27-250a4): if a given power (*dunamis*) can move a given weight over a given distance in a given time, it will move half the weight over twice the distance in the same time. As the balance beam turns, the smaller weight covers a proportionally greater distance in the same time; hence it is moved "by the same force" as the larger weight closer in. This amounts to understanding *ischus* as the

simply as a push or effort that is applied to a mechanical device. What is surprising is that the same "push" produces a quicker movement if it is applied farther from the center. As noted above, the author associates *ischus* with speed; thus the greater speed of the point on the larger circle is responsible for the greater effect produced by a force acting farther from the center.[21] But *ischus* itself still refers to the input force that moves the radius of the circle. Here it is relevant to note that the term *ischus* has connotations of physical strength, while the other principal term for force used in Aristotelian contexts, *dunamis*, refers primarily to a capacity or ability (it is derived from the verb *dunamai*, "to be able"). The emphasis on input force as effort is understandable given the author's concern to explain practitioners' knowledge. For what is remarkable about mechanical devices from the point of view of practical experience is the fact that the same input yields a greater effect.

In keeping with the non-technical role of *ischus* in the argument of problem 1, the author frequently substitutes *baros* or "weight" for *ischus* in applying the circular motion principle. This occurs already in problem 1, where the principle is applied to the original question about the balance:

> Accordingly, the reason (*aitia*) why, from the same force (*apo tês autês ischuos*), the point more distant from the center travels more quickly than the nearer point, and the greater line traces out the greater circle, is clear from what has been said. And why larger balances are more accurate than smaller ones is evident from the following. For the cord becomes a center (for this is stationary), and the parts on

"effective weight" exerted by a given *baros* at a certain distance from the center (cf. De Gandt, "Force," 116-118). On this interpretation, it is clear how a small weight at a large distance from the center can balance a large weight at a short distance: for the *ischus* exerted by both is the same. Yet this is not the argument that the author actually makes, either here or in the analysis of the lever (see section 5 below). The argument of problem 1 is based not on the principle of *Physics* VII, but on the deflection principle. And the *ischus* is not the *result* of the swifter motion of the larger circle, but the cause or source of its motion.

[21] Cf. pr. 17, 853a24-25 (the wedge "is stronger (*ischuei pleon*) because of its speed"); pr. 19, 853b18-22 (the effectiveness of the axe is due to the speed of its motion); pr. 22, 854a31-34 (the nutcracker, since its action does not involve blows, loses much of the *ischus* due to movement).

> each side of the beam become the radii. So (*oun*) from the same weight (*apo tou autou barous*), the extremity of the beam must be moved more quickly, the more distant it is from the cord; and some weights are not clear to sense perception when placed on small balances, while in large balances they are clear. (849b19-28)

The author identifies the cord of the balance with the center of the circle, and the parts on each side with the radii. Then he applies the circular motion principle, but with the substitution of weight (*baros*) for force (*ischus*): from the same weight (*apo tou autou barous*), a point farther from the center will be moved more quickly. The author imagines the same weight being placed on two balances, one of which has longer arms; the magnitude of the swing produced in the larger balance is larger because of the circular motion principle. Evidently he is not concerned to distinguish being moved "by the same force" and "by the same weight." In problem 3 the greater radius is said to be moved more quickly *by an equal weight* (*hupo tou isou barous,* 850a36-7; see section 5 below). Recall that the remarkable fact about the lever mentioned in the introduction (847a28-b15) is that it makes it possible to move a large weight (*baros*) by a small force (*hupo mikras ischuos*). Since problem 3 explains how a large weight (*baros*) can be moved by a small *weight*, the implication is that a small weight is associated with a small *ischus*. Finally problem 9 suggests a similar correspondence between *ischus* and *baros*, with a direct reference back to problem 1:

> Why is it that we move things that are raised and drawn by means of greater circles more easily (*rhaon*) and more quickly (*thatton*)? For instance, larger pulleys rather than smaller ones, and similarly for rollers. Is it because the longer the radius, in an equal time it is moved over more space, so that it will also do the same thing if an equal weight (*tou isou barous*) is on it, just as we also said that larger balances are more accurate than smaller ones. (852a14-20)

The remaining passages in which the circular motion principle is invoked are summarized in the following table:

Pr.	Ref.	Text
8	851b36-38	"for greater (circles) are moved, and move weights, more quickly (*thatton*) by an equal force (*hupo tês isês ischuos*)"
12	852b8-9	"the greater the radius, it is moved more quickly (*thatton*)"
13	852b15-18	"the (radii) of larger circles are moved more quickly (*thatton*), and more, from the same force (*apo tês autês ischuos*), than those of smaller circles; for by the same force (*hupo tês autês ischuos*), the extremity farther from the center changes position more quickly (*thatton*)"
14	852b27-28	"the farther it is from the center, everything is moved more easily (*rhaon*)"
15	852b33-34	"the greater radius always traces out a larger circle from an equal motion (*apo tês isês kinêseôs*)"
27	857a31-32	"the greater the (distance) from the center ... it changes position more"

The author consistently refers to "the same" or an "equal" *ischus* applied at different distances from the center; *ischus* is consistently used of input force or effort. Of the nineteen instances of the term, fourteen are expressions of agency (expressed with the prepositions *apo* or *hupo*, or the dative of agent). We may briefly consider two examples. (1) Problem 13 asks (852b11-13): "Why are larger handles more easy to move around a spindle than smaller ones, and in the same way thinner windlasses are more easily moved by the same force (*hupo tês autês ischuos*) than thicker ones?" The author applies the circle model and then goes on to state the circular motion principle at 852b15-18, as summarized in the table above. Here the point is that the same effort applied to two different devices (windlasses with longer or shorter arms) has different effects.[22] (2) Problem 18 asks why it is that the use of a compound pulley enables a large weight to be raised "even if the pulling force (*hê helkousa ischus*) is small (853a37)." The author continues (853a37-39): "Is it because the same weight is moved from a smaller force (*apo elattonos ischuos*) if the lever is used than from the hand?" Again we

[22] Evidently the thinner windlasses have longer arms.

have a comparison between the input force exerted in two different situations, in this case with and without mechanical aid.[23]

5. The Lever

I begin with a translation of the key passage of problem 3 (850a30-b6):

> (a) Why do small powers (*dunameis*) move great weights (*barê*) by means of the lever, as was also said at the beginning, even though the weight of the lever is taken in addition? It is easier to move a smaller weight, and it is smaller without the lever. (b) Is it because the lever is the reason, since it is a balance (*zugon*) with the cord (*spartion*) attached below and divided into two unequal parts? For the fulcrum (*hupomochlion*) is the cord (*spartion*): for both of these remain stationary, like the center (*kentron*). (c) And since the greater radius is moved more quickly by an equal weight (*hupo tou isou barous*); and there are three things involved in the lever, the fulcrum (cord and center), and two weights, the mover (*ho kinôn*) and the moved; so (*oun*) as the weight moved is to that which moves it, the length is to the length, inversely. (d) And it will always cause movement (*kinêsei*) more easily (*rhaon*), the more distant it is from the fulcrum. (e) The reason is the one stated before, that the line that extends more from the center traces out a greater circle, so that from the same force (*apo tês autês ischuos*) the mover (*to kinoun*) that is farther from the fulcrum will change position more.

[23] If any term is a candidate for expressing the idea of "effective weight" in the *Mechanica* (cf. n. 20 above) it is *rhopê* ("inclination"), not *ischus*. Thus at the end of problem 1 the author writes that "the magnitude of the inclination (*rhopê*) produced by the same weight (*hupo tou autou barous*) is much larger in larger balances" (849b32-34). Here *rhopê* refers to the visible "swing" of the balance, its deviation from equilibrium. The terms *rhopê* and the associated verb *rhepein*, "to incline," are closely connected with the balance in Greek literature of the fifth and fourth centuries BC (cf. LSJ s.v.). In Aristotle *rhopê* refers to the downward tendency possessed by all heavy bodies (e.g. *De caelo* 297a28). In the *Mechanica* the term refers either to a disturbance of the balance from a state of equilibrium (as in pr. 1), to something that produces such a disturbance (at 850a13-16 it is used a small weight added to one side of the balance), or to a downward tendency as in Aristotle (858a15, cf. 858a21). There is no evidence in the *Mechanica* for the idea that equilibrium is produced by equal *rhopê* on either side of the balance, as in Archimedes. Nor is *rhopê* consistently used of the effect produced by a weight on the balance: recall that mechanical problems are situations in which "things with a small *rhopê* move great weights (*barê*)" (847a22-23), implying that a small *baros* has a small *rhopê*.

The overall structure is the author's usual one of question (a), model identification (b), and explanation (c-e). Note first that the question (a) is phrased in terms of small "powers" (*dunameis*) moving large "weights" (*barê*); *dunamis* thus takes the place of *ischus* in the original statement of the lever's operation at 847a28-b15 ("at the beginning"). In (b) the lever is identified with a balance having unequal arms and supported from below.[24] In (c) we have a statement of the circular motion principle formulated in terms of weight (*baros*), an analysis of the lever as consisting of fulcrum and two weights (mover and moved), and a claim concerning the relationship of weights and distances: "as the weight moved is to that which moves it, the length is to the length, inversely." The "mover" slot of the lever model is here filled by a weight (*baros*). In (d) the author claims that the movement will be easier, the farther the mover is from the fulcrum. Finally, (e) cites the circular motion principle as the explanation: the same mover placed farther from the fulcrum will change position more "from the same force" (*apo tês autês ischuos*), and this is why it causes motion more easily. The substitution of *ischus* for *baros* in (c) is striking, and confirms the analysis of section 4 above.

Despite the author's claim to reduce the lever to the balance, the references to the "center" in both (b) and (c) signal the circle model, and it is in fact the circular motion principle that supplies the explanation (c, e). The notion of the equilibrium of forces on either side of the fulcrum plays no role in the argument. Nor is it clear that the statement of the relationship between weights and distances is meant to express an exact proportionality. The idea may just be that any change in the relationship of the weights is compensated by the inverse change in that of the distances: for example, if one of the weights is diminished, its distance from the fulcrum must be increased.[25] It is also not entirely clear whether the relationship between weights and distances is supposed to be an inference from

---

24) Problem 2 distinguishes between the cases where the balance is suspended from above the beam and from below.

25) The author uses the term *antipeponthen* (850b2), which expresses the idea of reciprocal proportion in Euclid. But the Euclidean way of putting the point would be to say that the weights "are reciprocally proportional" (*antipeponthasi*) to the distances

the circular motion principle as stated in the first part of (c), or a statement of an additional fact about the lever parallel to the remark that it involves a fulcrum and two weights. That it is a consequence of the circular motion principle is suggested by the particle *oun*, which frequently has an inferential sense (e.g. 849b24, 851a21, 851b4, 854a24, 854a29). But no argument for this is given. Insofar as there is an argument here it is to support claim (d), that "it (the mover) will always cause movement (*kinêsei*) more easily (*rhaon*), the more distant it is from the fulcrum." This is because, as is said in (e), the same mover placed farther from the fulcrum will change position (*metastêsetai*) more. So again we have the circular motion principle applied to explain the greater ease of movement produced by increasing the distance from the center. The same force, applied farther from the center, has a greater effect because of its quicker movement.

This emphasis on the effect of increasing the separation from the fulcrum is apparent in the citations of the lever principle throughout the text:

Pr.	Ref.	Text
4	850b14-15	"the mover of the weight always moves (*kinei*) more weight, the more distant it is from the fulcrum"
6	851b2-4	"the farther the fulcrum, the same power (*hê autê dunamis*) moves the same weight more easily (*rhaon*) and more quickly (*thatton*)"
16	853a11-12	"the longer the (distance) from the fulcrum, the more it must be bent"
20	854a12-13	"the longer the length of the lever from the fulcrum, the more easily (*rhaon*) ... it causes motion (*kinei*)"
22	854b7-8	"the more the lever extends from the fulcrum, it moves (*kinei*) more easily (*rhaon*), and more, from the same force (*apo tês ischuos tês autês*)"
29	857b15-16	"the more distant (the mover) is from the weight, the more easily (*rhaon*) it causes motion (*kinei*)"

(cf. Eucl. 11.34, 12.9, 12.15), not (as here, more literally) "what the weight suffers in relation to the weight, the length suffers in relation to the length, inversely."

Each of these statements explains the consequences of increasing the distance between the mover and the fulcrum. Doing so increases the ease of movement (3d, 6, 20, 22, 29), the speed of movement (6), or the amount of movement (3e, 22; cf. 16). The mover remains the same while its distance from the center is increased. The author evidently understands the quicker or greater movement of the *weight* to follow from the quicker or greater movement of the *mover*. Only in problem 4 is there any reference to the fact that the mover can move more weight if its distance from the fulcrum is increased.[26] Finally we should note that the correlations between distance from the fulcrum and effect are expressed not as exact ratios but rather using vaguer locutions such as "the more ... the more" (e.g. *hosôi* ... *tosoutôi* in the Greek), a further indication that establishing an exact proportion between weights and distances is not the author's concern.

The author is clearly aware of some sort of reciprocal relationship between weights and distances in the lever. And the fact that he attempts to reduce the lever to the balance suggests that he recognizes the balance with unequal arms as an illustration of this relationship. But it is really *only* an illustration. The analysis of the lever is based entirely on the circular motion principle, and the main point that is established in problem 3 and applied throughout the text is that the *same* input (whether force, power, or weight) will have a greater effect, the farther it is from the fulcrum.

6. Conclusion

There is much more of interest in this short text. But I hope that the analysis I have presented, however partial, at least points the way to a clearer picture of its place in the history of mechanics. If my analysis is correct, then it is a mistake to look to the *Mechanica* for an application of the principles relating force and movement in the *Physics* or for the development of innovative concepts of force. But the author succeeds remarkably well in devising an

[26] However, 850b14-15 might also be translated "the mover of the weight always moves (*kinei*) the weight more, the more distant it is from the fulcrum."

explanatory framework to address his main challenge: the ability of technological devices to contravene what seemed to be the obvious notion that one needs a large force to move a large weight. The three simple models of circle, balance, and lever supply the author with the conceptual tools to analyze a wide variety of mechanical devices. The flexibility in his use of terms such as "force," "weight," and "power" is matched by standardization in the terminology for the model slots, indicating the crucial importance of the latter. Despite its importance as an illustration of the ability of a small force to balance a large weight, the balance plays a relatively unimportant role in the author's reduction program. This reflects the fact that his argument for the circular motion principle begins not from considerations of equilibrium, but rather from an innovative analysis of circular motion. It is only after Archimedes that the exact proportionality between weights and distances on the balance beam in a state of equilibrium comes to be the core of theoretical mechanical knowledge.

The Simple Ontology of *Kalām* Atomism: An Outline

A.I. Sabra*
Harvard University

Abstract

This paper aims to present concisely the Islamic kalām atomism as an alternative philosophy to Hellenizing *falsafa*. Kalām is a theological-philosophical discourse which, first (in the third/ninth century) ventured to rival the *falsafa* represented early by al-Kindī (d.ca. 252/866), then by al-Fārābī and Avicenna in the fourth/tenth and fifth/eleventh centuries, and which eventually (in the sixth/twelfth century and after) appeared to be inclined to propose a mingling of the kalām discourse with *falsafa* in a series of varied "syntheses".—Focusing on the simple ontology of the basic kalām atomism, and noting the hybrid character of kalām, the aim of this paper is to help to clarify the inevitable problematic consequences of those late ventures of Islamic intellectualism.

Keywords

kalām, atomism, Abū Ḥāmid al-Ghazālī, al-Ashʿarī, al-Ījī, al-Naẓẓām, al-Muʿtazila

I

"... only the Exalted God and His actions exist."
The Ashʿarite Abū Ḥāmid al-Ghazālī, d. 505/1111[1]

* History of Science Department, Science Center 371, Harvard University, Cambridge, MA 02138, U.S.A. (sabra@fas.harvard.edu). Web Site, including a complete list of publications: http://www.people.fas.harvard.edu/~sabra.

1) Al-Ghazālī, *al-Mustaṣfā min ʿilm al-uṣūl*, Būlāq, 1322, vol. I, pp. 26-27: "[...] knowing/knowledge is the occurrence of forms, or similitudes, of things, in the mirror of reason (*al-ʿaql*); and reason is the innate disposition (*al-gharīza*) through which it is able to receive the forms; now, the soul, or the human reality that is endowed with this innate disposition to receive the intelligibles, [acts] like a mirror. [...] this analogy will help [the reader] to understand what knowing really is; for when the real intelligibles

1. Introduction

1.1. *Al-Kalām*, or what came to be known to its practitioners as "the science of *al-kalām*, *'ilm al-kalām*," is a theological-philosophical discourse which arose in Baṣra and Baghdad shortly before the time when the 'Abbasid rulers in Baghdad undertook to promote the astounding projects of translating Greek medicine, science, mathematics and philosophy into Arabic at the middle of the second/eighth century.—It should be noted immediately that at least some of the early 'Abbasid rulers were also supportive of the first large kalām 'school', the so-called Muʿtazila or Muʿtazilīs, who were known for championing the role of reason/*al-ʿaql* in theological thinking. In the fourth/tenth century a well-practiced Muʿtazilite, Abū l-Ḥasan al-Ashʿarī (d. 324/935), broke away from the Muʿtazila to found what eventually came to be the largest school of kalām known as the Ashʿarīs or Ashʿarites. Despite some significant doctrinal differences between the two schools, it is the opinion of the present writer (and others) that some specific and important Ashʿarite doctrines can and should be understood as natural, if not logical, developments of Muʿtazilite doctrines. And in accordance with this opinion, it is my assumption that a correct and complete understanding of Ashʿarism is not possible without grasping the earlier Muʿtazilite arguments, many of which have been preserved in al-Ashʿarī's own clear and precise accounts in his outstanding *Maqālāt al-Islāmiyyīn*/ *Islamic Doctrines*.[2]

1.2. The well-known third-/ninth-century *littérateur*, al-Jāḥiẓ (d. 255/868-9), who originated in Baṣra, was also a noted *mutakallim*, or practitioner of kalām, and the leader of a school of *kalām* known after his name. He wrote in his monumental book, called *Animals*

impress themselves upon the rational soul, they will be called 'knowledge'. And, just as the heavens and the trees and the rivers are conceived to be seen in a mirror as if they existed in the mirror, and as if the mirror contained them all, so the whole divine presence (*al-ḥaḍra al-ilāhiyya*) is conceived to be impressed upon the human soul. Now, the 'divine presence' is an expression for the totality of existing things, all of them being [part] of the divine presence — since only the exalted God and His actions exist (*idh laysa fī l-wujūdi illā Allāha taʿālā wa afʿālahu*)."

[2] Hellmut Ritter, ed., Abū l-Ḥasan al-Ashʿarī, *Maqālāt al-Islāmiyyīn*, 2nd ed. (Wiesbaden, 1963).

(*Kitāb al-Ḥayawān*), a sentence to the effect that the practice of kalām as an argumentative discourse on religion (called *kalām al-dīn*) would be less than perfect unless it was "brought to the level" of the discourse of philosophy (*kalām al-falsafa*). I have used that sentence as a motto for an article with the title "Kalām Atomism as An Alternative Philosophy to Hellenizing Falsafa."[3] — On this occasion I should like to focus on some features of the atomistic theory forming a basis of kalām as discussed by the early school of the Muʿtazila *and* as later developed by al-Ashʿarī and his early followers.

1.3. Even as a discourse *on religion*, kalām obviously inclined, right from the start, to use forms of arguments some of which were clearly employed by ancient (and modern) philosophers; and it is of course important to identify these forms, their sources and characteristics.—In the article just mentioned I briefly describe some of these forms. Now I am concerned only with a basic, though apparently overlooked, form of argument, namely the ontology that lays the foundation of a certain kind of "atomism" explored and generally chosen by the Muʿtazila *and* later developed and adapted by al-Ashʿarī and his followers.

2. The Ontology of Kalām Atomism

2.1. Ontology is the theory of what there is. My aim is to answer the question: what is it, for kalām, and, in particular, Ashʿarite kalām, that exists in the created World (*al-ʿālam*); or, more specifically, what is it that exists in the World according to the basic doctrine of kalām atomism known to the mutakallimūn (practitioners of kalām) as *daqīq al-kalām*, subtle kalām.—*My proposed answer to this question will be: not things, but events or occurrences* (ḥawādith) *in space and time; and this answer will imply that the atom itself is*

[3] This article is contained in James E. Montgomery, ed., *Arabic Theology, Arabic Philosophy: Essays in Celebration of Richard M. Frank* (Leuven, 2006), 199-272.—Hereafter, I shall refer to this article as *KAF.*

such an occurrence or event that is presupposed by all other events in the World.[4]

2.2. My argument can be summarised briefly as follows. By definition, events are characterised by four co-ordinates: three of them are those of space, and they attach to the atom called 'simple' or 'individual' substance (*al-jawhar al-basīṭ*, or *al-jawhar al-fard*), *or* to aggregates of atoms called 'composite substances', also by virtue of being occupiers of place; the fourth co-coordinate is the time at which a "quality" or "property" (*maʿnā*) or "accident" (*ʿaraḍ*) *momentarily* inheres, or comes-to-occur, in a simple atom or in an aggregate of atoms called 'body'.

2.3. Space and time, both being neither substances nor accidents, they are consequently conceived and clearly treated as relations or conventions.[5]

3. The Definition of Atom as a Postulate

3.1. The definition of atom as a simple, individual or indivisible substance is itself also a postulate asking us to *grant* these qualifications, just as in the way that Euclid's "definition" of point asks us to *grant* that a point is "that which has no part/*sēmeion estin, ou meros outhen*." That means that we are not at liberty to confer upon the atom other qualities such as size and shape.—As a substance, then, the atom is completely characterized as a place-occupying entity, and completely defined as a simple or indivisible entity that occupies place.

---

4) Note that the English verb 'to occur', like the Arabic verb *ḥadatha*, from which the singular noun *ḥadath* (*occurrence*) and the derived plural, *ḥawādith*, is intransitive, and so is 'to-come-to-be', *kāna*. By contrast, 'to create,' and the Arabic *khalaqa*, are transitive, implying a creator/*khāliq* and a created/*makhlūq*.

5) See *KAF*, 206-07, esp. 215-17; ʿAḍud al-Dīn al-Ījī, *Kitāb al-Mawāqif fī ʿilm al-Kalām* [hereafer: *Mawāqif*], ed. Ibrāhīm al-Dusūqī ʿAṭiyya and Aḥmad al-Ḥanbulī (Cairo, 1357/1938-39); on relativity of time, see 112.—See also: Al-Tahānawī, Muḥammad Aʿlā ... al-Fārūqī, *Kashshāf Iṣṭilāḥāt al-Funūn*, 3 vols, Dār Ṣādir, Beirūt, no date, vol. II:619-623, entry on time: *al-Zamān*, esp. 621, on "the mutakallimūn," particularly al-Ashʿarī and his followers.

3.2. The well-known debate, first initiated among the Muʿtazila, as to how many atoms are necessary to make up a "body" (*jism*), or composite *substance*, came to be considered by some leading Ashʿarites as a useless digression: we can regard two composed atoms as 'length', meaning that they can be imagined to be joined by a straight line, or else *call* them 'an aggregate of two atoms'; similarly, we can *call* three composed atoms a 'plane', meaning that they can be joined by three non-collinear lines, *or* 'a particular aggregate of three atoms'; and, again, we might *call* four atoms a 'body', *or* 'another particular aggregate of four atoms'.—*But a body must be defined as that which has length, breadth and depth*—these three characteristics being distinguished among themselves by means of the concept of 'right angle'.—For the influential fourteenth-century Ashʿarite ʿAḍud al-Dīn al-Ījī (d. 756/1355), the early well-known debate had been merely "a verbal [that is, useless] dispute."[6]

3.3. As for us, we can easily see the decisive Euclidean contributions to the definitions of atom and body.

4. Modes of Coming-to-be in Space: al-akwān

4.1. The behaviour of atoms *in space* is completely described in terms of exactly four accidents (*aʿrāḍ*. sing. *ʿaraḍ*) all of which expressed as distinct modes of "coming-to-be" (*akwān*, sing. *kawn*, from *kāna*, to-be or to come-to-be):– at any single moment, an existing atom either comes-to-be in-motion, or comes-to-be at-rest; and, simultaneously, it might come-to-be in-separation-from and/or in-contact-with another atom or group of atoms.

4.2. Thus motion involves two places and two times; and rest involves two times and one place; *and, therefore, both motion and rest involve a* coming-to-be *at every moment during their initially created existence.*

4.3. Al-Ashʿarī, in contrast to the Muʿtazila, and especially against the often misunderstood al-Naẓẓām (d. 230/845), did not regard the *initial* "creation" of an atom or body in the World as motion,

[6] See, in *KAF,* Marṣad (i) in Ījī's *Mawāqif* On Bodies, 259-271, especially the discussion on what he calls "the true [doctrine]," and note 32.

for the reason that initial creation is not preceded by the atom's or body's being-at-rest or in-motion.)

4.4. In regard to separation: when an atom A comes-to-be in-separation-from B, then B would have simultaneously come-to-be in-separation-from A. This illustrates the relativity of spatial locations.

4.5. In regard to contact, it may be noted that the term actually covered all cases involving touching (*mumāssa*) of substances—such as pressure, impact, collision, cohesion, etc.

4.6. To be noted, and kept in mind, is the fact that the four *akwān* are all explicated in terms of the basic concept of "coming-to-be" (*al-kawn*).—*Not to observe this is to miss the direction of argument in all of kalām ontology from the start. To complete the argument, we only need to turn to kalām explication of accidents, as* will now be shown.[7]

5. Motion as "Departure" (zawāl). Rest as a Coming-to-be (kawn). An Obvious and Definitive Rupture with Aristotelianism

5.1. Two assertions are especially important about the kalām doctrines of motion and rest; both assertions are shared by the Ashʿarīs and the Muʿtazilīs. First is the definition of motion *only in terms of place and time*, and in the total absence of all Aristotelian concepts like actuality and potentiality, and completion or perfection, and of any Neoplatonic concepts. Descartes would agree.

5.2. The definition of motion is consistent with al-Ashʿarī's explicit understanding of motion simply in terms of "departure" (*zawāl*) from one place to another—period.

5.3. The second assertion is the explication of rest *in terms of a 'coming-to-be'*—exactly as in the case of motion; which implies that to-be-at-rest at *p* at time *t* is not, of necessity (or, of itself) followed by a being-at-rest at *p* at the immediately following moment;

---

[7] On the *akwān* doctrine, and on motion, see *KAF*, pp. 209-11, and pp. 211-15; for the supporting quotations in English translation, see *KAF*, pp. 221 and following pages.

which in turn signals that a cause is *always* involved in *both* motion *and* rest, and in both "separation" and "contact".

—As the reader will have guessed, the cause, for the Ahʿarīs, is *always* meta-physical, being no other than God's *free* and *unmediated* action.—But we still have to introduce the role of accidents.

6. Accidents: What they Are, Generally[8]

6.1. Generally, accidents (*aʿrāḍ*, sing. *ʿaraḍ*) are occurrences that come to inhere in (*taqūmu fī*) a single or composite substance, only for a single moment.—*Note that it is through this instantaneous inherence or attachment that an accident acquires its momentary position in space, thereby gaining the qualification of being an event or occurrence in space as well as time.*

6.2. But, as a result of considering all accidents as occurrences *in a subject* (*fī mawḍūʿ*), that is, a place-occupying *substance*, it was inevitable for the mutakallimūn, both Muʿtazilīs and Ashʿarīs, to confront the following question with regard to motion as departure from one place to another: namely, *where* must the accident motion occur[9]—*at the first* or *the second place*?

6.2.1. That, we now know, was the kind of question that plagued all exclusive subject-predicate ontologies throughout the history of philosophy, until they were exorcized by Bertrand Russell. For al-Ashʿarī, the answer was: that *stepping into* the Mosque from the street is identical with *abandoning* or *departing from* the street.

7. Accidents as Defined by the Muʿtazila and the Ashʿarites

7.1. For the Ashʿarīs, accident is *both what comes-to-be and stands*, resides or inheres in (*qāma fī*) what occupies place. Whereas, for the Muʿtazilīs, accident is that which, *if* it exists, must inhere in what occupies place, their reason being that accident is, by itself,

---

[8] In Ījī's *Mawāqif*, the discussions of "accidents" (*al-aʿrāḍ*) occupy the pages *KFA*, 96-181; they have yet to receive the attention they certainly deserve.

[9] See al-Ashʿarī's *Maqālāt al-Islāmīyīn*, 353-56; *KFA*, 235-37; and Ibn Fūrak, *Mujarrad Maqālāt al-Ashʿarī*, ed. Daniel Gimaret (Beirut, 1987).

only a definite or determinate (*thābit*) "description," whether it does or does not exist.

This is not the place to comment in any detail on the important discussions in kalām about "whether the non-existent is a thing" (*hal al-maʿdūm shayʾ*, i.e. some-qualification-or-other, which seems to be a question about 'existence' rather than about 'thing'.) —However:

7.2. Explicating, briefly, the Muʿtazilite argument, we may say that a *'square circle'* and an *'oval circle'* both do not exist, and yet they are distinct one from the other by virtue of their "determinate" and "differing" accidents, namely 'square' and 'oval'.[10]

7.3. The falāsifa, for their part, are understood as holding that an accident is that which, *if* it has *external* existence, must inhere in a supporting subject (*mawḍūʿ*) also termed *maḥall muqawwim*, supporting substrate).

8. Characteristics of Accidents, Especially for al-Ashʿarī and His Followers

8.1. By definition, then, an accident, for al-Ashʿarī and his followers, does not inhere in another accident, but always in a substrate (*maḥall*) which itself occupies place; speed, for example, is not a quality of motion but of the body in motion.

8.2. Moreover, accidents do not migrate from one substrate to another; what may appear as a loco-motion (*intiqāl*) of one and the same accident is in fact a creation in some substance of another instance (*shakhṣ*) of the quality inhering in the other substance.

8.3. Thus *one and the same* quality cannot inhere in two substrates at once; the black inhering in one substrate is existentially *other than* the black inhering in the other.

---

[10] It may be helpful to note that (for the Muʿtazila) a *ṣifa thubūtiyya* is *either* a definite (*thābit*) attribute of the Thing (*al-dhāt*) of which it is an attribute, without including any additional *qualities* of the Thing, such as being a substance, or something else that exists; *or* the *ṣifa thubūtiyya* is a qualifying attribute (*ṣifa maʿnawiyya*) indicating a characteristic over and above the Thing, such as occupying-place (*taḥayyuz*), or coming-to-be (*ḥudūth*), or accepting accidents.

8.4. And, finally, an accident does not itself endure for more than one instant; only the free-acting Creator determines how long it persists, that is, how long it *continually comes-into-being*.

8.5. Accidents may of course belong to a living thing, and when they do they are the life itself and all the accidents/qualities that *accompany* the presence of life, including, e.g., willing, perceiving, believing, knowing, and the ability to perform other actions.

8.6. *All* remaining accidents are comprised of the four comings-to-be (*al-akwān*) plus the sensible qualities (*al-maḥsūsāt*) of place-occupying entities.

9. Two Historical Comments

9.1. As can be imagined, all these doctrines were bound to provoke, not trivial or "verbal" disputes, but substantial and long-lasting arguments between, on one hand, the mutakallimūn, both Ashʿarīs and Muʿtazilīs, and, on the other hand, the committed "falāsifa," who clung to positions received from Aristotle and his ancient supporting commentators and from the Neoplatonists, or, what is historically most important, the hopeful "conciliators" of kalām and falsafa.

9.2. The resulting debates in fact constituted the major and dominant part of what we often refer to as "*Islamic* philosophical thought" from, say, the time of Fakhr al-Dīn al-Rāzī (d. 606/1209) on.

II

10. God and Humans

So far in this account, God has been mentioned as the sole cause/creator of all events in a World of substances and accidents, including of course humans (*al-insān*) and what humans seem to be able to do. By introducing humans *as creatures under obligation to their creator* we come to a distinct and fundamental aspect of kalām referred to as a 'theology' concerned with what are called Traditions/*Samʿiyyāt*.—Under this aspect we meet, for example, the concept of origination or initial creation of the World *in time*, and we see ourselves as "actors" in a drama of birth, death and resurrection, always expressed in terms of right and wrong, punishment and reward, both measured and distributed by a just and compassionate God. Above all we come to the questions of God's existence and His attributes, being questions of *our knowledge* of His existence, His attributes and His demands, etc. At this point philosophy must make room for what is acquired from the Revealed Book and from the Last Messenger to whom The Book was first revealed; and, accordingly, kalām discussions have been bound to take the form of a hybrid discourse that is concerned with Traditions/*Samʿiyyāt* as well as rational arguments.

And yet, we should keep in mind that neither the "subtle kalām/*daqīq al-kalām,*" concerned with the behaviour of atoms and bodies in space, nor the "exalted kalām/*jalīl al-kalām,*" mainly concerned with God and His relations with humans, were ever completely detached from what we normally call rational thinking.—It should be remembered that both the Muʿtazila and the Ashʿarites were definitely and explicitly against *taqlīd*, or the slavish acceptance of authority. And one would therefore hope that the recent modern period in which kalām was mostly regarded as merely "apologetic discourse" is now definitely over.

III

The preceding is written as a philosophical argument clarifying a position that may be suggested by the following hypothesis consisting of thirteen propositions:

1. The world (*al-ʿālam*) is all that occurs (*kullu mā yaḥduth*).
2. Occurrences are events (*ḥawādith*) in space and time: all have space and time coordinates.
3. Occurrences consist of the coming into being of place-occupying entities called 'substances' (sing. *jawhar*), and of accidents (sing. *ʿaraḍ*) that may only reside (*taḥillu fī*) or inhere (*taqūmu fī*) in substances momentarily.

4. Substances are initially created (*tukhlaq*, *tubtada*ʿ, *tukhtara*ʿ) and subsequently re-created (*tuʿād*) at every moment of their span of existence.
5. Substances are of two types (*ḍarbān*): (1) "simple" or "individual" substances (sing. *jawhar basīṭ,* or *jawhar fard* or *munfarid*), called 'atoms' (sing. *al-juz*ʿ, or *al-juz*ʿ *alladhī lā yatajazza*ʿ: the part, or the indivisible part); and (2) composites of these, called 'bodies'.
6. Four accidents, known as *al-akwān*, the comings-to-be, completely describe the possible behaviour of substances *as place occupiers*: at every moment, a substance (atom or body) either (1) comes-to-be *in motion*, from one place to another, or (2) comes-to-be *at rest* in one and the same place; and, through motion and rest, a substance may (3) come-to-be *in contact with*, or (4) *in separation from*, another substance or substances.
7. Accidents acquire their *spatial positions* from their inevitable association with place-occupying substances.
8. Space and time, being neither substances nor accidents that *reside* or *inhere in* substances, *are* relations:
9. Space is an order of relations that happen to obtain between substance-occupied places (positions) at any moment.
10. Time is an arbitrary convention of correlating coincident or simultaneous events.
11. All occurrences are contingent, all of them being choices or free actions of a transcendent, unique and omnipotent Agent:
12. In the created world of *events*, there are no Aristotelian substances, no natures or natural powers or forces, no forms or essences or necessities; there is only one, meta-physical, cause which acts without means or intermediary agents:
13. "Creating," the act of a transcendent Being, "*is* that which is created" ("*al-khalq huwa l-makhlūq*").

The Footprints of "Experiment" in Early Arabic Optics

Elaheh Kheirandish*
Harvard University

Abstract

This study traces the early developments of the concept of experiment with a view of extending the subject in both content and approach. It extends the content of the subject slightly backward, prior to the methodological breakthroughs of the *Optics* of Ibn al-Haytham (Alhazen or Alhacen, d. ca. 1040), which are credited as a "significant landmark in the history of experimental science." And it extends the approach to the subject slightly forward, from the premise that early science was "largely carried out in books," to a close examination of the books through which the footprints of 'experiment' may be traced. The point of departure is the *Optics* of Aḥmad ibn ʿĪsā, a revealing text for the early developments of concepts such as 'demonstration' and 'experiment', and one through which some modern discussions are examined and extended with reference to this and other historical sources.

Keywords

Aḥmad ibn ʿĪsā, Ibn al-Haytham, al-Kindī, Qusṭā ibn Lūqā, Khāzinī, Fārisī, Çelebī, *iʿtibār* (examination, experiment), *muʿtabir/ṣaḥib al-iʿtibār* (examiner, experimenter), *miḥna/imtiḥān* (test), *tajriba* (experience), *unbūb* (tube), *nuḥās* (copper), *mithāl* (illustration), *ʿilla* (cause), *khalaf* (contradiction), *burhān* (demonstration), *ḥiss* (sense-perception), *qiyās* (comparison, syllogism)

Introduction

Pre-modern concepts of experiment have been variously associated with medieval science in general and the science of optics in par-

* Harvard University, 38 Kirkland Street, Cambridge MA 02138, U.S.A. (kheirand@fas.harvard.edu). An earlier version of this present paper, entitled "Greco-Arabic Optics and Early 'Demonstrative' Variations," was delivered at the Medieval Science Colloquium VII held on the occasion of John Murdoch's 80th birthday at Boston College in May 2007.

ticular. From pioneering works on the medieval 'origins of experimental science' to in-depth studies on the 'origin of the concept of experiment' itself, the early development of the so-called 'experimental method' is a recurring theme that has formed the basis of a spectrum of approaches to the subject.[1] At one end of that spectrum is the approach that "the contemporary notion that observation and experiment are necessary ingredients of a discipline called science is not a conception properly applicable to antiquity and the Middle Ages"; at another end, that "the rise of the experimental method," defined as "the most important idea of the past millennium" above many others, was "set in motion" about 1000 years ago; and somewhere in between, that the *confirmatory experiments* of medieval Arabic optics still differ from the *discovery experiments* of seventeenth-century optics.[2] The present study is aimed at tracing the early developments, rather than the origins, of the concept of experiment, with a view of extending the subject in both content and approach. As a continuation of my previous work on early Arabic optics, it extends the content of the subject slightly backward, from the conceptual and methodological breakthroughs of the *Optics* of Ibn al-Haytham (Latinized as Alhazen or Alhacen, d. ca. 1040), credited with a "significant landmark in the history of experimental science...operating consciously and systematically with a concept of experiment *as a distinct method of proof*, and not merely to perform or refer to experiments."[3] The approach to the subject

---

1) A.C. Crombie, *Robert Grosseteste and the Origins of Experimental Science*, 1100-1700 (Oxford, 1962). A.I. Sabra, "The Astronomical Origin of Ibn al-Haytham's Concept of Experiment," in *Actes du XIIe Congrès International d'Histoire des Science* (Paris, 1971), 133-136 (reprinted in Sabra, *Optics, Astronomy and Logic, Studies in Arabic Science and Philosophy* [Aldershot, 1994], VI).

2) The quotations are taken from: John Murdoch, *Album of Science: Antiquity and the Middle Ages* (New York, 1984), 3; Richard Powers, "Eyes Wide Open: When an Obscure Arab Scientist Solved the Riddle of Light, the Universe no Longer Belonged to God," in *The Special Millennium Issue of New York Magazine on the Best Ideas, Stories and Inventions of the Last 1000 Years* (New York, 1999), section 6; and A.I. Sabra, *The Optics of Ibn al-Haytham: Books I-III On Direct Vision, Translated with Introduction and Commentary*, 2 vols., (London, 1989), 2:18 (emphasis in the original).

3) Elaheh Kheirandish, *The Arabic Version of Euclid's Optics: Kitāb Uqlīdis fī Ikhtilāf al-manāẓir*, Edited and Translated with Historical Introduction and Commentary, 2

is itself extended slightly forward, from the premise that in antiquity and the Middle Ages "science was livresque. It was not just set down in books; it was largely carried out in books,"[4] to a close examination of the books through which the footprints of 'experiment' may be traced.

The point of departure is the *Book on Optics and Burning Mirrors* by Aḥmad ibn ʿĪsā, a text with particularly revealing conceptions and expressions related to the historical meaning and developments of concepts such as 'demonstration' and 'experiment'. This is an early Arabic 'composition' according to its full title, with a little known author and date and other curious features including uncommon references to ancient Greek authors.[5] If not as early as "before 250H" (864 CE), as contemplated by Max Krause presumably because of the dates of one of the more likely figures carrying the author's name, the text is among the earliest Arabic optical compositions. This is suggested by A.I. Sabra with reference to the text's non-standard vocabulary and absence of Arabic author citations. It would

---

vols. (New York, 1999); Elaheh Kheirandish, "The Many Aspects of Appearances: Arabic Optics to 950 AD," in *The Enterprise of Science in Islam: New Perspectives*, eds. Jan P. Hogendijk and Abdelhamid Sabra (Cambridge, 2003), 55-83. Quoted statement from Sabra, "The Astronomical Origin of the Concept of Experiment," 136. Cf. also Roshdi Rashed, "Experimentation," in *Geometry and Dioptrics in Classical Islam* (London, 2005), 1039-1045, which includes Ibn Sahl and other relevant cases and references to earlier works by M. Nazif (translated), in addition to M. Schramm cited by Sabra.

4) Murdoch, *Album of Science*, 3.

5) *Optics and Burning Mirrors composed by Aḥmad ibn ʿĪsā, in the Tradition of Euclid, on the Reasons for sight* (*Kitāb al-Manāẓir wa al-marāyā al-muḥriqa taʾlīf Aḥmad ibn ʿĪsā ʿalā madhhab Uqlīdis fī ʿilal al-baṣar*). Max Krause, "Stambuler Handschriften islamischer Mathematiker," *Quellen und Studien zur Geschichte der Mathematik, Astronomie und Physik*. Abteilung B, 3 (1936), 437-532 reported two Arabic manuscripts of the text; a third Arabic manuscript in Hebrew letters is reported by Y. Tzvi Langermann, "Arabic Writings in Hebrew Manuscripts: A Preliminary Relisting," *Arabic Sciences and Philosophy* 6 (1996), 137-160 and "Transcriptions of Arabic Treatises into the Hebrew Alphabet: An Underappreciated Mode of Transmission," in F.J. and S.P. Ragep, eds., *Tradition, Transmission, Transformation* (Leiden, 1996), 247-60. Partial transcription and French translation of the section on burning mirrors is included in Roshdi Rashed, *Oeuvres philosophiques et scientifiques d'al-Kindī*, vol. 1 (Leiden, 1997), 571-645. An edition and translation of the optics section with historical notes is under preparation by the present author.

still be *early* if it was a later, post-al-Kindī (d. ca. 870) 'compilation', as argued by Roshdi Rashed, who explains absent citations as the author's 'discretion' in naming his Arabic sources and overlaps with them as 'borrowings', rather than possible sharing of common prototypes.[6]

The few historical sources whose direct textual correspondence with this text could have provided leads for dating cannot presently have exact dates attached to them either. The pseudo-Euclidean *De speculis*, a text attributed to Euclid in its incomplete Arabic version, and one whose previously unidentified correspondences with Ibn 'Īsā's *Optics* I discuss in this paper, has an anonymous authorship.[7] The only known historical source citing an *Optics* (*Kitāb al-Manāẓir*) by an "Aḥmad ibn 'Īsā," which might provide documentary evidence for the text's latest, rather than earliest, possible composition date, is the medieval Arabic ophthalmologic compilation, *Light of the Eyes and the Compendium of the Arts*, the dating of which is itself dependent on the exact dates of its supposed author.[8]

---

6) On the suggested date of "schon vor 250h," see Krause, "Stambuler Handschriften," 513; on the references to ancient Greek authors, and lack thereof to Arabic authors, see Sabra, *Optics of Ibn al-Haytham*, 2:xxxvi-xxxvii, and n. 39 (see also lxiii, n. 93); on the argument in favor of a post-Kindī compilation, and discretion on naming Arabic sources, see Rashed, *Oeuvres philosophiques et scientifiques*, 1:58-59. On the suggestion of the date of Ibn 'Īsā's text as "very probably from the tenth century, since it borrows propositions from al-Kindī's treatise on burning mirrors," see Rashed, *Geometry and Dioptrics*, 1043.

7) On the pseudo-Euclidean *De speculis* in Arabic (*Kitāb al-Mir'āh li-Uqlīdis*), see Elaheh Kheirandish, *Arabic Version of Euclid's Optics*, 1:243, and "What Euclid Said to His Arabic Readers: The Case of the Optics," in Gérard Simon and Suzanne Débarbat, eds., *Optics and Astronomy* (Proceedings of the XXth International Congress of History of Science, Liège, 20-26 July 1997, vo. XII) (Turnhout, 2001), 17-28. The Latin version of the text (which is to be distinguished from the Euclidean *Catoptrics* extant in Greek and Latin, not Arabic) is published in *Alkindi, Tideus und Pseudo-Euklid: Drei optische Werke*, eds. Axel Anthon Björnbo and Sebastian Vogl, in *Abhandlungen zur Geschichte der mathematischen Wissenschaften* 26/3 (1911), 97-119; on its partial Arabic version, see A.I. Sabra, "A Note on Codex Biblioteca Medicea Laurenziana Or. 152," *Journal for the History of Arabic Science*, 2 (1977), 276-283, containing reference to Sebastian Vogl, "Über die (pseudo-)euklidische Schrift '*De speculis*'," *Archiv für die Geschichte der Naturwissenschaften und der Technik* 1 (1900), 419-435.

8) *Nūr al-'uyūn wa jāmi' al-funūn*, MS. Bibliothèque Nationale arabe 3008, 178 fols.; Arabic ed. M.Z. Wafai and M.R. Qal'ajī based on this and two other Arabic manu-

The present paper presents sections from the *Optics* of Ibn 'Īsā that contribute to the early history of optics in general and to the concept of 'experiment' in particular.[9] The opening sections are selected to discuss and extend present knowledge in the following directions: (1) Examination of some modern discussions of early experiments with reference to the works of Ibn al-Haytham and Ibn Sahl, and their extension in the light of Ibn 'Īsā's *Optics*. (2) Discussion of the occurrence and meaning of the term *i'tibār* in Ibn 'Īsā's *Optics* in the context of the work itself, and identification of specific components relevant to early experiments: comparative examination, use of instrument, generalized procedures, etc. (3) Documentation of selections from the *Optics* of Ibn 'Īsā and the two above-mentioned sources before and after it, and conclusions based on these and other historical sources.

Examinations and Extensions

In 1989, A.I. Sabra, "repeated and expanded the evidence" of his 1971 article, "The Astronomical Origin of Ibn al-Haytham's Concept of Experiment," which in his own words argued that: "in the

---

scripts (Riyāḍ, 1987), with introd. in English. For references to this text, see J. Hirschberg, *The History of Ophthalmology*, vol. 2, trans. from German Frederick C. Blodi (Bonn, 1985), 71-72, 86-87, 242; Sabra, *Optics of Ibn al-Haytham*, 2:lxvi and n. 103-106; and Rashed, *Oeuvres philosophiques et scientifiques*, 58, n. 61. All the above sources have the name Ṣalāḥ al-Dīn Kaḥḥal of Ḥamat (Ḥamawī) (ca. 1296) as the author. Tzvi Langermann, in his most recent articles, "Ibn Kamūna in Aleppo," *Journal of the Royal Asiatic Society* (*JRAS*), 3rd series, 17 (2007), 1-19, 18, n. 60 and "From my Notebooks: A (The?) Geniza Fragment on Optics," *Aleph* 8 (2008), 345-358, 351, n. 16, writes of the author (whom he refers to as Abū Zakarīyyā' Yaḥyā ibn Abī al-Rajā') that "his correct identity was established by Gregor Schoeler, 'Der Verfasser der Augenheilkunde K. Nūr al-'Uyūn', *Der Islam* 64 (1987), 89-97." Read in n. 60: Ṣalāḥ al-Dīn for Ńalāḥ al-Dīn, and in n. 60 and n. 16, *K. Nūr al-'uyūn* (as in Ullmann's and Schoeler's originals), and Riyāḍ for Riyadh.

9) The appearance of the concept of "experiment" and the term "*i'tibār*" and its distinct meanings in Arabic texts is discussed in Sabra, "The Astronomical Origin," and Sabra, *Optics of Ibn al-Haytham*, with reference to earlier works such as Matthias Schramm's *Ibn al-Haythams Weg zur Physik* (Wiesbaden, 1963), and more recently, by Rashed, *Geometry and Dioptrics*, who besides modern references (1040-1043, n. 10-13) includes Ibn 'Īsā's use of the term.

Optics of I.H. there appeared for the first time a distinct concept of experiment consistently associated with three cognate words, *iʿtabara*, *iʿtibār* and *muʿtabir*, which the Latin translation of the book rendered as *experimentare*, *experimentatio* and *experimentator*," and that

> the appearance of this concept, being essentially different from the Aristotelian and medical *empeiria* (almost always expressed in the Arabic literature by *tajriba*, experience), should be regarded not as a development within Aristotelianism or Galenism but as a "result of taking over into optics an idea [of testing] which had an established career in astronomy."

In the much more extensive treatments of the volumes on *The Optics of Ibn al-Haytham*, one that goes beyond both Ibn al-Haytham and his *Optics*, Sabra added that in his opinion, [1] "the experiments in the *Optics*, or most of them, are essentially different in form from both the repeated experiences of the physicians, and the 'comparisons' of the astronomers," and that [2] "a change in the meaning of the term was bound to happen when *iʿtibār* was transferred to an experimental discipline like optics."[10]

In 2005, Roshdi Rashed made remarks on 'experimentation' occasioned by a study of the works of Ibn Sahl, belonging to a generation before Ibn al-Haytham, with conclusions that would similarly benefit from the contextual meaning of both the concept and the terminology of the *Optics* of Ibn ʿĪsā. Those most relevant here are that [1] "the term *iʿtibār*, belongs neither to the Greek catoptrical traditions, nor to the Arabic catoptrical tradition," a discussion in which he includes "a compilator like Aḥmad ibn ʿĪsā [who] uses the term *iʿtabara* once in the general sense of 'to consider', in the expression 'to think and to consider' (*fakkara wa-iʿtabara*), and hence in a non-technical sense"; and [2] that some occurrences of the term *iʿtibār*, which "covers an entire spectrum of meanings," do "have a meaning in common: experimentation carried out with an instrument fashioned with this goal in view."[11]

---

10) Sabra, *Optics of Ibn al-Haytham*, quotations from 2:14 and 18.

11) Rashed, *Geometry and Dioptrics*, 1043 and 1044, respectively. On the works of Abū Saʿd al-ʿAlāʾ Ibn Sahl, see also Roshdi Rashed, "A Pioneer in Anaclastics: Ibn Sahl on Burning Mirrors and Lenses," *Isis* 81 (1990), 464-491.

Both Sabra and Rashed refer extensively to Ptolemy's *Optics* (2nd c. CE) and to a terminology corresponding to *iʿtibār* in the Latin version of the apparently lost Arabic original of that text.[12] What follows is an extension of the above arguments and sources through texts readily available in Arabic. The texts are discussed either for the first time or from previously overlooked angles to track the visible trails of early experiments from the identification of their key components in Ibn ʿIsā's own *Optics* to analyses of relevant concepts and expressions in other Arabic sources.

Identifications and Discussions

Comparative Examination: The text of Ibn ʿĪsā's *Optics* indicates that the intended meaning of the term *iʿitibār* in the single occurrence of its verbal form in that text is a *comparative* form of examination in both its immediate and larger contexts. The immediate context of the verb *iʿtabara* suggests a meaning distinct from the preceding term *fakkara* ('thinks' or 'considers'). The employment of two adjacent terms with the same meaning would itself be hard to explain if different connotations were not intended. The larger context of the joint expression and the collective meaning of the four cases presented also point to the same direction: the case in which the concept and term appear (case 4) involves conclusions based on a physical examination with the aid of an instrument such as a tube, conclusions that are *comparable* to those in the preceding demonstrations (cases 1-3) and linked to them through the presence of a terminology to that effect (case 4).

From this passage in Ibn ʿĪsā's *Optics* it seems that a concept appearing with the terminology of *iʿtibār* is still used in a context not so "different from the comparisons of astronomers" (as described by Sabra with reference to Ibn al-Haytham), in the sense that the reference points for comparison are previous results. Nor is it easily

---

12) Albert Lejeune, *L'Optique de Claude Ptolémée dans la version latine d'après l'arabe de l'émir Eugène de Sicile*, (Leiden, 1956, repr. 1989) 31; A. Mark Smith, *Ptolemy's Theory of Visual Perception: An English Translation of the Optics with Introduction and Commentary* (Philadelphia, 1996), 1-270.

distinguishable from the repeated 'tests' (*imtiḥān*) of early catoptrics (as described by Rashed with reference to Ibn Lūqā), insofar as its verifications are repeated, even if with the same instrument. But despite the limited scope of the passage at hand, the contextual meaning of the verb *i'tabara* does not allow its interpretation to be reduced to a more general sense of the term as 'consideration' (as suggested by Rashed with reference to Ibn 'Īsā), a sense made further superfluous by an adjacent synonym. It rather seems that a more specific usage of the term here would be consistent with both the astronomical sense of "testing by comparing one observation with another" in Ptolemy's *Almagest* [*peira* and *synkrisis* in Greek, and *miḥna/imtiḥān* and *i'tibār* in Arabic], and the most common occurrences of such a concept in Ptolemy's *Optics* (with whatever lost terminology in either Greek or Arabic) in the specific sense of involving instrumentation.[13] As we shall see, there is yet another ancient source whose surviving Arabic text is promising for uncovering links in these and other connections.

Use of Instruments: the use of an instrument such as a tube (*unbūb*) in Ibn 'Īsā's *Optics* to demonstrate the form and function of visual radiation seems, surprisingly enough, linked to the Euclidean, rather than Ptolemaic, tradition, indicating earlier usages of at least the instrumental component of the concept of experiment. Reference to the same apparatus (tube on eye surface) in the same context (eye curvature limiting rays) is readable (though without the specific term *i'tibār*) in the badly damaged Arabic manuscript of the pseudo-Euclidean *De speculis*, a text titled as Euclid's own in Arabic, and one that reveals a version of the visual-ray hypothesis strikingly close to both al-Kindī and Ibn 'Īsā. The latter's specification of the material of the tube as copper (*nuḥās*) or another texture may seem like a sign of physical considerations in 'testing' (*i'tabara*) previous results through an actual instrument, as opposed to an imaginary tube in a 'thought experiment'. Either way, it is related to theoretical

13) Sabra, *Optics of Ibn al-Haytham*, 14-18 (14-16 on the terminology of the *Almagest* of Ptolemy; 17-18 on the *Optics* of Ptolemy and Ibn al-Haytham); Rashed, *Geometry and Dioptrics*, 1042-44 (1042-43 on the *Optics* of Ibn Lūqā and Ibn Īsā, 1044 on the *Optics* of Ptolemy and Ibn Sahl).

considerations in blocking vision by a non-transparent texture such as copper (occurring also in following propositions), as a rationale behind that demonstration.

There is still more to early experiments than comparative examination and use of an instrument, two components stressed, respectively, by Sabra and Rashed. Before getting to the cases of 'generalization', 'contradiction', and 'sense-perception' of the three demonstrations not previously discussed with reference to Ibn 'Īsā or others (cases 1-3), this would be a good place to correct some published speculations.

Speculations about Identity (Euclid and 'others'): The only known historical source referring explicitly to Ibn 'Īsā's *Optics* by author and title begins its formulation of the Euclidean visual-ray hypothesis by the statement that "Euclid and other(s) said in the Book of *Optics*...." The statement is then followed by the three main components of that hypothesis whose variations from the main corpus of Euclidean optics are almost identical in at least three Arabic sources: the pseudo-Euclidean *De speculis*, the *Optics* of Ibn 'Īsā, and the *Rectification* of al-Kindī (his *De aspectibus* is extant only in a Latin version).[14] Among important questions raised by the passage itself, the most immediate with regard to its opening lines is: who is meant by the expression 'other(s)' next to Euclid's name? The suggestion that Ibn al-Haytham is meant is downright false, given the substantially more advanced visual theory of his *Optics*. A consideration of Ibn 'Īsā or al-Kindī in that connection has its own difficulties, given the presence of the former's name elsewhere in the text and the absence of the latter's name in that connection altogether. The statement in a passage in Ibn 'Īsā's own *Optics* referring

[14] On al-Kindī's *De aspectibus*, see *Alkindi, Tideus und pseudo-Euklid*, eds. Björnbo and Vogl, 1-70. A new edition of that Latin text and an Arabic edition of the *Rectification* (*Taqwīm al-khaṭā' wa al-mushkilāt allatī li-Uqlīdis fī Kitābihi al-mawsūm bi al-Nāẓir* [*i.e. al-Manāẓir*]), with a French translation are included in Rashed, *Oeuvres philosophiques et scientifiques*, 161-358. For other studies of al-Kindī's *Optics* and their ancient background, see David C. Lindberg, *Theories of Vision from Al-Kindi to Kepler* (Chicago, 1976); and Peter Adamson, "Vision, Light and Color in al-Kindī, Ptolemy and the Ancient Commentators," *Arabic Sciences and Philosophy* 16 (2006), 207-236.

to "Philosophers, among them and with them Euclid," is no less problematic for similar speculations concerning al-Kindī.[15]

By contrast, it is rather likely that the person meant by 'other' in these cases is the author of the pseudo-Euclidean *De speculis* or its source, that is, the author of the Greek text rather than the Arabic version quoted. This is indicated by the context of both quoted statements: in the case of *Light of the Eyes*, by the context of the statements just before it that mentions Democritus, Epicurus, Empedocles (Duqlis), Hipparchus, Galen, and Plato by name and their visual theories, and in the case of Ibn Īsā's *Optics*, by references to a number of Greek authors throughout the text (Euclid, Anthemius, Archimedes, Aristotle, the Sophists, Hippocrates and Galen), this time without a single reference to an Arabic author.[16] But the interpretation of the terms 'others' and 'philosophers', cited along with Euclid, as references to the pseudo-Euclidean author or his source must remain tentative, however plausible, without more evidence to raise it from the status of speculation, to one of identification.

Speculations about Dating (Ptolemy's *Optics* in Arabic): Significantly, the old question of the date of the lost Arabic translation of the *Optics* of Ptolemy, the text that forms the basis of its only extant version in Latin, must itself remain open-ended. It may not be set as "between the 70's of the 9th and the 10th century," solely on the basis of Ibn Sahl's citations of Ptolemy's *Optics* in his works of 983-985, together with the absence of any knowledge of it in

15) On the association of Ibn al-Haytham by the editors of *Light of the Eyes*, see *Nūr al-ʿuyūn*; Arabic ed. Wafai and Qalʿajī, 55, n. 3; on the association of Ibn ʿĪsā, see Sabra, *Optics of Ibn al-Haytham*, lxvii, and n. 106 (where mention is made of explicit references later in that text); on the association of al-Kindī with Ibn ʿĪsā's passage, see Rashed, *Oeuvres philosophiques et scientifiques*, 59, and n. 62. For the Ibn ʿĪsā passage, see sec. IV below. The latter's name appears in both long form (Aḥmad ibn ʿĪsā) and short form (Aḥmad) in a different passage in *Nūr al-ʿuyūn* (ed., 480-481).

16) Sabra, *Optics of Ibn al-Haytham*, 2:xxxvi, n. 39 and lxvii, n. 105 provide the full lists; the latter includes reference to *Aetius Arabus*, 202-03 [ed. Hans Daiber (Wiesbaden, 1980)], a text that has a passage almost identical to the first one mentioned above, to the exclusion of the version of the visual-ray hypothesis in *De speculis*, making its addition by the author of *Light of the Eyes* a reasonable act.

the works of al-Kindī and Ibn Lūqā in the 870s.[17] The failure of al-Kindī and ibn Lūqā, and of Ibn ʿĪsā as well, to cite Ptolemy's *Optics*, in the vast and poorly connected environments where those contemporaries lived and worked, does not demonstrate that the Arabic translation was not made until *after* their time: the translation might have been made, but not have been *accessible* except in the cases of Ibn Sahl and Ibn al-Haytham. Neither can the absence of the vocabulary of *iʿtibār* in early optical and catoptrical texts accessible to us be supposed to extend "down to the beginning of the tenth century," nor can it be related to the dates of Ibn ʿĪsā who has that vocabulary as "very probably from the tenth-century." First because Ibn ʿĪsā's source is not considered to be Ptolemy's *Optics* as judged on internal grounds; and second, because the term *iʿtibār* had been available in texts as early as the Arabic *Almagest*. Rashed's explanation of the overlaps of Ibn ʿĪsā's text with the optical works of al-Kindī as borrowings without citations is itself subject to further investigation, especially when clues from common Greek sources have not been exhausted.[18]

These standing questions concerning the Euclidean and Ptolemaic optical traditions are not the only cases for which the evidence of early Arabic optics, and of the *Optics* of Ibn ʿĪsā in particular, can positively contribute to reinterpretations and reformulations. The cases to be focused upon in the remainder of the present paper are those specifically related to 'demonstration' and 'experiment'. These are adopted from the opening sections of the optics part of Ibn ʿĪsā's text, a part made up of definitions and

---

17) Rashed, *Geometry and Dioptrics in Classical Islam,* 1045: "As far as the latter [Arabic translator of Ptolemy's *Optics*] is concerned, all we know for the time being is that he lived between the 70's of the 9th and the 10th century. Ibn Sahl cites the *Optics* in his work of 983-985 ... [and] examination of the optical works of al-Kindī and Ibn Lūqā shows, contrary to what has been asserted, that they did not know Ptolemy's *Optics*..."

18) Ibid., 1043: "The former [*iʿtibār*] belongs neither to the Greek catoptrical traditions, nor to the Arabic catoptrical traditions down to the beginning of the tenth century. What is more, this absence is still observed in the work of such tenth-century catoptricians as ʿUṭārid, and a compilator like Aḥmad ibn ʿĪsā... very probably from the tenth century, since it borrows propositions from al-Kindī's treatise on burning mirrors."

propositions on the subject of vision, the main subject of optics at the time. The selections are in the form of textual summary and historical commentary on four successive demonstrations for one and the same proposition: these are labeled as demonstrations involving 'causal explanations' (*'illa*) (case 1), contrary 'indication' (*dalīl*) (case 2), 'sense-perception' (*ḥiss*) (case 3), and comparative examination (*i'tibār*) (case 4), the last representing a phase earlier than its more established meaning as 'experimentation'.[19] What follows is as much a reflection on the nature and courses of transmission of the Euclidean and the Ptolemaic traditions, as it is on the methodological shifts from geometrical and logical illustrations to physical and experimental demonstrations.[20]

Documentation and Conclusion

From Ibn 'Īsā, *Optics*: MS Laleli 2759; MS Ragip Pasa 934; *Nūr al-'uyūn*, ed., 1987 (henceforth cited as NU).

- *Definitions*: Variation on the Opening Euclidean Definitions.
- *Translation*:

> The philosophers, and Euclid among them and with them, said that: [1] The eye spreads out (*yanbathth*) from its viewing [organ] (*nāẓir*) a luminous power (*quwwa nūriyya*) that affects that part of the illuminated atmosphere that it encounters, namely air when it is illuminated, in a gathering of illumination.... [2] And the

19) The format is adopted with some adjustments from Kheirandish, *The Arabic Version of Euclid's Optics*. The full texts of some preliminary samples and titles appear in Elaheh Kheirandish, "Science and Mithāl: Demonstrations in Arabic and Persian Scientific Traditions," in *Sciences, Crafts, and the Production of Knowledge in Eastern Islamic Lands (ca. 184-1153 AH/800-1740 CE)*, eds. Najm al-Din Yousefi, Elaheh Kheirandish, Carol Bier, *Iranian Studies*, 4 (2008), 465-489. Modifications and references to the corresponding sections in *Light of the Eyes* (*Nūr al-uyūn*, ed. 1987: NU) are added here.

20) On Euclidean optics, see *Euclidis Optica, Opticorum recensio Theonis, Euclidis Opera Omnia,* eds. J.L. Heiberg et H. Menge, vol. 7 (Leipzig, 1895); on Ptolemaic optics, see Lejeune, *L'Optique de Ptolémée*; Smith, "Ptolemy's Theory of Visual Perception"; Sabra, *Optics of Ibn al-Haytham;* and A.I. Sabra, "Ibn al-Haytham's Criticisms of Ptolemy's *Optics*," *Journal of the History of Philosophy* 4 (1966), 145-149.

shape enclosed by that illumination is an even cone (*makhrūṭ*), its pointed end being at the viewing [organ] and its extremity, in touch with the viewed object. [3] That upon which this luminous ray (*al-shuʿāʿ al-nūrī*), namely the illumination spread out of sight, falls, sight sees and perceives it, and that upon which the ray does not fall, it does not.... (MS Laleli 2759, fols. 26a-27a; MS Ragip Pasa 934, fols. 5b-7a; NU, 55).[21]

- *Proposition*: Enunciation on Visual Radiation and Perception
- *Textual Summary*: [1] Spreading out of luminous power to facing objects in continuous conic form, with apex next to the eye and base next to the object. [2] Straight-line correspondence of effective points on the eye's surface and facing points on visual object. [3] Limited scope of visual radiation due to the eye's spherical configuration, and its obstructive curvature. (MS Laleli 2759, 27a-27b; MS Ragip Pasa 934, fols. 7a-7b; NU, 55-65).
- *Demonstrations*: Here, a number of cases are given, which will now be discussed in detail.

Case 1: Illustration Involving Causal Explanation

Textual Summary: [1] The cone TAB (see figure 1) is drawn with its apex

T at the center of the eyeball and base AB at the visible object. [2] The effective area of vision is represented by the arc GDE, cut off by the cone. Any point on AB is said to be seen through every point on GE, from which a straight line may be drawn to the visible point. In the case illustrated by the figure, point W on the axis of the cone is visible from all points on GE, and thus [3] W will be seen more clearly than points close to A and B because it is reached through more lines than any of those points. For example, GZ being drawn tangent to the circle at G, it

[21] This is a translation of a preliminary transcription. The Arabic text corresponds closely to a passage in the pseudo-Euclidean *De speculis*: MS Biblioteca Medicea Laurenziana Or. 152, fol. 104a, see "Supplementary Passages" in Kheirandish, *The Arabic Version of Euclid's Optics*, 1:243.

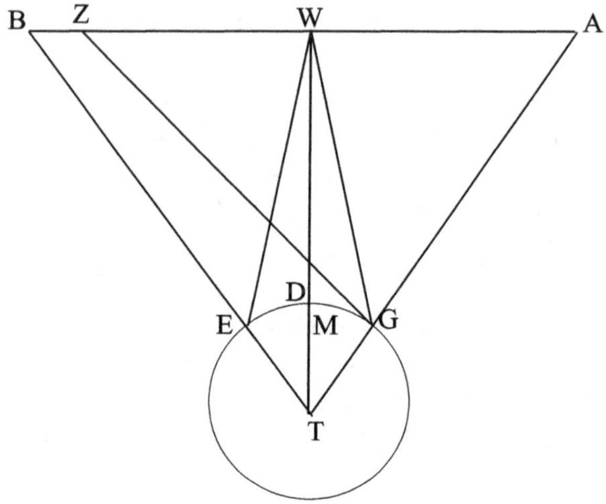

Figure 1

> is clear that no point on segment ZB will be visible from G.... (MS Laleli 2759, fols. 27a-29b; MS Ragip Pasa 934, fols. 7a-9b"[22]; NU, 56-57).

Historical Commentary: This demonstration is the first of four successive demonstrations for one of the opening propositions of Ibn ʿĪsā's *Optics*. Similar to the three cases following it, it involves an illustration (*mithāl*), in this case providing a geometrical proof for the components of the stated enunciation of the opening proposition [1-3 above]. The demonstration continues with a part not explicit in the enunciation that [a] 'because' upon all the points closer to the extreme point on the object (B), fewer of the rays fall than do on the central point (W), then the area around the apex of the perpendicular (WT), upon which the rays originating from arcs GD and ED fall from beginning to end, is subject to a more accurate vision due to what falls upon it. Providing a 'causal' explanation, a standard feature for geometrical demonstrations, is then followed by a generalized premise that [b] this is the case for *all* bodies, because that upon which the ray shines and covers all of it (namely

[22] Textual summary quoted from Sabra, *Optics of Ibn al-Haytham*, 2:25-26, where it is part of his discussion of "punctiform analysis" of light radiation with reference to treatments by al-Kindī and Ibn al-Haytham.

the region around perpendicular TW), has the (most) accurate vision, and that is what we wished [to demonstrate].

Ibn ʿĪsā's treatment of this case has features that are certainly comparable to al-Kindī, in both his *De aspectibus* and *Rectification* (*Taqwīm*), extant in Latin and Arabic respectively. The explanation of 'accuracy of vision' (*ṣidq al-ruʾya*) based on a *causal* premise (*ʿilla*), in this case the *amount* of falling rays, is a common feature of late treatments as well.[23] But the explicit formulation of a *generalized* premise, here that this is the case for *all* bodies, is one of the key features of the experimental treatments of Ibn al-Haytham (see case 5).

Case 2: Illustration Involving Contrary Indication

Textual Summary: If the ray were not to spread out of the arc to perceive the object through a power as described, then only a certain amount of the object, and only that amount, would be seen. But we have seen that sight does perceive half of the heavens. And this is an *indication* (*dalīl*) that from sight a luminous ray issues spreading out in the illuminated atmosphere in the form of a cone, as we described in the preceding (*awwal*) proposition. The demonstration continues with an analogy with touch, taste, hearing and smell, stating that if sight were not connected continuously with its object, its vision would be nullified. But in the case of sight, if a radial power (*quwwa*) would not spread out by which sight would see what it encounters, perception of half of heaven would not occur.... (MS Laleli 2759, fols. 29b-30b; MS Ragip Pasa 934, fols. 9a-10b; NU, 57-58).

Historical Commentary: The case of demonstrations involving contradictions is clearly rooted in the Greek classical tradition, a case for which the classic work of Árpád Szabó is a useful starting

23) Kheirandish, *The Arabic Version of Euclid's Optics*, vol. 2, 30-34 is devoted to relevant treatments. A proposition in al-Kindī's *Taqwīm* is most closely comparable to case 1 above in both content and figure, where line GZ is drawn correctly as a tangent to the circle at point G in its single known manuscript: Qum MS. Marʿashī Najafī 7580, fol. 70; Rashed's ed., 167.

point for the distinctions between three stages of Greek demonstrations: 'illustration through visualization' (physical representations and superimpositions), 'anti-illustrative and anti-empirical tendencies' (generalizations and abstraction), and 'indirect demonstrations' (self-contradictions and non-visual conceptualizations).[24] The breakdown of the so-called logical display (*deiknymi*) into a sequence of verbal statements, concrete examples, and demonstrations by example, corresponds to many Arabic optical demonstrations as well. It would help to consider that the divisions of a Euclidean proposition in Arabic (as provided in a passage in Yaʿqūbī's *Ta'rīkh*), also includes the case of contradictions, a division not found in its Greek version (as provided by Proclus). The third division of a proposition in Arabic is not only labeled 'contradiction' (*khalaf*); it is defined as the "contrary of illustration, and the extension of enunciation to what is not possible."[25] It is, therefore, not surprising that the second case of demonstration in Ibn 'Īsā's text is not accompanied by a figure in any of the known manuscript copies of the text. While the source of the analogy of vision with external senses of hearing, touch, smell and taste remains unclear beyond the expected natural philosophical traditions, later application of contradictions is clearly identifiable in experimental treatments by Ibn al-Haytham (see case 6).

---

24) Árpád Szabó, "The Development of Mathematical Demonstration and of the Concept of Evidence: The Transformation of Mathematics into Deductive Science and the Beginnings of its Foundation on Definitions and Axioms," *Scripta Mathematica* 27 (1964), 27-139.

25) On the Euclidean tradition in geometry, see T.L. Heath, *The Thirteen Books of Euclid's Elements*, 3 vols. (New York, 1956), 1:129-131; includes a passage from Proclus with the original Greek terms: Proclus, *A Commentary on the First Book of Euclid's Elements*, Trans. with Introd. and Notes by Glenn R. Morrow (Princeton, 1970), 159-161: enunciation, exposition, specification, construction, proof and conclusion: *protasis, ekthesis, diorismis, kataskene, apodeixis, sumperasma*; Yaʿqūbī, *Ta'rīkh*, ed. M.Th. Houtsma (Leiden, 1883), 136: *khabar, mithāl, khalaf* (contradiction), *tartīb, faṣl, burhān, tamām.*

Case 3: Illustration Involving Sense-Perception

Textual Summary: This demonstration offers a proof that involves sense in the form of the obstruction of an object by a physical obstacle represented geometrically by a horizontal line (see figure 3). The demonstration begins by considering, opposite to the exposed sensitive arc of the eye (GD) and parallel to the object seen (AB), a line (HT) equal in width to the width of the exposed arc (GD) and by explaining that the part actually seen of the object (AB) goes beyond what is 'blocked' in the wording of the proposition, i.e., many times more than the amount equal and parallel to the assumed line (WZ) for a 'reason': it demonstrates, this time through a visual as well as logical display, that in this case, the reason, is the visual configuration and radiation specifications posed in the given assumptions. The proposition ends by stating that it would become clear (*qad ẓahara*) to whoever looks into geometry that what we say is true (*ḥaqq*) and obvious [not hidden (*lā khaffā'*)]. (MS Laleli 2759, fols. 30b-32a; MS Ragip Pasa 934, fols. 10b-12a; NU, 58-59).

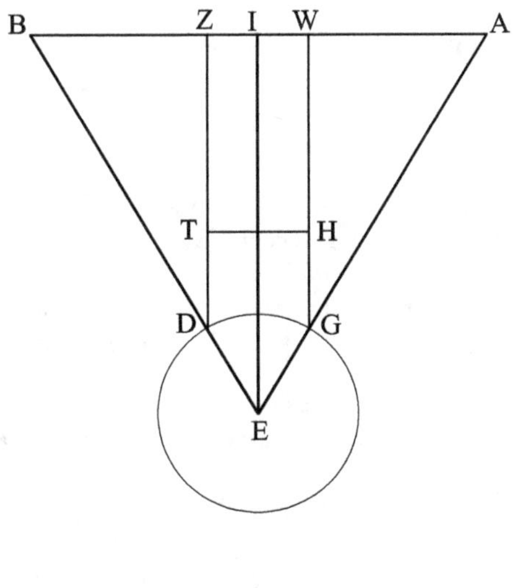

Figure 3

Historical Commentary: The third of the four demonstrations does not appear in this form in sources presently known. The expression 'illustration seen by sense' (*mithāl yurā ḥissan*) is itself rare other than comparable cases in Ibn 'Īsā's own *Optics*, such as the closing propositions uniquely carrying his name in *Light of the Eyes*. Expressions comparable to 'sensory illustration' (*mithāl ḥissī*) do exist, however, in the form of 'explanations involving sense' (*bayān ḥissī*) in contemporary works such as that by Qusṭā ibn Lūqā.[26] But conceptions such as experimental proof of enunciations by physically blocking areas within facing visual regions may not appear again until Ibn al-Haytham.[27]

Case 4: Illustration Involving Comparative Examination

Textual Summary:

> To illustrate the 'natural' tendency of the 'luminous ray' to spread out (*yanbathth*) in the shape of a cone, one end of a tube [is placed] at the surface of the eye: the tube will restrict the passage of the luminous power issuing from points on that surface, but as the light comes out of the tube at the other end it fans out to assume the natural conic shape. Thus the visible area at a distance from the far end of the tube will be larger than the defined straight lines drawn from the center of the eye or from points on the visually effective area and grazing the far edge of the tube.[28] (MS Laleli 2759, fols. 32a-33b; MS Ragip Pasa 934, fols. 12a-13a; NU, 59-60).

Translation (see figure 4):

> We also (*ayḍan*) see this in another way. So we consider circle GID as the sphere of the viewing [organ] (nāẓir) with center E from which the ray (*shu'ā'*) issues, and arc GD as the exposed part (*ẓāhir*) of the eyeball...; and we consider (*nufriḍu*) a tube, GHDT, in copper or some other [material], and we mount it on arc GD,

---

26) For Ibn Īsā's name next to the expression in *Nūr al-'uyūn*, see Paris MS, fol. 32b-33a, ed., 476-477. For Ibn Lūqā's text, see Rashed, *Oeuvres philosophiques et scientifiques*, 572-645, at 589.

27) "It is evident from this experiment (*i'tibār*) [with equal scripts on equal paper strips] that of the visible objects facing the eyes and binocularly perceived, the clearest is that which lies where the two axes meet, and the closer the object is from this point, the clearer (*abyan*) it will be"; see *Kitāb al-Manāẓir*, bk. III, 2, sec. 61, ed. Sabra, 366; tr. Sabra, 1:245.

28) This textual summary is quoted from Sabra, *Optics of Ibn al-Haytham*, 2:26, where it is part of his discussion of "punctiform analysis" of light radiation with reference to treatments by al-Kindī and Ibn al-Haytham.

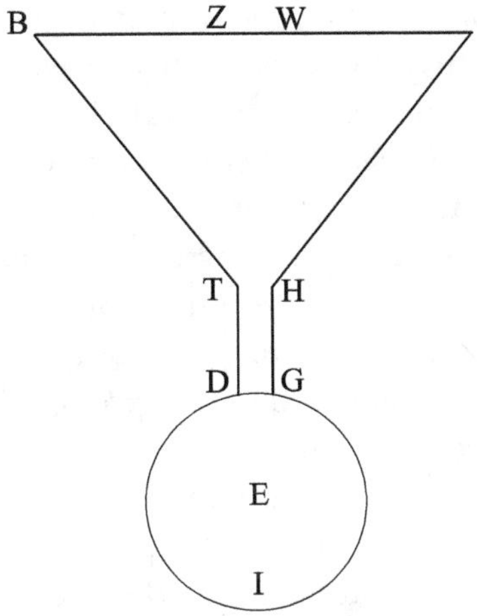

Figure 4

and also consider the object, line AB, parallel and opposite to it. And we take on line AB, line WZ equal to the magnitude of tube's width, line HT, and parallel to it. So from this, it is necessary that since arc GD is the viewing [organ], then it would perceive only line WZ of line AB, and nothing else, that is, if the eye were to see only by its fitness, not a power (*quwwa*) issuing from it. As such (*fa kadhālika*), this would be a correct comparison/syllogism (*al-qiyās*), because what preceeded (*taqaddama*) in our description was that if a ray did not issue from arc GD as we described it [i.e. power spreading out conically], then objects would rather be seen by themselves or by a cause (*'illa*) not outside of them. Then we had restricted that (arc GD) (*ḥaṣarnāhā*) by tube HGDT, because nothing would be seen through the tube in the right or north, or up or down, only that opposite to it, that is magnitude WZ. But we had [still] seen from arc GD through tube HGDT all of it (AB). So from this proposition it is also (*ayḍan*) necessary, for whoever considers (*fakkara*) and examines it (*i'tabara*) that if it were the case that arc GD, which is the viewing [organ], were to see things only by its soundness or by a cause (*'illa*) not surpassing that (*tajāwazahā*), then it would not see from the viewed object, upon this restricted viewing, more than its magnitude. But we [still] see with this restriction on it, many times more of its magnitude. So it is necessary from this that from sight a luminous ray spreads out in the illuminated air, and that whatever upon it this ray falls is seen, and whatever upon it, it does not fall is not seen... And this would be sufficient for whoever would look, however little, into geometry.[29]

---

[29] This is a translation of a preliminary transcription, slightly modified from an earlier form in Kheirandish, "Science and Mithāl," and supplied with some Arabic expres-

Historical Commentary: The last demonstration must be distinguished from the previous three in more than one way. The ways most relevant to the present discussion are: (a) the use of the term *iʿtibār* in the expression "whoever considers and examines (*iʿtabara*) it"; and (b) the use of an instrument such as a tube (*unbūb*). Regarding the first point, the nature of the employment of the term *iʿtibār* indicates that it is used in the context of a comparative examination, one that takes into account not only the cases before it, as suggested by the statement "from this proposition it is *also* (*ayḍan*) necessary...," but also the expression *al-qiyās*, whose context may be interpreted as comparative. As for the second point, namely the use of an instrument such as a tube, the procedure involves physically placing a tube on the sensitive arc of the eye to determine, through a comparative test, that the same assumption of a diverging visual power holds with a tube restricting the passage. Because the part of the object seen is much larger than the portion restricted by the narrow tube, Ibn ʿĪsā argues that the visual emission must be a power that spreads out from the end of the tube in conic form to touch physically a larger part of the object.

Do these four treatments come from previous sources? The question cannot be positively answered in each case. In the third and fourth cases, the exact meaning and application of 'sense-perception' and 'examination' cannot be entirely clear without any knowledge of earlier sources. And in the first two cases, alternative demonstrations prove statements that are treated as self-evident definitions in Greek. But with the increasing generalizations, contradictions, sense-perception and instrumentation of experiments, questions of development can be more directly posed than those of origins.

Two cases from the *Optics* of Ibn al-Haytham (numbered 5 and 6 after Ibn ʿĪsā's four cases) provide some means of comparison and evaluation. Figure 5 corresponding to both is a reconstruction.

---

sions. The use of a tube in copper also occurs with reference to a mirror a few folios later (Ragep pasa fol. 34a; Laleli, fol. 14a).

Case 5: Rectilinear Vision through Experimental Verification

(Ibn al-Haytham, *Optics*: Book I. Chap. 2, Sec. 5-7, Sabra, tr. vol. 1, 7-8; ed., 64-65).
Text:

If a survey is made of all visible objects at all times, and if they are experimentally [*iʿtibār*] and accurately examined, they will be found to be uniformly as we have described them, with no variation or change. This therefore proves that for every seen object that exists with the eye in the same atmosphere, and is not perceived by reflection, there exists between each point on the seen surface of the object, and a certain point or multiplicity of points on the surface of the eye, a straight line or lines which are not interrupted by any opaque body.

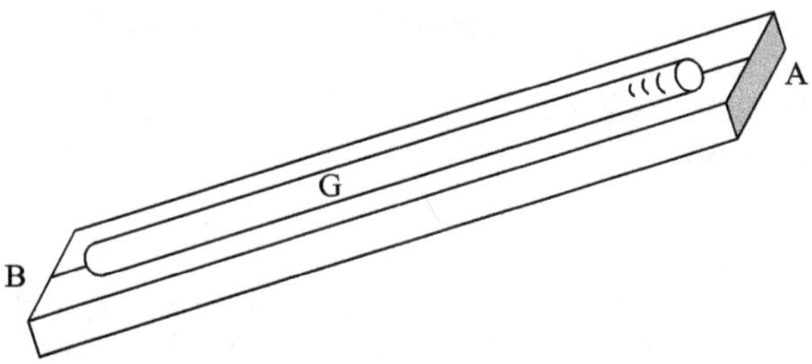

Figure 5

An accurate experimental examination of this fact may be easily made with the help of rulers and tubes [*unbūb*]. Let the experimenter who wishes to make such an examination [proceed as follows]. [Cf. figure 5] Take a very sound and straight ruler and draw along the middle of its surface a straight line parallel to its sides. Take a hollow cylindrical tube, very straight in length, perfectly round and ending in parallel circles; let its thickness be the same throughout and let it be fairly wide but not wider than the eye socket: draw on its outer surface a straight line extending from a point on the circumference of one base to the opposite point on the other side; and let this tube be a little shorter in length than the ruler. Divide the line along the middle of the ruler into three parts, and let the intermediate part be of the same length as the line on the surface of the tube; the remaining parts on either side may be of any length. Attach the tube to the surface of the ruler.... The tube should be so closely and firmly fastened that it cannot be loosened or displaced.

When the instrument has been perfectly prepared and the experimenter wishes to examine the perception of visible objects by sight, he should aim at one of these objects, put one end of this ruler close to the lower eyelid of one eye and the other

> end close to the surface of the object, cover the other eye, and while in this condition, look through the opening in the tube; he will see that part of the object which is opposite the opening of the tube at the end of the ruler. If he covers the opening of the tube with an opaque body, that part of the object will be screened off which he has seen through the opening. Upon removing the cover, he will perceive the same part as he did at first. If, by means of the opaque body, he covers any part of the opening, then there will be screened off only that portion of the visible part situated opposite the covered part of the opening, namely the part that lies on a straight line with the eye and the screening body-this straightness being secured by the ruler and the straightness of the tube.... When the cover is removed, the eye will again perceive that portion of the visible object. That is always found to be so, with no variation or change.

Case 6: Rectilinear Vision through Experimental Contradiction

(Ibn al-Haytham, *Optics*: Book I. Chap. 2, Sec. 8-9, Sabra, tr. vol. 1, 8; ed., 65-66).
Text:

> ...If it were possible for sight to perceive an object existing with it in the same atmosphere through non-rectilinear lines, then it would perceive part of the object opposite the tube's opening after the opening has been obstructed. But we find, when such an object is experimentally [*i'tabara*] examined and observed in the manner we have described that it ceases to be visible upon closing the opening. It follows from this experiment, with a necessity that dispels doubt, that sight does not perceive any visible object existing with it in the same atmosphere, this perception being not by reflection, except through the straight lines alone that can be imagined to extend between the surface of the object and the surface of the eye.

The cases above allow for several comparisons between the methodologies of Ibn 'Īsā and Ibn al-Haytham, despite the much richer, and apparently later, treatments of the latter. We have seen that some components of early experiments such as the presence of 'examination' and 'instrumentation', or the use of 'generalization' and 'perception', are identifiable in both cases, however different their form and capacity. But there are important distinctions between these treatments starting with objectives. It may not be immediately clear that the respective examinations (*i'tibār*) of these two authors, clearly representing different stages in the earliest phases of experimentation, were presented to verify different, if not opposite, assumptions. In the case of Ibn 'Īsā, who was a proponent of the visual-ray hypothesis, the examination was meant to verify the assumption of outward radial powers according to some of its Euclidean versions. In the case of Ibn al-Haytham, who refuted that hypothesis in all its variations, it was directed to the one-to-one correspondence between facing sensory and visible points, an assumption he is credited for raising to the status of the principle called the "punctiform analysis of light radiation" and associated with later theories.[30] The clearly different objectives of the two treatments, otherwise appearing to draw different results from one and the same procedure, is not their only outstanding distinction. Constancy and consistency, two components of later experiments, are quite pronounced in Ibn al-Haytham's 'uniform' results and repetitions 'without variation or change', respectively.

It is appropriate to close with a few comparable cases whose previously unknown or unreported texts, in both Arabic and Persian, deserve some priority with regard to the direction of further study on the present subject.

The concept of an 'experimenter' (*mu'tabir*), for example, a concept that appears throughout Ibn al-Haytham's works, is one that occurs elsewhere, though in a much more limited and isolated form, in the context of optics. This is in a work by 'Abd al-Raḥmān Al-Khāzinī (ca. 1100) other than his so-called 'Tested Tables' (*Zīj al-Mu'tabar*), a treatise preceding the first book of his *Zīj al-Sanjarī*, similarly astronomical under the self-explanatory title of "Points on the Manner of [Comparative] Examination (*i'tibar*) of the Positions of the Planets...." With constructs such as the 'person examining' (*ṣāḥib al-i'tibār*) and his 'sense of vision' (*ḥiss al-baṣar*), this treatise contains material that would be of much interest to the history of optics as well as astronomy.[31] When it comes to more mature

[30] See Lindberg, *Theories of Vision*, 26-30; Sabra, *Optics of Ibn al-Haytham*, 2:23-28; and more recently, Adamson, "Vision, Light and Color in al-Kindī," 211-12.

[31] *Risāla fī Nukut kayfiyat i'tibār mawāḍi' al-sayyārah*. A copy of the Vatican manuscript was provided to me by the late David Pingree for "its technical material on

concepts of 'experiment', however, comparisons are most appropriate with Kamāl al-Dīn al-Fārisī (d. ca. 1318), best known for his monumental commentary on Ibn al-Haytham's *Optics*. In addition to various occurrences of experimentation (*i'tibār*) in the known works of this outstanding scholar, a treatise on the manner of solar propagation in the atmosphere contains the same term not only in the context of repeated examinations (*i'tibār ākhar*), but also with reference to astrolabes (*usṭurlāb*) and the agreement of their results (*tawāfuq*), as well as terms such as observation (*raṣad*) and comparison (*qiyās*).[32]

For some revealing passages on the alternative cases of experience (*tajriba*) and testing (*imtiḥān*), we can turn to works as late as the early tenth/sixteenth century. One of these is an anonymous Persian composition dedicated to the Ottoman Sultan Selim I (r. 918/1512-927-1520), another, is a shorter Arabic text carrying the name of Sa'd ibn 'Īsā (better known as Sa'dī Çelebī: d. ca. 945/1529).[33] Interesting from the present standpoint is that the author, who may

---

optics, but representing the concerns of the astronomical observers pretty fairly," according to a handwritten note he attached to it.

32) *Risāla fī kayfiyya ḍaw' al-shams fī al-hawā',* Ayasofya MS 4855 (dated 721H), 178b-199a, 178b, 183b. On the optical works of Fārisī, see Sabra, *Optics of Ibn al-Haytham*, 2:lxiv-lxxiii; and A.I. Sabra, "The Commentary That Saved the Text: The Hazardous Journey of Ibn al-Haytham's Arabic Optics," *Early Science and Medicine* 12 (2007), 117-133.

33) Both texts are extant in apparently unique manuscripts in Istanbul. The Arabic text, *Risāla fī 'ilm al-manāẓir* (*Treatise on the Science of Optics*), has the name Sa'd ibn 'Īsā in the colophon (MS and author information kindly provided by Himmet Taskomur). The supposed author or owner/transcriber, Sa'dī Çelebī, the mufti and book collector of Constantinople, is different from Mirem Çelebī (ca. 1525), included in the *Mathematicians, Astronomers and Other Scholars of Islamic Civilization and their Works (7th-19th c.)*, by Boris Rosenfeld and Ekmeleddin İhsanoğlu (Istanbul, 2003), 319, no. 940. On the Persian text, *Risāla fī Sabab ru'yat al-ashyā' wa bayān al-madhāhib fīhi wa bayān ru'yat al-ashyā' fī al-marāyā al-maṣnū'a*, (*Treatise on the Cause of the Vision of Objects and the Explanation of Traditions* [*on Vision*] *and the Vision of Objects in Constructed Mirrors*), see also Kheirandish, "The 'Manazir' Tradition through Persian Sources," *Les sciences dans le monde iranien*, eds. Z. Vesel, H. Beikbaghban and B. Thierry de Crussol des Epesse (Tehran, 1998), 125-45. On the period and the region, see Tofigh Heidarzadeh, "Patronage, Networks and Migration: Turco-Persian Scholarly

be the same person given the close overlaps of these two texts, presents the striking narrative of a 'high standing' mirror sent by 'European scholars' to 'scholars in Khurasan' for the explanation of its rationale in the manner of 'experience and testing' (*tajriba va imtiḥān*).[34] Regardless of the exact author and date of such a significant exchange, the text is of great historical interest for the later phases of instrumentation and experimentation in both Islamic and European lands, especially in the light of the scarcity of evidence with respect to the much-debated role of optical instruments in fifteenth and sixteenth-century Europe.[35]

Finally, what does this all mean with respect to the development of the concept of experiment in general and its early instances in particular? It may be helpful to keep in mind the distinctions that have been made with reference to Ibn al-Haytham's much more advanced use of the concept of experiment between, not only his own *confirmatory experiments* and the *discovery experiments* of seventeenth-century optics, but also the 'experimental' and 'logical' confirmations by him and people before him.[36] In the case of Ibn 'Īsā it may be said that his demonstrations are such that all including those involving instrumentation (*unbūb*) and 'examination' (*iʿtibār*) need qualifications as 'experiment' or 'experimentation', justifying the presentation of these expressions in quotation marks throughout this paper. From the standpoint of the 'scientific method' at

Exchanges in the 15th and 16th Centuries," *Archives Internationales d'Histories des Sciences*, 55 (2005), 419-434.

34) "... Scholars (*ḥukamā*) from Europe (*farangistān*) had constructed (*sākht*) a high standing (*'aẓīm al-sha'n*) mirror (*mir'āt*) and sent to scientists (*'ulamā*) in Khurasan in the manner (*sabīl*) of 'experience' (*tajriba*) and 'testing' (*imtiḥān*) for the explanation (*bayān*) of its rationale" (*ḥikmat*)."

35) Mark A. Smith, "Reflections on the Hockney-Falco Thesis: Optical Theory and Artistic Practice in the Fifteenth and Sixteenth Centuries," *Early Science and Medicine*, 10 (2005), 163-185.

36) The pairs 'confirmatory and discovery experiments' and 'experimental proof' (*ithbāt bi al-iʿtibār*) and 'proof by syllogism' (*ithbāt bi al-qiyās*) are used with reference to Ibn al-Haytham, respectively by Sabra, "iʿtibār/experimentatio," *Optics of Ibn al-Haytham*, 2:18, and Rashed, "Experimentation," *Geometry and Dioptrics in Classical Islam*, 1040, n. 10, in translated quote from M. Nazif (Cairo, 1958).

large, supposed 'experimental' demonstrations expressed and transmitted in *form* through a clearly bookish mode, may then be said sometimes, but not always, to cross over the textual line, in *content*.

The "Experience-Based Medicine" of the Thirteenth Century

Michael McVaugh*
University of North Carolina

Abstract
We should not assume that medieval physicians did not take pains to found their practice upon evidence. Academic physicians at Montpellier ca. 1300 were cautious about accepting textbook claims for the powers of drugs, and tried to verify each drug's physiological effects before using it; yet they were also flexible, ready to believe that powerful new medicines might be discovered empirically that were unknown to their authorities or superficially inconsistent with existing knowledge. Likewise, physicians were careful to observe their patients closely and to try to identify the condition from which each was suffering, and when they were unsure of the nature of an illness, they feared to administer medicines lest their known effects might be harmful to the patient. Anticipating today's "evidence-based medicine," the physician's practice involved the conscientious use of current best evidence.

Keywords
evidence-based medicine, drug testing, *empirica*, *proprietas*, experience, diagnosis, therapeutics

I.

At the beginning of the twenty-first century, the phrase "evidence-based medicine" is capable of arousing fierce debate within (and

* Department of History, CB #3195, University of North Carolina, Chapel Hill, NC 27599-3195, U.S.A. (mcvaugh@email.unc.edu). I gratefully acknowledge the very helpful comments and contributions by Drs. Walton O. Schalick III and Fernando Salmón in response to earlier versions of this paper.

without) the medical community.[1] It has been defined by its supporters as "the conscientious, explicit and judicious use of current best evidence in making decisions about the care of individual patients."[2] Their best-evidence ideal is typically to be able to assess a case against hundreds of comparable cases that have been analyzed probabilistically in clinical trials in order to determine statistically the risk or benefit of a given treatment. Those with reservations are sometimes skeptical about the extension of probabilities to the treatment of individual patients or groups of patients (e.g., the critically ill), pointing out, for example, that difficulties in the identification of diseases and syndromes can make it impossible to decide what the "comparable case" would be.[3] Critics are also often unhappy with the insistence that other kinds of information—patient narratives, case histories—must be considered as "weak" evidence compared to the conclusions derived from clinical trials.

Ultimately what is in question here is the meaning of "conscientious, explicit and judicious use" and "current best evidence." If "evidence" can be construed to include all a practitioner's hard-won experience, one could say, with Edmund Pellegrino, that

> evidence-based medicine is the newest name for a very old idea. The best physicians have always sought the best empirical evidence available on which to base diagnoses and therapeutic decisions.[4]

---

1) The intensity of feeling on this subject was first made fully clear to me by the exchange of letters between J.K. Aronson and Druin Burch in the *Times Literary Supplement*, 15 and 22 February 2008.

2) David L. Sackett, William M.C. Rosenberg, J.A. Muir Gray, R. Brian Haynes, and W. Scott Richardson, "Evidence Based Medicine: What It Is and What It Isn't," *British Medical Journal*, 312 (1996), 71-72, at 71.

3) For example: "There are a number of difficulties in the conduct of randomized trials in the critically ill. These include difficulties in the definition of diseases and syndromes, a heterogenous [*sic*] population of patients undergoing a variety of therapeutic interventions, and outcomes that may not be able to discriminate between beneficial and risky therapies," in Paul C. Hébert, Deborah J. Cook, George Wells, and John Marshall, "The Design of Randomized Clinical Trials in Critically Ill Patients," *Chest*, 121 (2002), 1290-1300, at 1290 [abstract].

4) Edmund D. Pellegrino, "Medical Ethics and Virtue Ethics," *Theoretical Medicine and Bioethics*, 23 (2002), 397-402, at 397.

The proponents of evidence-based medicine react indignantly to this idea that their approach might be "old hat." At their most extreme, they come close to suggesting that the only believable medical evidence is produced by systematic research designed to elicit hard facts, as though medicine should ideally be a Comtian positivist science.[5] Perhaps not surprisingly, one of the leaders of the movement, David L. Sackett, has made it explicit that he locates its philosophical origins in mid-nineteenth-century Paris.[6] Yet if we take literally the definition given above, it is difficult not to feel that Pellegrino must be right and that evidence-based medicine was surely practiced before 1850. It would not merely seem condescending, it would seem ahistorical to posit that the pre-modern practitioner must have been ignorant and unthinking and did not appreciate the importance of determining the effectiveness of different therapies. As a matter of fact, some clinician-historians have already been brought to recognize medieval practice as in some sense a precursor of today's "evidence-based medicine,"[7] and indeed if we go back to the earliest moment in the history of academic medical communities, we will already see there what I think can be recognized as "the conscientious, explicit and judicious use of current best evidence in making decisions about the care of individual patients."

That earliest moment is the thirteenth century, when the institutional development of European universities began—in the case of Paris, development around a core of instruction in the liberal arts; in the case of Montpellier, around a medical faculty that existed as early as 1220 and by 1300 was arguably the most important center of medical teaching in Europe. During that century, the intellectual resources readily accessible to medical masters had multiplied:

---

5) Trisha Greenhalgh, "Narrative Based Evidence in an Evidence Based World," in *Narrative Based Medicine: Dialogue and Discourse in Clinical Practice*, ed. Trisha Greenhalgh and Brian Hurwitz (London, 1998), 247-265, at 247-249.

6) Sackett et al., "Evidence Based Medicine," 71.

7) D. Craig Brater and Walter J. Daly, "Clinical Pharmacology in the Middle Ages: Principles that Presage the 21st Century," *Clinical Pharmacology and Therapeutics*, 67 (2000), 447-50; Walter J. Daly and D. Craig Brater, "Medieval Contributions to the Search for Truth in Clinical Medicine," *Perspectives in Biology and Medicine*, 43 (2000), 530–540.

by century's end, they were trying to absorb most of the works of Galen as well as the great textbooks of Avicenna and Rhazes, all available to them in Latin. The discovery of this medical literature had explosively extended the horizon of academic medicine in all directions, theoretical and practical, but for our purposes what matters is that the new Greco-Arabic literature had brought to their attention a number of new drugs that they might be able to use in their practice, as well as new uses for familiar drugs. One consequence of this was to draw the masters' attention back to classes of medicines with well-recognized if dangerous powers—laxatives, opiates, and caustics: could they be used as widely as the new texts claimed? Another was to make them wonder whether the texts' claims for the powers of unfamiliar medicines were warranted, and whether their promise that *new* medicines could be created to meet particular conditions was justified. The Montpellier masters' problem around 1300 is perhaps not unfamiliar to us today: they had to decide how best to evaluate a mounting number of competing claims for the effectiveness of new drug therapies.

Three of the general questions that they raised about medical evidence seem to me in principle not so very different from questions that concern the medical profession of today:

> How far can we trust existing scientific reports about the properties and effects of medicines? Should those reports be subject to modification by our own controlled experience?
>
> How can we best evaluate that science of medicinal properties and modify or extend its conclusions, as necessary?
>
> How can physicians determine whether medicines whose physiological effects on the body have been established are *clinically* effective, and against what conditions?

From their response to those questions we may be able to judge just how conscientious medieval physicians were about using the best current evidence in deciding to treat their patients.

We are fortunate that two members of the Montpellier medical faculty at that moment who were sensitive to the problem happened to be articulate enough to let us see their response to it. One of these was Bernard Gordon, who began to teach there in 1283, composed his first written medical work in the mid-1290s, and his last in 1308. The other was Arnau de Vilanova, who had studied medicine there about 1260 and returned to teach in the school from about 1291 to 1301, during which time he too completed a number of medical treatises. It is not surprising that their works often address common issues, and those issues include their response to the opportunities for a new drug therapy.

II.

One such opportunity was presented to them by works from the Arabic medical tradition, arguing that individual medicines could be combined according to quantitative rules so as to produce a new compound medicine with predictable new properties, tailor-made for a particular patient. Galen had laid particular emphasis on the power of medicines to act on the qualitative nature of the body (its complexion) through their own primary qualities of hot or cold (and dry and moist), a power that he discussed at length in his works *De simplici medicina* and *De complexionibus*; both these books were part of the "new Galen" coming into popularity in the later thirteenth century that teachers were using to interpret basic texts like Hippocrates' *Aphorisms*. Knowing whether a specific medicine would heat or cool the body, and to what degree (Galen recognized a scale of four degrees of hot or cold in medicines), was essential if the physician wanted, say, to cool a fevered patient; but its hotness or coldness was also linked in a more or less regular way to specific physiological or clinical effects, like tightening passages within the body (*constrictio*) or bringing local infections to a head (*maturatio*): an aperitive medicine, for example, capable of opening and loosening the body, was naturally hot and moist. Once the physician has determined the nature of the medical problem to be corrected, he administers a drug whose established primary quality of hot or cold, or higher-order physiological effect, will counteract the specific problem he has identified. He may have learned the medicinal properties of drugs by a lifetime of testing, or he may simply accept textbook opinion as to their properties; but once the primary qualities of his medicines have been established experimentally, the ensuing treatment should be straightforward. What later Arabic authorities had proposed, particularly Alkindi and Averroes, was that the amount of hotness in a medicine depended on its degree and on the dose used of it, and that computations based on those two numbers would allow the physician to determine the medicinal degree of any mixture of medicines—and, conversely, would allow him to create a new medicine of any degree his patient might need.[8]

Here was a highly promising approach to refining the first stage in medical treatments, tempering a patient's complexion. Alkindi's work describing the necessary calculations had been admired by Latin physicians—at a distance—for fifty years before Arnau de Villanova mastered the technique and explained its arithmetic in his *Aphorismi de gradibus*, towards the end of his Montpellier teaching. But Arnau recommended restraint in making up these new medicines, warning of the variability of the ingredients and the unknown synergies that they might display when mixed, and Bernard Gordon elaborated on these uncertainties in 1303 when he outlined a shorter version of the technique. Everything depended upon knowing the qualitative nature of the individual components, and that could only be arrived at by testing them sensorily. In fact, Galen had already set out six rules in *De complexionibus* for determining whether drugs were hot or cold,[9] and Bernard now adapted them

---

8) The issues are outlined in the introduction to Arnau de Vilanova, *Aphorismi de gradibus*, in *Arnaldi de Villanova Opera Medica Omnia* [hereafter *AVOMO*], 2 (Barcelona-Granada, 1975).

9) Galen, *De complexionibus* III.5; *Opera* (Venice, 1490), 2:20v–21r. His rules were summarized in the 1290s by the former Paris master Jean de Saint-Amand as follows: "ut cibus et medicina quae offerunt sint munda ab omni qualitate extranea... ut illa offerantur corpori cujus complexio est singularis simplex... ut fiat differentia inter operationem per se et operationem per accidens ... ut quantitas medicinae mensuretur secundum quantitatem dispositionis corporis ... ut medicina sit simplex singularis... ut sit corpus existens purae qualitatis." See Georg Matern, "Die drei Bücher des Galen über die Temperamente," Inaug.-diss. (Berlin 1894), 23. Avicenna's rules in *Canon* II.i.2 (Venice, 1507; fol. 82r–v) obviously derive from Galen's, though they have been slightly altered and expanded: "ut sit medicina vacua a qualitate acquisita; ut illud in quo experitur egritudinem habeat singularem; ut medicina sit experta iam in

to his present purpose: the drug to be tested should not have any accidental properties, should not be artificially warmed or cooled; it should be homogeneous; it should be tested on a human body neutral in temperature between hot and cold; and its primary effect on the body should be carefully distinguished from any secondary ones.[10] And he concluded with a further rule of his own to be followed generally in testing medicines, whether familiar or unfamiliar:

> If we wish to try a certain medicine out on the human body, we should try it out first on birds, then on dumb animals, then in hospitals, then on lesser brethren (*fratribus minoribus*), and so on in turn, because it could kill if it turned out to be one of the poisons.[11]

Indeed Bernard, more than Arnau, reveals a certain skepticism as to whether the properties of the ingredients could ever be known with sufficient precision to make the system work with any confidence. Both evidently admired the scientific beauty of the Arabic calculations and appreciated their apparent power intellectually, but that did not translate into an immediate uncritical commitment to an apparently exciting new therapeutic option.[12]

---

duobus contrariis ita ut cum utrique eorum simul conferat non iudicetur quod sit contraria complexioni alicuius ipsorum; ut sit virtus que est in medicina opposita et quod equatur ei de potentia egritudinis; ut observetur tempus in quo apparet eius impressio et operatio; ut observetur processus operationis eius secundum assiduationem; ut experimentum sit in corpore hominis."

10) Luke E. Demaitre, *Doctor Bernard de Gordon: Professor and Practitioner* (Toronto, 1980), 127.

11) J.L. Pagel, "Ueber die Grade der Arzneien," *Pharmaceutische Post*, 28 (1895), 65-67, 131-133, 142-144, 180-182, 221-225, 257-262, at 143: "Et possumus addere 6[am] condicionem gratia doctrinae: quod si aliquam medicinam volumus experiri in corpore humano, quod prius experiamur in avibus et postea in brutis animalibus et postea in hospitalibus et postea in fratribus minoribus et postea in aliis per ordinem, quia si forte esset de genere venenorum interficeret."

12) I know of only one case where the techniques can be shown to have been applied to a recipe, and there they were used to characterize the nature of an existing recipe rather than to direct the preparation of a new one: see Michael R. McVaugh, "Quantified Medical Theory and Practice at Fourteenth-Century Montpellier," *Bulletin of the History of Medicine*, 43 (1969), 397-413, esp. 404-405, 411-413.

It is not surprising that Arnau and Bernard should have been so hesitant to trust in the predictability of the powers of these new medicines, for even the old familiar ones could not be counted on to behave consistently, and physicians invoked a string of causes that could affect normal medicinal activity and that needed to be borne in mind when administering them to a patient: the season and region in which they were collected; their freshness; the part used; their purity—all these factors could strengthen or weaken the characteristic effect of the drug. This problem was most serious in the case of medicines whose effects on the body were most powerful, like laxatives and opiates, and for this reason practitioners were reluctant to use opiates, since the distance between a therapeutic and a fatal dose was so narrow: a sample of opium might easily kill if it provèd fresher or purer than expected. The surgeon Henri de Mondeville studied medicine at Montpellier about 1300, perhaps indeed with Arnau, for a number of his ideas repeat Arnaldian positions. Henri acknowledged the powers of opium, mandrake, henbane, and solatrum, but warned that they should be given dry, not fresh; that some parts of the plants were safer than others; and that "nunquam nisi pauca et multum repressa et in mortis articulo ministrentur."[13] Indeed, Bernard Gordon recommended avoiding the use of opiates altogether.[14]

The alternative was to test a sample first, to see whether its action was too strong or too weak. Bernard explained how to test the effectiveness of a theriacal medicine (one acting as an antidote to poison:

> take two pheasants, cut off their crests, apply a poison to the wounds (or administer it orally) and wait until they begin to stagger. Then put theriac on the crest wound and in the drink of one of them: if this one lives and the other dies, the theriac is good.[15]

---

13) Julius Leopold Pagel, *Die Chirurgie des Heinrich von Mondeville* (Berlin, 1892), 524.

14) Demaitre, *Bernard de Gordon*, 125-126.

15) J.L. Pagel, "Ueber den Theriak nach einer bisher ungedruckten Schrift des Bernhard von Gordon," *Pharmaceutische Post*, 27 (1894), 260: "Accipiatur gallus silvester qui fasianus appellatur, quia (et) ille est melior propter majorem siccitatem, tunc

This was one way in which the variability of the familiar and the mystery of the unfamiliar could be brought under control.

If the effectiveness of these active pharmaceuticals was variable, that could be explained; but the very nature of that effect was not always a subject of agreement. Clinical results—especially of drugs whose physiological effects were subtler than purging or narcosis—were notoriously difficult to interpret, as Galen himself had recognized:

> If someone is treated with different medicines, and improves or worsens as a result, it is not easy to decide which of these helped or harmed him; for example, if the patient sleeps smoothly and after his sleep is anointed with an epithimum, then given a cataplasm or a clyster, and finally given some dish to eat, after which he has a sudden bowel movement, it isn't easy to decide which of these things helped him or harmed him.[16]

The primary quality of a particular medicine, hot or cold, was ordinarily uncontroversial, but its specific physiological action within the body could be a matter of debate: the regular association of primary with secondary qualities appeared to break down in practice, since different observers came to different conclusions.

Disagreements among the authorities of the new Greco-Arabic literature made this problem particularly visible in the later thirteenth century. Roger Bacon may not have practiced medicine, but he was an acute and knowledgeable student of the world of academic medicine, and in the 1260s he was already identifying this as one of its major problems:

---

accipiantur duo et abscinantur cristae et tunc in locis abscisis apponatur toxicum vel detur per os et tunc dimittantur donec incipiant titubare; tunc alteri ponatur tyriaca supra cristam et cum hoc detur sibi in potu, tunc si ille vivat et alter moriatur bona est tyriaca." I have slightly modified the translation in Demaitre, *Bernard de Gordon*, 127.

16) Galen, *Comm. Hippocrates, Aphorisms* I.1; in *Articella* (Venice, 1523), [II] fol. 2v (*translatio antiqua*): "Si quis enim diversis medicaminibus medicinatur, si quolibet diversorum iuvetur sive disturbetur, quod horum iuverit sive nocuerit non est facile videre; verbi gratia, si suaviter infirmus dormierit et post somnum epithimate se inunxerit, deinde cataplasmaverit seu clysterizaverit et post hoc quali quid comederit, et ex hoc spontanea solutio evenerit, non facile est videri quod eorum iuverit seu nocuerit."

> [different] writers say that the same simple medicine purges different things, that is, contrary humors; thus Haly says that senna purges yellow bile, Avicenna (in his chapter on fumitory) that it purges adust humors, the Latins that it purges melancholy. Likewise the Latins say that rhubarb purges bile, but the Greeks and Aristotle that it purges phlegm.[17]

Bacon went on to describe an Indian form of rhubarb unlike the commoner European types which strengthened the body *contra legem aliarum medicinarum, et hoc scitur per experientiam*;[18] that exemplified the problem perfectly. However neat the Galenic-Avicennan model taught in the schools, there were innumerable exceptions to it, as experience showed. Should perhaps the regularity, the lawfulness of medicine itself be questioned?

Some thoughtful people were certainly convinced that it ought to be. We know this from the reaction of a Dominican friar, Nicholas, who lived in Montpellier c. 1250-70, apparently studying arts and medicine there, before returning to his native Poland.[19] He took away with him a fierce hostility towards the scholastic learning that was gaining hold at the school, expressed in his work *Antipocras*, from the 1270s. In this long poem he attacked the rational, cause-based medicine of the medical faculty and declared that what the physician needed for healing was instead a firsthand knowledge of the powers given by God to natural things, the kind of direct knowledge that came easily to simple villagers. William Eamon and Gundolf Keil have demonstrated that the *Antipocras* was influenced

---

17) Roger Bacon, *De erroribus medicorum*, in *Fratris Rogeri Bacon De retardatione accidentium senectutis*, ed. A.G. Little and E. Withington (Oxford, 1928), 151: "Auctores eandem medicinam simplicem dicunt purgare contraria, i.e. contrarios humores, ut Haly dicit quod sene purgat coleram rubeam, et Avicenna capitulo de fumo terre quod purgat humores adustos, Latini quod melancoliam. Similiter dicunt Latini quod reu purgat coleram, sed Greci et Aristoteles quod fleuma." The disagreement evidently struck Bacon as of particular importance, because he called attention to it again later in the work, 164.

18) Ibid., 163.

19) William Eamon and Gundolf Keil, "'Plebs amat empirica': Nicholas of Poland and His Critique of the Medieval Medical Establishment," *Sudhoffs Archiv*, 71 (1987), 180-196. André Gouron, "Deux universités pour une ville," in *Histoire de Montpellier*, ed. Gérard Cholvy (Toulouse, 1984), 103-125, at 114-115, touches on the role of the Dominican *studium* in education there.

by the *De mirabilibus mundi* of pseudo-Albertus Magnus very early in the latter work's circulation, and have proposed that this work was available to him at the Dominican community at Montpellier. The *De mirabilibus* acknowledged the reality of apparent "marvels" in nature, but concluded nevertheless that they were rational phenomena, even though some—like magnetism, or the power of amulets to cure disease—had hidden causes and could be known only by experience. Nicholas, however, took a different view of such marvels ("empirica," to him) and asserted that their cause was transcendent, divine, certainly neither elemental nor material and therefore inherently outside all scholastic system, Galenic explanatory theory notwithstanding.[20]

A way out of this problem—how to find a place for the unpredictable element within medicine while maintaining its essentially lawful and causal character—was soon available in Avicenna's treatise *De viribus cordis* ("On cardiac medicines"), translated for the first time from Arabic into Latin by Arnau de Vilanova, probably at Barcelona c. 1280.[21] It is tempting to wonder whether these intellectual tensions were part of the reason why Arnau chose to translate this particular work. In the 1260s he himself had been a medical student at Montpellier, and he must have been aware at that time of this issue within medicine; indeed, he could perfectly well have encountered Nicholas and discussed it with him, could conceivably even have read the *De mirabilibus*, for he also studied theology for a time with the Montpellier Dominicans.[22] Avicenna's treatise deals with the many medicines that can be shown to have an effect upon the heart, and recognizes that in many cases the

---

20) Nicholas of Poland, *Antipocras*, ed. Ryszard Ganszyniec, *Brata Mikołaja z Polski Pisma Lekarskie* (Poznań, 1920), 45, lines 12-13: "'Plebs amat empirica,'" 187: "Hoc in sermone sine cause cognitione / Monstro posse bene curam fieri, Galiene." Cf. Eamon and Keil, "'Plebs amat empirica'."

21) The text is being edited for Arnau's collected medical works by Kristin Peterson. A portion of the work was also attached to book IV of Avicenna's *De anima* in its twelfth-century Latin translation: Avicenna, *Liber de anima*, ed. S. Van Riet (Louvain and Leiden, 1968), 98*-99*. An English translation of the Arabic is available in *Avicenna's Tract on Cardiac Drugs*, ed. Hakeem Abdul Hameed (Karachi, 1983), 11-66.

22) Francesco Santi, *Arnau de Vilanova: L'obra espiritual* (Valencia, 1987), 84.

effect is an unexpected one—is not, for example, obviously reducible to the action of the primary qualities, hot and cold. To explain this, Avicenna developed an idea that was rapidly adopted by Western physicians once *De viribus cordis* entered into circulation. If substances are truly mixed together into a new substance, he proposed, not just finely divided and intermingled, the complexion of the new compound will be a product of that of its ingredients; but the new complexion also prepares the way for the mixture's acquisition (from God, or the stars) of a totally new property (*proprietas*; Ar. *khāss*), specific to that substance, which cannot be related to the properties of the starting points. So it is with the magnet: it possesses the *proprietas* of being able to attract iron, which is new and unpredictable and cannot be rationalized as originating from hotness or coldness. This concept made it possible for non-Galenic properties of clinical activity to be attributed to medicines without having to relate them to hotness or coldness; and it also implied that familiar substances might be discovered to have new properties that were not a part of their nature as understood by contemporary science.

It is important to recognize that Avicenna did not put *proprietas* forward as a quality of bodies distinct in kind from hotness or coldness or their higher-order correlatives, as Nicholas's *empirica* were. *Proprietas* was no different at all from other qualities of bodies;

> The common folk suppose that every action manifested by a body arises from hotness or coldness or wetness or dryness… . And when an action can be assigned to none of these, and the reason for its production by that body is not obvious to sense, they imagine that all its principles are unknown, but this is not so.[23]

In fact, Avicenna went on, our knowledge of the primary qualities is merely of the active power they are observed to display; fire is called hot because of the combustive power we recognize in it, just

---

23) Avicenna, *De viribus cordis* I.10; MS Paris, BN lat. 6949, fol. 91rb: "Subponitur apud [vulgares] quod omnis operatio procedens a corpore fiat a caliditate vel frigiditate vel humiditate vel siccitate… . Cum ergo actio nulli attribuitur predictorum, nec est sensui manifesta ratio processus eius ab illo corpore, putant quod omnia principia eius ignota sint, cum tamen non sit ita."

as the magnet is observed to have an attractive power. We are equally ignorant of the deeper reasons why hot things burn and why magnets attract, though ultimately both are powers implanted by God. *Proprietates* are simply one aspect of the nature of things, and the reason the vulgar think they are different is because they are uncommon. If fire were rare and appeared only occasionally and irregularly and magnetism were a phenomenon that we encountered daily, he argued, we would think fire's peculiar power to grow in size while corrupting other things was utterly miraculous, far more remarkable than magnetism, and we would no doubt worry about our inability to explain its cause; but it is magnetism that happens to be rare, and we are accustomed to fire, so we treat magnetism as a miracle and take fire for granted. Yet both are equally natural properties of bodies. In the context of the later thirteenth century, *De viribus cordis* should certainly be seen as providing an implicit answer to radical empirics like Nicholas. It was widely read and accepted in the later Middle Ages, in part no doubt because it provided a rationale for accepting and incorporating unpredicted and unexpected medicinal effects into the body of scholastic medicine.

This brief sketch suggests some of the problems that confronted academic physicians and their students at the end of the thirteenth century as they tried to interpret the new Greco-Arabic science of drugs in the light of their own experience. And the Montpellier response shows us a community fully accepting the principle of a science of pharmacy, and acknowledging the value of published data, while remaining conscious that that work was not necessarily sound and might be contradicted by their own experience, and ready to give credence to that experience even if it was not in accord with theory.

III.

How did this prudential principle work itself out in real life? How did the Montpellier physicians go about deciding when the evidence available to them indicated that the current body of scientific knowledge was inadequate? These issues were very much on Arnau de Vilanova's mind when he drew up a commentary on

Hippocrates' first medical aphorism ("Vita brevis, ars vero longa, tempus acutum") for his students at the University of Montpellier in 1301.[24] The *Aphorisms* as a whole (seven books) were a foundational work in the medical curricula at every thirteenth-century medical school; every student would have begun his studies with "Vita brevis … ," which continues "experimentum fallax, iudicium difficile" (ἡ δὲ πεῖρα σφαλερή, ἡ δὲ κρίσις χαλεπή), and every teacher, not just Arnau, would have had the opportunity to make plain to his students why experience can be so misleading and interpretation so problematic.

Arnau's instructions to his students in his "Vita-brevis" commentary show a medical teacher consciously and conscientiously at work, explaining how judiciously to apply the best of current medical science to particular patients. We might outline his advice as follows: A physician's knowledge of current medical science, from its worst to its best, comes from a range of sources. First, most obviously, from the written references available to him, which he and his colleagues recognize as the foundation of their knowledge: the Greeks, Hippocrates and Galen; the Arabs, especially Avicenna and Rhazes; and a few Latins, like Petrus Hispanus and Gilbertus Anglicus. All these writers offered long lists of remedies for different conditions. Yet despite their fame, despite their authority, you cannot automatically assume that each medicine they pass on will be therapeutically effective in a given case. Often they prove to offer contradictory advice. You have to consider whether a given remedy is consistent with "science," with the broader foundations of the medicine that you have studied as developed by Galen as well as of the natural philosophy worked out by Aristotle. If it is not scientifically plausible, even though you cannot put your finger on the reason for your mistrust, you should not risk its use until you can resolve your doubts.

---

24) Fernando Salmón and I are preparing an edition of this commentary for Arnau's collected medical works. The version printed in sixteenth-century editions of his writings is incomplete and often somewhat corrupt, and my citations of the commentary will therefore be from a manuscript copy, editorially emended, with a supplementary reference to the printed version when possible.

You also have to determine whether the medicine's supposed effect is consistent with your own experience, and in the process you may be able to refine the content of "best scientific knowledge."

> For example: Dioscorides and Galen and Avicenna all agree that rhubarb tightens the belly and is good for dysentery and ruptures and so forth, yet with all of us experience has found that it loosens the belly, and physicians are more likely to use it to loosen than to tighten. However, [those authorities] were speaking of young, fresh rhubarb, which is extremely bitter, like *sorba*; we on the contrary are using aged rhubarb, whose bitterness and intense sharpness have been reduced so that its weak power to attract yellow bile can produce its effect.[25]

And if in the end you feel confident of the medicine's property, you will still need to decide how it is to be modified for the particular patient in question, for the vehicle in which it is administered and the dose that is given are not the same for everyone:

> There is one rule for using mild things on a temperate body, another in distemperate ones, another in a bilious one, another in a phlegmatic one, and further, in each of these instances there is one rule for children, another for youths, another for the old.[26]

Sometimes your sources will explain how the medicine is to be adapted to specific situations, but not always, and then you must cautiously determine this for yourself.

---

25) Arnau de Vilanova, *Repetitio super Vita brevis*; MS Munich, CLM 14245 (hereafter Munich), fol. 21r; Arnaldi de Villanova *Opera* (Venice, 1509), fol. 276va: "Verbi gratia: Dyascorides et Galienus et Avicenna concorditer ponunt quod reubarbarum stringit ventrem et confert dissinterie et rupturis etc., tamen aput nos omnes experientia invenit quod laxat ventrem et potius communiter medici utuntur eo ad solvendum quam stringendum. Sed illi loquuntur de reubarbaro existente in sua iuventute vel recentia quod est ponticissimum ut sorba; nos autem utimur reubarbaro inveterato et antiquato in quo remissa ponticitate et intenso acumine debilis eius potentia in attrahendo coleram rubeam potest manifestare suum effectum."

26) Ibid., Munich, fol. 19r: "Aliud est enim canon operandi cum dulcibus in temperato corpore, aliud vero in distemperatis et distemperatorum, aliud in colerico, aliud in flegmatico, et in hiis etiam alius in puero, alius in iuvene, alius in sene; similiter quoque in aliis causarum et corporum differentiis."

Thus far the use of medical authorities. But not all knowledge of medicinal activity has to come from books. Avicenna's discussion of *proprietas* underlines this:

> for since a knowledge of *proprietates* cannot come from reasoning, only from *experimentum* or a revelation [*revelatio*], and *experimentum* and revelation are possible for both the common people [*vulgo*] and the learned, it is possible that a knowledge of *proprietas* may come to the common people before it comes to others.[27]

Indeed, writers like Galen and Rhazes acknowledge that they have taken some of their remedies from a popular tradition. Therefore, exactly like the remedies of the learned, popular remedies should be considered seriously to see, first of all, whether they are consonant with reason. If not—Arnau gives as an example of a misguided popular remedy, the vulgar belief that by bandaging a split in a tree, a fracture miles away can be healed—it is mere superstition; but if it seems plausible, it should be investigated, in order to find out what the causal factors might be that facilitate its activity.

> For example: a physician comes into a region where the common people maintain it as established [*pro experto*] that cabbage softens the bowels and prevents constipation. He should see what the climate of that region is, and from that he can infer the complexion of its inhabitants; then he should see whether they eat it out of desire or out of necessity [?]. Then, regarding the cabbages, [he should decide] what element predominates in their composition, which he can judge from their form: if the region's cabbages have broad, round, white, soft leaves, it is certain that water predominates in them. Also whether they are cooked slowly, and whether they are drained or not, because if the region is moist and the inhabitants eat a great deal, their bodies will be humid. And if the cabbage is as has been described and is eaten, unstrained, just after it has been boiled, all these things

27) Ibid., Munich, fol. 19v; *Opera*, fol. 276ra: "Nam cum noticia proprietatum non possit haberi per rationem sed tantum experimento et revelatione, et experientia casualis et revelatio sint communes vulgo et sapientibus, possibile est ut proprietatis noticia primo habeatur a vulgaribus quam ab aliis." In the *Medicationis parabole* that Arnau had finished in the previous year he also alluded to the role of *revelatio* in medical knowledge, explaining in his commentry to that work that the word can of course refer to knowledge communicated by God, but, in this context, refers to knowledge gained by the physician from his observations. See Joseph Ziegler, "Steinschneider (1816-1907) Revised: On the Translation of Medical Writings from Latin into Hebrew," *Medical Encounters*, 3 (1997), 94-102, esp. 96.

> together ensure that it will powerfully soften the belly; but this effect would not be expected in the inhabitants of a different region unless all the same factors concurred.[28]

On the other hand, if a remedy is found to be unchanging in its action on different populations, a *proprietas* has been discovered:

> the Arabs have determined that *polimonium* repels the poison of scorpions in whatever manner and whenever it is administered, and the physician can determine the same thing in every region, merely adapting the method of administration to its quantity and circumstances, with respect to the complexion of the body and of the medicine.[29]

By such methods popular discoveries can be tested, confirmed, and brought into the network of rational causation.

Indeed, there seems to have been a growing feeling in the thirteenth century that the apparently miraculous was likely to be a natural phenomenon;[30] no doubt the success of Avicenna's concept of *proprietas* was partly due to the fact that its insistence on the naturalness of the apparently unique conformed so well to this mood. So we need not be surprised at this particular scholastic teacher's openness to empirical discovery and to knowledge that

---

28) Ibid., Munich, fol. 20r; *Opera*, fol. 276ra–b: "Verbi gratia: vulgus regionis quam ingreditur medicus habet pro expertus quod caules mollificant et prohibent constipationem. Attendere debet tunc qualis sit complexio regionis, quia ex hoc elicet complexionem habitatorum; deinde utrum ex parte consuetudinis habeant copiose refici vel ex parte vivere. Deinde ex parte caulis quod elementum dominatur ibi in commixtione eius, quod per formam perpendit: nam si caules illius regionis habuerint folia lata et rotunda et alba et mollia, certum est quod aqua dominatur in eo. Similiter an leniter ibidem coquantur, cum excolatione vel sine, nam si regio fuerit humida et habitatores multe comestionis, corpora erunt humida. Et si caulis fuerit qualis est dictus et sumatur post primam ebulitionem non excolatus, omnibus concurrentibus potenter mollificat ventrem, nec inhabitantibus alterius regionis expectandus est ille effectus nisi predicta concurrunt."

29) Ibid., Munich, fol. 20r; *Opera*, fol. 276rb: "De polimonio experti sunt arabes quod pellit venenum scorpionis quocumque modo et quandocumque propinatur, et experiri potest medicus in simili casu in qualibet regione, variato tamen modo administrandi et quantum ad quantitatem et quantem ad concomitantiam, secundum diversum respectum complexionis corporis et complexionis medicine."

30) Eamon and Keil, "'Plebs amat empirica,'" 191-192.

arises outside existing rational structures; it is broadly in line with wider intellectual trends and more narrowly, perhaps, with issues that had been fought out at Montpellier in the years when Arnau and Nicholas of Poland were both studying there. Like Nicholas, Arnau could accept the reality of apparently inexplicable cures, but unlike Nicholas he believed firmly, following Avicenna, in the ability of medical science to bring them into a rational schema of nature.

There is thus a definite line to be drawn: Arnau thinks that unexpected discoveries in medicine are possible and even likely, and that they should be pursued, but he does not believe that *all* reports of new phenomena need to be taken seriously. There is of course no inconsistency here with his willingness to believe that unknown *proprietates* are hidden in nature waiting for the *vulgus* or the *sapiens* to discover them, with his expressed conviction that evidence is the test of truth. Reports of unforeseen and unpredictable properties of natural things deserve to be carefully considered so long as they do not contradict what we might call the natural order, the most basic principles of nature that are the foundation of Aristotelian-Galenic science and medicine. Arnau knew that for some claims evidence simply could not be forthcoming. No more than a modern researcher would he have thought it worth the trouble to undertake a clinical trial of the popular treatment of fractures mentioned above, by experimentally binding up split trees.[31]

---

31) In this connection, see the discussion of Arnau's position by Sebastià Giralt in Arnau de Vilanova, *Epistola de reprobacione nigromantice ficcionis, AVOMO* 7.1 (Barcelona, 2005), esp. 83-91. Lea T. Olsan suggests the existence of a spectrum of medieval opinion as to the status of claims for empirical therapies (*empirica*), from Nicholas's readiness to accept them as manifestations of God's absolute power to Guy de Chauliac's pronounced suspicion of them because of their irrationality in Lea Olsan, "Charms and Prayers in Medieval Medical Theory and Practice," *Social History of Medicine*, 16 (2003), 343-366, esp. 349-351. Arnau would fall in the middle of this spectrum, open to admitting their reality if they were not utterly inconsistent with reason. Guy's attitude is touched on in Michael McVaugh, "*Incantationes* in Late Medieval Surgery," in *Ratio et Superstitio: Essays in Honor of Graziella Federici Vescovini*, ed. Giancarlo Marchetti, Orsola Rignani, and Valeria Sorge (Louvain-la-Neuve, 2003), 321-345.

For Arnau and his students, then, scientific and empirical remedies both have a possible role in medical practice, though each has to be thought through and tested before it can actually be used. Totally new medicines, on the other hand, medicines that have no scientific or evidential authority behind them, must *not* be administered to a patient, with one exception: substances whose possible effects on the human body have not yet been determined must not be taken internally nor applied externally, but they can be worn as amulets, and quickly removed if need be.[32] Arnau's openness to the use of such therapies, for whose usefulness there would seem to be no evidence whatsoever, is actually based on the best scientific knowledge of his day. Amulets had of course long been recommended for the treatment of particular illnesses, as far back (at least) as Galen, but late thirteenth-century Montpellier seems to have paid particular attention to the treatise *De phisicis ligaturis* (also referred to as *De incantatione*) by Costa ben Luca, which presented a general discussion of the theoretical basis for their activity—Arnau makes use of it in the "Vita-brevis" commentary, actually citing it in the present context.[33] In his work, Costa expresses no particular skepticism about the efficacy of amulets (or charms or incantations), but from the outset he indicates that he believes that their action is on the mind rather than the body—that they exemplify the Galenic

32) Arnau de Vilanova, *Repetitio*, Munich, fol. 19v: "Nam si nullam, nullo modo debet experiri a parte intra nec a parte extra re alterata, sed propter rei alterationem debet corpori applicare secundum formam a sapientibus traditam in phisicis ligaturis, specialiter Aristotelis et Galieni, et Dyascoridis constat taliter ut si voluntas affuerit cito possit elongari vel amoveri a corpore." The very similar attitude of Arnau's Montpellier contemporary, Bernard Gordon, is made clear by Demaitre, *Doctor Bernard de Gordon*, 129-130.

33) Ibid., fol. 32r; *Opera*, fol. 279va: "Et ideo medicus prudens semper dicit infirmo et aliis quod ordinet usum talis aut talis antidoti ad talem vel ad talem preparationem introducendam in egro, ad hoc ut ipsum egrum et alios retineat semper in spe ulterioris et necessarii operis, ut animo non fatigentur aut perturbentur; et interim si apparuerit effectus sanationis, commendabitur medicus de recta administratione preparatoria corporis et adiutrice nature. Talis enim est cautela prudentis artificis, ut ait Costa ben Luce tractatu de phisicis ligaturis." The text is edited and analyzed by Judith Wilcox and John M. Riddle, "Qustā ibn Lūqā's *Physical Ligatures* and the Recognition of the Placebo Effect," *Medical Encounters*, 1 (1995), 1-50.

realization that the mind has the power to heal the body: an amulet appears to exert an effect because the patient believes that it is doing so, and his body responds accordingly. An alternative explanation might of course be that the amulet possesses a specific quality that is directly responsible for its effect on the body; Costa acknowledges that some authorities believe this and does not completely rule out the possibility, but his own conclusion is that insofar as they are successful, such therapies act as what we would today call placebos. The alternative—that amulets contained a *proprietas* that worked on the body, rather than on the mind—was accepted by some Latin philosophers (e.g., Albertus Magnus), but practicing physicians and surgeons must certainly have recognized the placebo effect in their own practice.

One who obviously did recognize it was Henri de Mondeville. He pointed out in his *Chirurgia* that Costa, Avicenna, and Galen had all shown how patients recovered more rapidly when things like incantations or amulets were employed as complements to the appropriate medicines, and he clearly concurred with their views. It seems quite likely that Henri absorbed this attitude from his earlier studies at Montpellier in the days when Arnau was teaching there. Arnau, as we have just seen, knew Costa's *De incantatione* and thought that it helped him understand the way in which amulets could be used. He did not spell out his thought-processes, but if he shared Henri's understanding, we can recognize why his assessment of the possible role of amulets is scientifically valid. If they are not taken internally or rubbed onto the patient's skin, they cannot harm him; if he believes they will help him, they may well do so; and if he happens to believe that they are injuring him, they can easily be withdrawn.[34]

---

34) Pagel, *Die Chirurgie*, 331. Mondeville's Avicennan reference is to *Canon* II.ii.124. The Galenic placebo he adduces was first described by Qustā ibn Lūqā, but his editors have not been able to identify a specific source for it in Galen's works: Wilcox and Riddle, "Qustā ibn Lūqā's *Physical Ligatures*," 45. On Mondeville's attitude see also McVaugh, "*Incantationes*," 339-344. Historians often scornfully depict Mondeville as obsessed with getting high prices for his services by, e.g., using more expensive remedies on richer patients without providing any reason to think that the cheaper remedies given to the poorer clients were any less effective. It is tempting to wonder, however, whether Mondeville might have recognized empirically what twenty-first century

Arnau's colleague at Montpellier, Bernard Gordon, might at first appear far less flexible in his attitude to such therapies. In his *De decem ingeniis curandorum morborum* (1299) he lumped prophecy and divination together with amulets and *empirica* as procedures that were inconsistent with the true canons of the medical art and therefore should not be used. Yet perhaps it should be noted that he does not actually say they are ineffective. Only a few years later, in fact, in his great *Lilium medicine* (completed in 1305), he showed himself quite willing to recommend amulets as therapies for a variety of conditions. Luke Demaitre has suggested that this was because Bernard sometimes despaired of curing certain illnesses by Galenic methods. He wrote of epilepsy that he had found it virtually incurable by rational means ("nescio causam, Deus scit"), but rather than restrict himself to academic science he went on to prescribe amulets and charms ("probatum frequenter," he said of one of the latter) for the disease. In the end Bernard reveals himself as willing, like Arnau, to explore all the resources available to him to identify the best possible therapeutic program for his patients.[35]

Surely it is hard to deny that this self-conscious program for evaluating claims of medicinal activity whose elements I have just set out is "evidence-based" in the sense of the definition we encountered at the beginning: it involves a conscientious use of the best evidence available (in 1300) to decide on the effects of drugs. It is cautious in accepting claims for their effectiveness, and can

---

medicine has just come to appreciate: that of two placebos which both show a positive effect, the more expensive one will be the more effective. See Rebecca L. Waber, Baba Shiv, Ziv Carmon, and Dan Ariely, "Commercial Features of Placebo and Therapeutic Efficacy," *Journal of the American Medical Association*, 299 (March 5, 2008), 1016-1017. Qustā's treatise is also analyzed by Béatrice Delaurenti, *La puissance des mots: 'Virtus verborum': Débats doctrinaux sur le pouvoir des incantations au Moyen Âge* (Paris, 2007), 176-183. Her focus is on its treatment of verbal charms or incantations rather than on its discussion of physical entities like amulets, but she still recognizes this text as the source of what she calls the Middle Ages' "argument de la confiance."

35) Demaitre, *Doctor Bernard de Gordon*, 157-159. Delaurenti, *La puissance des mots*, 298-299, follows Demaitre, concluding "sans doute Bernard de Gordon considère-t-il que les incantations qu'il prescrit ont une certaine efficacité thérapeutique." Nowhere in his works that I can see, however, does Bernard indicate that he is familiar with Costa ben Luca's work and its rationale for the use of such techniques.

envisage testing them against controls; and, while it seeks to find a theoretical justification for medicinal activity, it is quite prepared to make use of experimentally confirmed remedies even when the reason for their effectiveness is not understood.

IV.

But if the medieval physician was conscientious in his efforts to determine what active properties were truly characteristic of his drugs, was he equally conscientious when he tried to incorporate his knowledge and experience of drug activity into his clinical practice? When the necessity of therapeutic intervention required him to administer a medicine, how was he to decide what physiological effects were desirable in a given patient, and which medicines could be expected to manifest those effects? Or, starting from the other direction, how could he be sure, not just what action a particular drug had on the body, but whether that action would counteract the cause of his patient's illness? What did he actually do? Was he dogmatic about his approach, casual about his observations, uncritical about his assumptions—as our modern condescension to the remote past might lead us to assume—or did he in fact display "the conscientious, explicit and judicious use of current best evidence in making decisions about the care of individual patients"?

As we have seen, one caveat sometimes expressed about today's "experience-based medicine" is that its effectiveness depends on clearly defining and distinguishing among diseases and syndromes so as to treat its patients appropriately. How well do medieval physicians stack up in this respect? It is certainly true that they often had difficulty in identifying or defining particular diseases, and consequently in distinguishing between treatments that would be beneficial or ineffective in a given case. For they needed to be confident that the known effects of their drugs would address the cause and manifestations of the particular illness their patient was suffering from, and diagnosis was a subject on which there was very little consensus in 1300. Their sources described individual illnesses as well as medicines, no doubt, but the descriptions they provided were highly variable—there was naturally no reference standard, no International

Classification of Diseases to consult—and they were often impossible to match with the condition of a particular patient. Arnau underscores precisely this difficulty in his commentary. In their day, he says, Galen and Hippocrates were already pointing out how much disagreement there was about illnesses and their symptoms, how many there were and of what kind; "and this disagreement among scholars has *still* not come to an end, the specific cause of this or that symptom is *still* not known; we don't know exactly what *lepra* is, or exactly what all the prognostic signs signify, and so on."[36] When the physician is not sure what it is that he is treating, what the essential symptoms and above all the cause of the illness is, he cannot be sure whether its outcome is due to his treatment or not, no matter how much experience he has with his medicines; as a perplexed Montpellier master (Cardinalis) put it in the generation before Arnau, "if a physician gives violet sugar to someone with a choler-based fever, he won't know what cured him if he recovers, and he won't know what killed him if he dies."[37] Is it then with regard to pathology and nosology that medieval physicians can be faulted for failing to base their actions upon "best evidence"?

And yet they did not fail to study their patients closely: they recognized, as Galen had done before them, how important it is to do so, and they were just as sophisticated about the matter as he had been.[38] Arnau tells his students what to look for when they are called to a sickbed. First, what was the patient's normal state of

[36] Arnau de Vilanova, *Repetitio*, Munich, fol. 18v: "Patet multitudo desiderabilium et cognoscendorum quantum ad ea que ibi naturaliter insunt humano corpori, sicut sunt morbi et accidentia eius, in quorum numero et determinatione quam discordes et diversificati fuerunt investigatores diversorum temporum patet per Ypocrate et Galieno in regimine acutorum—et *adhuc* investigantium discrepantia non cessat, *adhuc* cuiuslibet accidentis causa specifica determinata non est, nec declarari determinate scitur quid sit lepra, nec de omnibus signis pronosticis quem determinate significent, et sic de aliis" (my emphases, of course).

[37] Cardinalis, *Comm. Hipp. Aph.* I.1, MS Cues 202, fol. 107rb: "In febricitante de colere gratia exempli utitur medicus zuccara violata et ideo si evadat nescit cui attribuat evasionem; et si moritur nescit cui attribuat."

[38] On Galen's methods, see the brilliant chapter by Luis García-Ballester, "Galen as a Clinician: His Methods in Diagnosis," in *Aufstieg und Niedergang der Römischen Welt*, ed. W. Haase and H. Temporini, II, 37.2 (New York, 1994), 1636-1671.

health, his complexion and constitution? Then, what changes have occurred in his normal functions, what unnatural changes can be seen in his body? The patient's own perceptions are important, and the physician should question him carefully about his condition. He should then take his pulse and palpate the parts of his body, sniff his breath, and examine his appearance, and finally study the external indices to his state—urine, feces, saliva, and sweat.

It is the next stage in diagnosis that is the difficult one: somehow matching the accidents that the physician has collected (signs and symptoms) to an illness entity defined both as to cause and as to locale. Certain widely accepted diseases can be relatively easily identified: the fevers caused by the different humors; distemperancy in a particular part of the body due to the presence there of excessive hot or cold matter. In such cases it is simply a matter of deciding on a remedy suited to the known cause, and adapting it to the age and condition of the individual patient. An actual case shows us how Arnau's mind works: a patient has twisting pains in his belly at a spot near the navel and kidney, but has no fever. The latter feature shows that the cause must be an accumulation of cold matter at that spot, so the physician responds with a program of heat to counteract it—first, mild heat applied topically; then, to be safe and to rule out the possibility of constipation, clysters; then hot baths; finally, if these treatments fail, a medicine to be taken internally: perhaps some kind of opiate in a decoction of hot wine, which Galen teaches will not only heat, it will also, by its *proprietas*, prevent the further accumulation of material.[39]

However, it seems clear from the direction taken by Arnau's discussion that he thinks the physician is more likely than not to find it difficult to make sense of the symptoms he has discovered and to determine the cause that has produced them. When he is in this situation of uncertainty, he should make a list of the symptoms observed, and try to think whether they are independent or interdependent; whether they indicate that a specific member is affected; and what they suggest the material cause in the body may be—a vapor, perhaps, or a humor. If this still does not enable him to identify the illness, he should consult a colleague, who may be able

---

39) *Repetitio*, Munich, fol. 24r-v; *Opera*, fol. 278rb–va.

to interpret the symptoms for him and suggest a medicine that will address their cause.[40] (The case of the twisting belly pains, above, had in fact arisen in this way; a bewildered colleague had reported the symptoms to Arnau, who was able to interpret their cause to his friend.)

If after studying the symptoms and consulting other physicians you still cannot decide, what do you do? You administer medicines—it is important to reassure the patient that something is being done (this is another medieval appreciation of the placebo effect)—but you must be sure that they will do no harm, being drugs of temperate complexion, neither hot nor cold, and drugs whose *proprietas* is to conserve the spirits or purify the body; these can be mixed together if desired, or combined with sugar, which is also temperate.[41] Having done so, you wait; you hope that the patient will recover; and if he does, you wonder—like Cardinalis in the example quoted above—whether you can be sure that it is what you did that made him well again.

Are these problems, this uncertainty about the world of illnesses, the result of an inattention to medical evidence, or a dependence upon the wrong evidence? Were they caused by the physician's failure to study the patient closely enough and to ask the right questions? The medieval physician evidently collected information about his patient's condition diligently, judiciously, critically—should he have done more? If he had looked more carefully, could he have determined that his patients were suffering from Addison's disease, say, or Bright's disease, or Crohn's disease, and responded appropriately? The very way in which I have chosen to phrase the question should indicate that he could not have done so, and why. The ABC's of disease today are not immediately and unequivocally given by experience acting upon individual patients, they have been built up

---

40) *Repetitio*, Munich, fol. 31v; *Opera*, fol. 279rb.

41) Alternatively, a potentially useful hot medicine can be tempered by mixing it with a corresponding cold one, and the combination can then be safely administered. For example, *melissa* strengthens the principal members, but it is hot and dry in the second degree; *poma maciana* when perfectly mature strengthen the same members and spirits by their *proprietas* and are of the opposite property. Hence these two drugs can be combined into an appropriate compound, mixed with something temperate like sugar. *Repetitio*, Munich, fol. 32v; *Opera*, fol. 279vb.

out of centuries of discrimination acting upon experience to gradually construct disease entities that today most physicians (but not all) will ordinarily be able to recognize in most (but not all) of the same patients. We should not imagine that "evidence" ought immediately to have led the medieval physician to construct our modern nosology.

Today the rhetoric about "evidence-based medicine" is toning down: its methods tend to be presented as a way of minimizing risk rather than establishing certainty.[42] At least some of its proponents accept that the data from today's randomized trials are, like all medical evidence, shaped to some degree by the thought of the researchers who gather them.[43] So in this sense there is no reason to treat "evidence-based medicine" as something fundamentally new, born out of mid-nineteenth-century positivism: it is, as Pellegrino wrote, what the best physicians have *always* tried to practice, back to the thirteenth century, as we have seen, and no doubt before, even though their methods of collecting and interpreting the evidence have changed. In fact, emphasizing the word "evidence" in this way may be misleading, suggesting as it does a realm of hard medical facts.[44] Evidence is not all there is to medicine, after all; data only become evidence at the moment when they are interpreted by experience and expectations. Taking the long view of medicine's history, it might be better to call that perennial ideal of "the conscientious, explicit and judicious use of current best evidence in making decisions about the care of individual patients" an "*experience*-based" medicine. In the light of his experience, the best physician is forever acting judiciously on evidence, in every age trying to decide conscientiously what is good evidence, what is the best evidence—and what is not evidence at all.

---

42) Harry M. Marks, *The Progress of Experiment: Science and Therapeutic Reform in the United States, 1900-1990* (Cambridge, 1997), 244.

43) Greenhelgh et al., eds., *Narrative Based Medicine,* 250. Marks, *Progress of Experiment,* 231-238, shows the inherent difficulties in interpreting modern medical research as disinterested objective science whose rationality is obvious.

44) See the discussion of the term by Erica Zarkovich and R.E.G. Upshur, "The Virtues of Evidence," *Theoretical Medicine and Bioethics,* 23 (2002), 403-405.

The Intellect Naturalized: Roger Bacon on the Existence of Corporeal Species within the Intellect

Yael Raizman-Kedar*
University of Haifa

Abstract
In this paper I challenge the claim that Bacon considered the operation of *species* as limited to the physical and sensory levels and demonstrate that in his view, the very same *species* issued by physical objects operate within the intellect as well. I argue that in Bacon the concept of illumination plays a secondary role in the acquisition of knowledge, and that he regarded innate knowledge as dispositional and confused. What was left as the main channel through which knowledge is gained were *species* received through the senses. I argue that according to Bacon these *species*, representing their agents in essence, definition and operation, arrive in the intellect without undergoing a complete abstraction from matter and while still retaining the character of agents acting naturally. In this way Bacon sets the intellect as separate from the natural world not in any essential way, but rather as it were in degree, thus supplying a theoretical justification for the ability to access and know nature.

Keywords
Roger Bacon, species, intellect, abstraction, illumination

The works that Bacon sent to Clement IV in 1267, with the ambitious goal of proposing a thorough revision of Christian learning, present the reader with a seemingly contradictory picture.[1] On the

---

* Philosophy Department, University of Haifa, Mount Carmel, Haifa 31905, Israel (yraizman@gmail.com).

[1] These works were the *Opus majus*, *Opus minus* and *Opus tertium*; the *De multiplicatione specierum* (henceforth DMS); and possibly some works on astrology and alchemy—perhaps even the *De speculis comburentibus*. See Alistair C. Crombie and John North, "Bacon, Roger," in *Dictionary of Scientific Biography*, vol. I (New York,

one hand, they comprise a detailed and well-elaborated account of sense perception; yet on the other hand, they display a profound belief in the crucial role of divine illumination in human knowledge.

In trying to make sense of Bacon's understanding of how the human mind attains knowledge of the natural world, some modern interpreters have chosen to lay stress on the so-called "mystical" aspects of Bacon's thought. French and Cunningham, for instance, call attention to Bacon's "illuminationist" understanding of how knowledge and wisdom are acquired. They stress the primacy of the "mystical intellectual grasp of things" in his writings[2] and, out of loyalty to their thesis that Bacon's thought is to be understood first and foremost as the philosophy of a Franciscan friar, dub Bacon a "Dionysian."[3] According to their description, "Bacon takes the Aristotelian account of how perception and understanding work and transforms it, without comment, into an essentially Neo-Platonist one."[4] Maurer, too, is convinced that Bacon thought science was acquired through an inner illumination, by which God revealed the principles of philosophy, and which would afterwards be completed by experience.[5] Maurer interprets Bacon's return to Oxford after his Parisian period, during which he lectured on Aristotle, as a turn away from scholasticism towards Augustine and his ideal of wisdom.[6] In discussing Bacon's earlier works, Crowley notes the secondary role played by illumination and stresses that Bacon there adheres to the teachings of Aristotle and never consciously departs

1970), 377-385, 378; David C. Lindberg, *Roger Bacon's Philosophy of Nature: A Critical Edition, with English Translation, Introduction and Notes, of* De multiplicatione specierum *and* De speculis comburentibus (Oxford, 1983), xxv; Jeremiah M.G. Hackett, "Roger Bacon: His Life, Career and Works," in *Roger Bacon and the Sciences: Commemorative Essay*, ed. Jeremiah M.G. Hackett (Leiden, 1997), 9-24, 22.

2) Roger French and Andrew Cunningham, *Before Science: The Invention of the Friars' Natural Philosophy* (Brookfield, Vermont, 1996), 243.

3) Ibid., 239. Such a label runs counter to the rarity of references to Dionysius in Bacon's writings.

4) Ibid., 239.

5) Armand A. Maurer, *Medieval Philosophy* (New York, 1962), 129-130.

6) Ibid., 127.

from them; but even he agrees that in his mature writings Bacon takes a noticeable turn towards illumination, a turn which Crowley attributes to the influence of Avicenna.[7] Rega Wood takes a similar stance, arguing that in his work after 1250 Bacon departs from the view that all science comes from sense and memory and appeals instead to internal spiritual experiences as producing scientific knowledge.[8] Hackett too stresses the prominence of divine illumination in Bacon. In seeking to account for the lack of a theory of abstraction in Bacon's mature writings, he claims that such a theory becomes futile given his "strong doctrine of illumination in which the *dator formarum* illumines the mind of the individual when the appropriate physio-psychological state has been engendered."[9]

In his pivotal work on intelligible *species*, Leen Spruit joins the crowd. In arguing against the notion of intelligible *species* in Bacon, he effectively sets up a barrier between sensory information and intelligible content in Bacon's epistemology. The process of the multiplication of *species*, Spruit declares, exerts no influence upon conceptualization.[10] When he refers to *species* as present in the rational soul and as originating from the cogitative faculty, Bacon merely qualifies them as instruments of soul, without presupposing any abstraction or impression from them.[11] Spruit claims further that it would be a logical fallacy on Bacon's part to accept that the corporeal *species* can be received into the strictly spiritual soul; and even if presumably abstracted, they would still be incompatible with

---

7) Theodore Crowley, *Roger Bacon: The Problem of the Soul in his Philosophical Commentaries* (Louvain/Dublin, 1950), 178-179. Crowley qualifies that Bacon did not follow Avicenna in making the active intellect a giver of forms.

8) Rega Wood, "Imagination and Experience in the Sensory Soul and Beyond: Richard Rufus, Roger Bacon and their Contemporaries," in *Forming the Mind—Essays on the Internal Senses and the Mind/Body Problem from Avicenna to the Medical Enlightment*, ed. Henrik Lagerlund (Dordrecht, 2006), 28-57, 56.

9) Jeremiah Hackett, "Roger Bacon," in *A Companion to Philosophy in the Middle Ages*, eds. Jorge J.E. Gracia and Timothy B. Noone (Malden, MA, 2003), 616-625, 623.

10) Leen Spruit, *Species Intelligibilis from Perception to Knowledge,* vol 1. *Classical Roots and Medieval Discussions* (Leiden, 1994), 154.

11) Ibid., 154.

the central role that Bacon assigns to innate ideas and divine illumination.[12]

Spruit here is in fact arguing against Katherine Tachau, who asserted that Bacon "evidently believed that the process of multiplication continued through the intellectual powers as well."[13] Like Tachau, Mark Smith, writing in general about the Perspectivists' psychology of perception, sees no problem in accepting the existence of intelligible *species* in Bacon.[14] My argument in this paper could serve to substantiate the position presented by Tachau and Smith, although it differs from theirs in some significant aspects, most notably, in taking issue with their views on abstraction in Bacon.[15]

The problem of the reception of *species* into the intellect in Bacon's thought appears to stand at the heart of the issue. If indeed we could establish a direct link between the *species* in the faculty of cogitation, the highest faculty of the sensitive soul, and the *species* inhering in the rational soul, and if we could solve the theoretical difficulties involved in establishing such a linkage, then we could justly stress the Aristotelian tendency in Bacon's thought. Then, of course, we would still have to give an account of the exact place and function of illumination and innate knowledge in Bacon's epistemology. However, if indeed *species* retain a place in the material and physical arena, and the possibility of their reception into the intellect is denied, then consequently the intellect would have to rely on either innate knowledge, divine illumination, or both to receive intelligible content.

---

12) Ibid., 154.

13) Katherine H. Tachau, *Vision and Certitude in the Age of Ockham—Optics, Epistemology and the Foundations of Semantics 1250-1345* (Leiden, 1988), 11.

14) A. Mark Smith, "Picturing the Mind: The Representation of Thought in the Middle Ages and Renaissance," *Philosophical Topics* 20 (1992), 149–170, 158.

15) Smith attributes abstraction to Bacon in A. Mark Smith, *Alhacen's Theory of Visual Perception—A Critical Edition, with English Translation and Commentary, of the First Three Books of Alhacen's De Aspectibus, the Medieval Latin Version of Ibn al-Haytham's Kitab al- Manazir* (Philadelphia, 2001), vol. 1, xcvii. Tachau ascribes abstraction to Bacon in *Vision and Certitude*, 22 and 26.

A major difficulty in opting for the first view, is that, without presupposing any abstraction mechanism, Bacon held firmly to the corporeal qualification of *species* and defined its being as corporeal, material, and natural: "... a species of corporeal and material things will always have material and corporeal existence...."[16] So he wrote, and added: "Therefore, I say that species have material and natural existence in the medium and in sense ...."[17] Moreover, Bacon believed that every *species* represents its agent "in nature, definition, specific essence, and operation."[18] Thus, once a *species* loses its inherent corporeal nature, it can no longer accurately represent the natural world. Bacon devotes a whole chapter (3.2) of the *De multiplicatione specierum* to counter the claim raised by Aristotle, Averroes, and Avicenna that *species* of corporeal agents receive spiritual being in media and in sense.

Defending the Aristotelian thesis gets even more complex since Bacon held a weak notion of the intentional existence of *species* in media and sense. A comparison to the view of Thomas Aquinas may help clarify his notion. According to Pasnau, the intentional existence of the *species* of P, as defined by Aquinas, means that the *species* is received by a patient without the patient actually becoming P.[19] As Pasnau illustrates, when something is made hot, the heat exists naturally in the recipient; but if someone merely thinks about heat, the intellect is not thereby made hot.[20] Although Aquinas saw no incompatibility between intentionality and physicality, he frequently used the words "intentional" and "spiritual" as synonyms.[21] Bacon seems to hold to a different understanding of the adjective "intentional." He writes about the "intentional and false" (*intentionale*

16) Bacon, *Perspectiva* 1.6.4, in David C. Lindberg, *Roger Bacon and the Origins of Perspectiva in the Middle Ages: A Critical Edition, with English Translation of Bacon's Perspectiva with Introduction and Notes* (Oxford, 1996), 88-89.

17) Bacon, *Perspectiva* 1.6.3, in Lindberg, *Roger Bacon and the Origins*, 83.

18) Bacon, *DMS* 1.1, ed. and trans. David C. Lindberg, *Roger Bacon's Philosophy*, 7.

19) Robert Pasnau, *Theories of Cognition in the Later Middle Ages* (Cambridge, 1997), 63.

20) Ibid., 33.

21) For example, in his commentary on the *De anima* II.24.45-56 he writes: "... for in the sense object it has natural existence, whereas in the sense it has intentional or spiritual existence." Quoted in Pasnau, *Theories of Cognition*, 40.

et vanum) existence of *species* and contrasts it with "real existence" (*rem veram*).[22] When specifying how a *species* differs from its issuing agent, he clings to the *species'* incompleteness. In all other respects, a *species* resembles its agent. The generation of *species*, as opposed to the complete generation characteristic of natural processes, is an incomplete one, in which "incomplete matter and incomplete form and an incomplete composite come into being…."[23] Compared to "real" beings, a *species* is so weak and deficient that it cannot be reckoned among the things of this world.[24] Yet this understanding of "intentional" as "incomplete," or having less being, does not make *species* either purely spiritual or separate from matter.

While Normore interprets Aquinas's *esse spirituale* as the presence of a form without the matter which usually accompanies it,[25] Bacon could never have accepted that a form might exist apart from its appropriate matter.[26] As Bacon explains it, the term "spiritual" which Aristotle, Avicenna, and Averroes used for sensory *species* "should be understood entirely to refer to insensible being," but there is no contradiction between insensibility and either corporeality or materiality. [27] Following this assertion, Pasnau infers that Bacon thought of *species* as "literal likenesses of external objects," which "make their recipients actually like their agents." "In Aquinas' terminology," Pasnau concludes, "they have natural existence."[28] This is not entirely accurate, since Bacon considered only the *species* of qualities like

---

22) Bacon, *Questiones supra librum de causis* 9, eds. Robert Steele and Ferdinand M. Delorme, in *Opera hactenus inedita Rogeri Baconi*, fasc. 12 (=*OHI* 12) (Oxford, 1909–1940), 41.

23) Bacon, *DMS* 1.2, in Lindberg, *Roger Bacon's Philosophy* 31. On intentional existence in Bacon see also Tachau, *Vision and Certitude*, 12.

24) Bacon, *Communium naturalium*, 1.1.2.3, *OHI* 2, 23: "… et quanto minus esse potest de natura specifica agencium est in eis; propter quod non vocantur res, set magis similitudines rerum."

25) Calvin Normore, "The Matter of Thought," in *Representation and Objects of Thought in Medieval Philosophy*, ed. Henrik Lagerlund (Aldershot, 2006), 117-134, 130.

26) See the justification of clause (3) in this paper.

27) Bacon, *DMS* 3.2, in Lindberg, *Roger Bacon's Philosophy*, 193. As Tachau (*Vision and Certitude*, 15) notes, Averroes did not in fact assert that light has spiritual existence in its medium. What he actually wrote was that it is not a body (*corpus*) but an *intentio*.

28) Pasnau, *Theories of Cognition*, 65.

heat, light, and colour to be literal likenesses. He stressed that for other things, *species* cannot realize themselves as fully as their agents, but remain in a state of a being which is incomplete.[29] Yet this being would still be material; and, in Bacon's terminology, it would have natural existence just the same.

Bacon had limited his discussion of the natural being of *species* to *species* in material substances, such as the transparent medium, the sense organs, and the brain. On how *species* are received within the spiritual substance of the rational soul, he promised that "later we will investigate the species of corporeal things as they exist in the soul and intellect ...",[30] but he leaves this promise unfulfilled. I shall here try to fill the gap which Bacon left open, by arguing that—in the sense defined—the *species* in the rational soul remain natural, adding the note that in Bacon natural and intentional existence are not mutually exclusive.

My thesis in this paper is that according to Bacon (1) *species* do arrive in the intellect through the senses; (2) moreover, they do so without being abstracted from matter and while remaining corporeal; (3) *species* must arrive into the intellect as corporeal, since they represent not the form alone but the whole composite of matter and form; (4) corporeal *species* can be received into the rational soul since the soul itself is not purely spiritual but comprises matter and form, for up to a point, the rational soul operates naturally; (5) the function of both innate ideas and divine illumination in Bacon is relatively minor and consists in enabling the rational soul to recognize valid arguments and stabilize cognitive habits; (6) this seemingly awkward treatment of cognitive content by Bacon stems from his view of representation, which requires a strong linkage between mind and nature.

My thesis should be qualified as applying only to Bacon's mature writings. The exact chronology of Bacon's works is much debated, and for the present paper I use the rather loose division into "early" writings and "late" or "mature" ones. The commentaries on the various *Physics* and *Metaphysics* books, as well as on the *De causis*, all

---

29) Bacon, *DMS* 1.1, in Lindberg, *Roger Bacon's Philosophy*, 13-15.

30) Bacon, *DMS* 3.2, in Lindberg, *Roger Bacon's Philosophy*, 191-193.

belong to the early phase of Bacon's career, in which he lectured at Paris. The three *opera* (*majus*, *minus* and *tertium*), the *Communium naturalium*, the *De signis*, and the *Compendium studii theologiae* were all written after 1250, that is after his return to Oxford and his supposedly crucial encounter with Grosseteste's writings. For various reasons, but mostly in light of the development of Bacon's thought on abstraction, I would date his questions on the *De causis* to the end of the early phase rather than to its midst as Rega Wood has suggested.[31]

The three main features in which the two periods of Bacon's career differ are his theory of the agent intellect, the development of his ideas on abstraction, and his theory on innate knowledge.[32] I will here sketch briefly the essentials of the developments in his theory of the agent intellect and then expand on abstraction, and innate knowledge in the appropriate sections.

Bowman, Dales, and others recognize three phases in Bacon's developing views on the agent intellect.[33] In the earlier commentaries, such as the questions on the *Metaphysics lambda*, Bacon teaches that the agent intellect is a part of the soul. In the second phase, following Avicenna and Alfarabi,[34] the agent intellect becomes a separate intelligence, placed outside the soul, yet it is still a created being. This second phase appears in writings such as the questions on the *De causis,* and the various questions on the *Physics*.[35] The third phase characterizes Bacon's mature writings and is the one presented in the *Opus majus*, where the active intellect is identified

31) Wood, "Imagination," 28-57, 53.

32) Other aspects can be added to the list, most notably the new stress on the importance of mathematics and experimental science for *scientia*, which characterized his mature period of writing. However, these aspects are irrelevant for the thesis I am advocating here.

33) See Leonard J. Bowman, "The Development of the Doctrine of the Agent Intellect in the Franciscan School of the Thirteenth Century," *Modern Schoolman* 50 (1973), 251-279, at 257 ; Richard C. Dales, *The Problem of the Rational Soul in the Thirteenth Century* (Leiden, 1995), 76-77; Crowley, *Roger Bacon*, 182-191; Wood, "Imagination," 54.

34) See Wood, "Imagination," 54.

35) Spruit, *Species Intelligibilis*, 154-155; Wood, "Imagination," 54.

with God.[36] This last phase rested on the argument that if the agent intellect knows all things and is always *in actu*, then it cannot be a part of the rational soul but God alone, since only God fits this description.[37] Bacon also referred to the analogies Aristotle had used to illustrate the relation between the agent and possible intellects: the artificer to his materials, the light of the sun to colors, and the sailor to the ship. All these, he concluded, entail an essential separation between the active element and the passive one.[38]

Given these changes in Bacon's thought, I shall limit my argument to the correct reading of Bacon's mature works alone, although at times will seek clarifications from some of the earlier texts, and especially from the *De causis*, which I consider to be a key text in the development of Bacon's thought. I will now attempt to justify each step in my argument as I stated it above.

1. Through the Senses to the Intellect

Several texts show that Bacon thought of the intellect as the receiver of *species* coming from the senses. For instance, in the *Communium naturalium* he explains foolishness as resulting from the inability of *species* to enter the intellect:

> For the truth is that the powers of the sensitive soul are in the service of the intellect, and when they are injured ... then the operation of understanding is hindered, because there is an error in perception and the *species* cannot reach the intellect according to their proper and true being; on the contrary, the being of the species is destroyed so that the intellect cannot be informed....[39]

---

36) Crowley, *Roger Bacon*, 165-166. See also Bowman, "The Development," 257.

37) Bacon, *Opus majus* 2.5, in *Opus Majus of Roger Bacon*, ed. Robert B. Burke, 2 vols. (Philadelphia, 1928), 44-45; see Bowman, "The Development," 257-258.

38) Bacon, *Opus majus* 2.5, in Burke, 44-45. In the next two pages (46-47) Bacon brings up some more considerations supporting his view that the agent intellect is in fact God. Crowley (*Roger Bacon*, 187) speculates that the decisive factor was the influence of Grosseteste and Adam Marsh, and Bacon himself claims he was influenced by William of Auvergne.

39) Bacon, *Communium naturalium* 1.4.2.3, *OHI* 2, 288: "Nam verum est quod virtutes sensitive deserviunt intellectui, et quando leduntur... tunc impeditur operacio intelligendi, quia error est in sensitiva et non potest species venire ad intellectum

Species coming from the senses are thus essential and indispensable for the proper functioning of the intellect. When they cannot arrive in the intellect, or when they get distorted on their way, the intellect cannot function properly. In the *Opus majus*, Bacon portrays the rational soul as added to the sensitive soul from without and united to it "primarily and immediately" through the cogitative faculty. Then he turns to speak of *species*: "and species are produced in the rational soul by the cogitative faculty."[40]

In the *De causis* Bacon invokes the principle that whatever the inferior power can do, a superior power can do even more effectively:

> ... the sensitive power, which is lower than the intellective soul, can have within itself the species of sensible things, through which it understands [those sensible things]; therefore in a similar manner ... the intellective soul knows corporeal sensible things, since it is more powerful than the sensitive soul .... And this is how [the intellect] holds the species of sensible things within itself.[41]

Since the intellective soul is more powerful than the sensitive, it follows that if the former can be united with sensible *species* and holds them in itself, then so can the latter.

2. Abstraction Abandoned

In his earlier writings Bacon indeed described *species* as being abstracted from their material conditions on their way to the intellect.[42] In his mature writings, however, not only is there no such

---

secundum rectum esse et veritatem; immo destruitur esse speciei ut intellectus non possit informari ...."

40) Bacon *Perspectiva* 1.1.5, in Lindberg, *Roger Bacon and the Origins*, 17. Also in the *DMS* 1.2, in Lindberg, *Roger Bacon's Philosophy*, 43, Bacon states that *species* of singulars as well as of universals are repeated in medium, sense, and intellect.

41) Bacon, *De causis* 14, *OHI* 12, 72: "... virtus sensitiva, que est inferior anima intellectiva, potest apud se habere species sensibilium per quas cognoscit ipsa, ergo similiter .... anima intellectiva cognoscit res corporales sensibiles cum sit potentior quam sensitiva .... et ita habet species rerum sensibilium apud se."

42) In the *Quaestiones supra libros quatuor physicorum Aristotelis*, *OHI* 8, 31, he describes the phantasms as being abstracted from their material conditions, and in the *De*

description, but Bacon denies abstraction—at least in the sense of the separation of form from matter—outright. On the reception of *species* into the sensitive soul, he writes:

> ... as regards what Aristotle says in *De anima*, book II, that sense receives the species of sensible things without matter, it is replied that he there uses "without matter," that is, "immaterial" to mean "insensible" rather than "spiritual."[43]

It is not true that the senses receive the sensible *species* without their matter; Aristotle's intention had been misunderstood. Alongside the abstraction of form from matter, Bacon also denies that universals are created from particulars, which was normally considered an act of abstraction; indeed he declares that view to be superfluous and false:

> ... a universal issues into being through the operation of nature .... But the apprehension of the mind contributes nothing to the operation of nature; therefore it is most vain to say that the mind makes a universal.[44]

At the mature stage of his thought, Bacon held that universals were not products of rational processes, nor were they exclusively received by the intellect from some innate or supernatural source. On the contrary, Bacon thought universals were a part of nature and were received into the intellect through the faculty of estimation. He denied that the rational soul was the cause of universality; for even if it did not apprehend particular things, they would still resemble one another.[45] The universal within the intellect is not the actual universal but its likeness or *species*. The intellect does not need to

---

causis 10, *OHI* 12, 59-60 he says the same of the *species* in memory and phantasy. In both cases the task of abstraction is performed by the agent intellect "beaming forth." Abstraction of *species* is also mentioned in the *Questiones supra primum Metaphisice*, *OHI* 11, 12.

43) Bacon, *DMS* 3.2, in Lindberg, *Roger Bacon's Philosophy*, 193.

44) Bacon, *Communium naturalium* 9.57, in *Three Treatments of Universals by Roger Bacon—A Translation with Introduction and Notes*, ed. and tr. Thomas S. Maloney (Binghamton, NY, 1989), 98.

45) See Dorothy E. Sharp, *Franciscan Philosophy at Oxford in the Thirteenth Century* (Oxford, 1930), 145; Maurer, *Medieval Philosophy*, 135.

abstract general concepts from many particulars, because it receives the *species* of universals ready-made.[46]

Bacon replaced the theory of abstraction with the principle of incorporation. According to this principle, to express the information it carries, a *species* must be incorporated within the right kind of medium. Thus, for example, the *species* of the proper sensibles are expressed within the external senses, the *species* of substances and complexions by the faculties of estimation and cogitation, and the *species* of universals by the intellect. A *species*' content is not extracted but rather expressed when it is incorporated within a suitable matter.[47]

The absence of abstraction from Bacon's mature writings has been noted by both Spruit[48] and Hackett, who wrote: "*Opus maius* V contains the physiological and mathematical basis for Bacon's theory of mind .... Most significant is what is missing: there is no theory of abstraction."[49] To ground my argument on more than just Bacon's silence, I propose to examine how abstraction in Bacon gets pushed aside by the full implementation of the principles that he had laid down for the propagation of *species*.

In the questions on the *De causis* the abstraction of *species* is treated, confusingly, alongside an argument that no such abstraction is necessary. Bacon begins his argument by distinguishing *species* having determinate dimensions from *species* having indeterminate ones. He rules that *species* of the first kind cannot be received into the soul, while *species* of indeterminate dimensions, arising in prime matter, "are not incompatible with the simplicity of the soul."[50]

---

46) In developing his notion of universals as issued from the things themselves Bacon has been influenced by Alhacen. According to Alhacen the universal form arrives into the eye together with the particular form. However, unlike Bacon Alhacen required repeated encounters with the universal form, as well as certain workings of abstraction so that the universal form would be impressed in one's imagination. Bacon went a few steps further. For Alhacen's account see his *De aspectibus* II.4.16 in Smith, *Alhacen's Theory*, 518.

47) For a description of the principle of incorporation see A. Mark Smith, "Getting the Big Picture in Perspectivist Optics," *Isis*, 72/4 (1981), 568-589.

48) Spruit, *Species Intelligibilis*, 154.

49) Hackett, "Roger Bacon," 623.

50) Bacon, *De causis* 10, *OHI* 12, 72, "... tales non repugnant simplici anime."

Bacon's use of Averroes' notion of indeterminate dimensions here is interesting. Averroes observed that the four elements share the attribute of divisibility as a common property, while divisibility, in turn, presupposes the possession of quantity. The forms of the four elements impose upon prime matter a set of determinate quantities, but prior to such an imposition, prime matter possess an indeterminate three-dimensionality only. Averroes identified this indeterminate extension with the corporeal form.[51] Bacon, in turn, considered that *species* representing the forms which inhere in prime matter, namely *species* of universals and substances, possess an indeterminate extension in the same way as prime matter does prior to the imposition of a substantial form. He assumed that this kind of indefinite extension can be received in the rational soul.

Yet there is a difficulty here: prime matter is not an active nature and therefore cannot issue its own *species*.[52] Bacon therefore explains that the *species* of universals are produced in the singular *species*, for a singular *species* cannot exist without the universal one and vice versa.[53]

The *species* of both universals and singulars are received in the intellect together:

> But from whichever particular arrives one species of a universal with the species of the particular, the species of universals are multiplied accordingly in the soul, and consequently the soul becomes stronger and more capable of bringing us to understand universals; and that is why the universals are called the objects of the intellect. But this is [done] by using an epithet instead of a name (*antonomasiam*), and without excluding singulars...[54]

---

51) See Arthur Hayman, "Aristotle's 'First Matter' and Avicenna's and Averroes' 'Corporeal Form,'" *Harry Austryn Wolfson—Jubilee Volume, On the Occasion of his Seventy-Fifth Birthday*, vol. 1 (Jerusalem, 1965), 385-406, 403.

52) In the *Quaestiones supra librum de sensu et sensato* 8, *OHI* 14, 29 Bacon lays down the principle that not only light but every active nature issues its own *species*. In the *DMS* 1.2, 41 he states that "the species of prime matter is not generated, but [only] of specific matter."

53) Bacon, *DMS* 1.2, 43.

54) Bacon, *Communium naturalium* 1.23.4, *OHI* 2, 104. "... set a quolibet singulari venit una species universalis cum specie singulari, et ideo multiplicatur species universalis in anima, et ideo fit fortior at potentior ad hoc ut per eam intelligamus

Universals are called the object of the intellect because they can be realized only by the intellect and not by the senses. Yet that does not exclude *species* of particulars from the intelligible realm. Both *species* of particulars and those of universals eventually reach the intellect, the difference being that *species* of particulars are also dealt with at an earlier stage, at the sensory level. It quickly becomes clear that it is not only the *species* of universals and substances—representing qualities which inhere in prime matter—that are devoid of determinate quantity. The group of *species* with fixed dimensions remains empty:

> Yet quantity, since it is bound up with matter, is not an active nature and so does not multiply its species; and because of this, its species is not under quantitative dimensions but only concerns a body inasmuch as it is a substance and therefore is received into the intellective soul.[55]

Species of material things do not represent the quantities of their issuing objects. According to Bacon, quantities and many other visual qualities, do not send forth *species*. Size, magnitude, distance, and the like are attributes that the intellect deduces from the arrangement of the *species* of light and color on the lens: "only the species of color and light are [actually] arranged on the pupil and not the species of magnitude itself."[56] That arrangement reflects the original proportions within the issuing object, since the propagation of *species* occurs in straight lines according to the laws of geometry.

Bacon's argument seems to be this: if *species* were endowed with fixed dimensions, they could not find their way into the intellect.

---

universale, et ideo universalia vocantur objecta intellectus. Set hoc est per antonomasiam non per exclusionem singularis…"

55) Bacon, *De causis* 10, *OHI* 12, 73: "Quantitas autem, cum materie debeatur, non est natura activa, ideo non multiplicat sui speciem, et propter hoc species non est sub dimensionibus quantitativis, set solum refertur ad corpus in quantum substantia est, et ideo in anima intellectiva recipitur."

56) Bacon, *DMS* 1.2, in Lindberg, *Roger Bacon's Philosophy,* 40-41. This explanation of visual sensation was derived from Alhacen. For the details of Alhacen's account see Smith, *Alhacen's Theory*, lxviii–lxxiii. See also Abdelhamid I. Sabra, "Sensation and Inference in Alhacen's Theory of Visual Perception," in *Studies in Perception*, ed. Peter Machamer and Robert Turnbull (Columbus, 1978), 160–85.

But since the range of qualities represented by *species* does not include quantitative ones, and since dimensionality is a quantitative property, *species* do not have to resemble the dimensional properties of their agents and therefore can be received into the intellect.

In the *DMS* Bacon makes his point again: a *species* is an incomplete entity and therefore does not exist of itself but always in something else. It is only through that something else, which becomes its material cause, that a *species* comes to resemble its agent. It is therefore only through that very material cause that place, position, and dimensions can be ascribed to *species*.[57] Thus a *species* need not be abstracted upon its reception into the intellect. It receives dimensions from its host; and by its very nature, it suits the intellect, which is itself devoid of spatial characterization.

3. *Species* Represent Matter and not Form Alone

One explanation for Bacon's dispensing with abstraction is his conviction that representation by form alone does not suffice to assure accurate correspondence between mind and nature. In several places he insists that a *species* represents the composite of form and matter rather than the form alone:

> ... it has been proved above that the species of a corporeal substance is the similitude of the whole composite, and that when a species is generated not only is its formal being produced in the medium, but also material being and true matter having incomplete being.[58]

In arguing for the view that a name is principally imposed to signify the composite of matter and form rather than the form alone, Bacon makes an interesting statement about the things that are best comprehended by the intellect:

> Just as each thing bears itself in relation to [its] being, [it presents itself] to an intellect in the same way, as Aristotle says in the first [book] of the *Metaphysics*.

---

57) Bacon, *DMS* 3.1, in Lindberg, *Roger Bacon's Philosophy*, 181.

58) Bacon, *DMS* 3.2, in Lindberg, *Roger Bacon's Philosophy*, 191. See also *DMS* 1.2, 29-31.

> But a composite has a more complete [type of] being than form, since it has the being of matter in addition to the being of form. Because of this it can more truly and properly be understood, as far as it is concerned, and therefore be named.[59]

Since a composite has a more complete being than that of a separate form, a *species* representing both the formal and material aspects of its agent would be more perfect than a *species* representing its formal aspect alone. A composite *species* therefore, not only provides the most faithful representation, but given its greater perfection is also more readily and quickly grasped by the intellect. This principle interlaces well with Bacon's insistence in the *Opus majus* that "spiritual facts are discovered through corporeal effects,"[60] and that since our intellect is associated with continuity and time we grasp quantities and bodies directly, but not the forms of incorporeal things. Even if incorporeal forms were available to our understanding, Bacon proceeds, we would not perceive them, "owing to the more vigorous occupation of our intellect in respect to bodies and quantities."[61]

4. The Rational Soul Includes Matter and Form

Yet if the intellect can receive *species* of both matter and form, it cannot be defined as purely spiritual. Bacon refers in some places to the respects in which the intellective soul can be called corporeal and discerns different degrees of corporeality or materiality. For example, in his second set of questions on the *Metaphysics* he distinguishes between a body (*corpus*) and two kinds of corporeal things (*corporeum* and *corporale*). He classifies the intellective soul as *corporalis* because it is conferred on a body, yet he notes that it remains separate and extrinsic from that body.[62] Once again, in his questions

59) Bacon, *Compendium of the Study of Theology* 3.76, ed. and trans. Thomas S. Maloney (Leiden, 1988), 81. Compare this statement with Aquinas: "… the more immaterially a thing receives the form of the thing known, the more perfect is its knowledge," *Summa Theologiae* 84.2. The disparity between Aquinas and Bacon is striking.

60) Bacon, *Opus majus* 4.3, in Burke, 124.

61) Bacon, *Opus majus* 4.3, in Burke, 126.

62) Bacon, *Questiones altere super libros prime philosophie Aristotelis* (*Metaphysica I-V*) in Maloney, *Three Treatments*, 56.

on the *Metaphysics*, he writes that the intellective soul can be called material since it is born with matter.[63]

Bacon seems to have found a satisfactory answer to his doubts about the ontological status of the intellective soul in the doctrine of universal hylomorphism. According to this theory, all substances except God consist of matter and form, whereas only God is entirely immaterial. Substances such as angels and the human soul are distinguished as spiritual matter from the matter of physical, sensible objects.[64] As a staunch supporter of this doctrine, Bacon held to the view that every finite being is essentially composite and so a simple finite being is a contradiction in terms. As receptors of divine light, both a human soul and an angel ought to be material, since matter is the principle of receptivity.[65] Accordingly, Bacon emphasized that the rational soul is not purely spiritual but composed of both matter and form:

> ... but I hold for certain that the soul is composed from matter and form, just like the angels; for the question is the same concerning the angels and rational souls.[66]

Rational souls are separate from sensible and corporeal matter but not from spiritual matter. Bacon then links the materiality of *species* with the materiality of the soul:

> And therefore this is as false as to maintain that a species has spiritual being ... since an angel and soul, although spiritual substances, nevertheless exist in matter, since they are composites of true matter and form ....[67]

---

63) Bacon, *Questiones supra undecimum prime philosophie Aristotelis, OHI* 6, 18.

64) Paul Vincent Spade, "*Binarium Famosissimum*," *The Stanford Encyclopaedia of Philosophy* (Fall 2003 Edition), ed. Edward N. Zalta, URL = <http://plato.stanford.edu/archives/fall2003/entries/binarium/>.

65) Crowley, *Roger Bacon*, 82-83.

66) Bacon, *Communium naturalium* 1.4.3.4, *OHI* 6, 291: "...Set ego teneo pro certo quod anima est composita ex materia et forma sicut angeli; eadem enim est questio de angelis et de animabus racionalibus." A similar statement appears in the *DMS* 3.2, in Lindberg, *Roger Bacon's Philosophy*, 191.

67) Bacon, *DMS* 3.2, in Lindberg, *Roger Bacon's Philosophy*, 191.

Considered in this way, the reception of *species*, which have no fixed dimensions and which represent both matter and form, within the intellective soul, which is likewise dimensionless and which is also a composite, becomes plausible.

Yet there seems to be another difficulty with the reception of *species* into the intellect, a difficulty which has to do with the status which Bacon had given to *species* as "natural." Bacon defines the intellect as a "rational substance" and deems it rational because it operates using deliberation and free will.[68] By contrast, Bacon characterizes natural action by the attributes of uniformity and necessity:

> … for only an agent that possesses free will and acts by deliberation can, for its part, act difformly. But a natural agent possesses neither will nor the ability to deliberate, and therefore it acts uniformly …[69]

The will as a basis for rational action is free to choose its course of action. Natural action is instinctive, immediate, and inherently deterministic.

Within the context of his classification of signs to the categories of "natural" and "given by soul," Bacon notes that some signs given by soul should also be considered natural. He provides the example of a dog's bark, which is natural because it is instinctive, automatic, immediate, and given "as if suddenly in the absence of perceptible time, and with a certain natural instinct and impetus …."[70] Bacon feels a need to explain in what sense a soul can be said to act naturally. Turning to Aristotle, he writes:

---

68) Bacon, *Communium naturalium* 1.2.4.6, *OHI* 2, 108: "intellectus autem hic vocatur substancia racionalis, que operatur ex deliberacione racionis et eleccione voluntatis."

69) Bacon, *DMS* 1.1, in Lindberg, *Roger Bacon's Philosophy*, 19.

70) Bacon, *De signis* 8, in K. Margareta Fredborg, Lauge Nielsen and Jan Pinborg, "An Unedited Part of Roger Bacon's *Opus Majus*: De Signis," *Traditio* 34 (1978), 75-136, at 83: "…quasi subito per privationem temporis sensibilis et quodam instinctu naturali et impetu naturae …"

> But in the second [book of the] *Physica* [nature] is considered more broadly than [in the *De anima*] as a force acting without deliberation whether in animate or inanimate things. Thus a soul can be included under nature in one way.[71]

Nature and reason are opposed. Yet even though the intellect as a whole is called "rational substance," in fact only some of its operations are truly rational. According to Bacon, any act—even if performed by the rational soul—which does not make use of deliberation should be considered natural:

> … the rational substance acts in many ways without deliberation and the choice of the will, therefore with regard to these actions it is said to act through nature ….[72]

Is the reception of *species* into the intellect a deliberative or natural act? Bacon's answer is clear-cut:

> For the rational soul does many things as by nature and many without deliberating.... And just, insofar as concerns the practical intellect, it has natural affections of this kind, in resemblance to animals, so likewise it has many natural cognitions and apprehensions and reflections [which it performs] without the exercise and deliberation of reason and without full use of it.[73]

The intellective soul does not always act by fully using its rational capabilities. In fact, many of its cognitions and apprehensions are received without exercising reason or use of its free will at all. In receiving *species*, the rational soul is passive; it is not at liberty to choose which of the apprehensions to accept and which to reject.

---

71) Bacon, *De signis* 13, in Fredborg et al., "An Unedited Part ," 85. Translated by Maloney, 1988, 139.

72) Bacon, *Communium naturalium* 1.2.4.6, *OHI* 2, 108: "…substancia racionalis multa agit sine deliberacione et eleccione voluntatis, ideo respectu illarum accionum dicitur agens per naturam…."

73) Bacon, *De signis* 12, in Fredborg et al., "An Unedited Part," 85: "Anima enim rationalis multa facit ut natura et multa ut non deliberans …. Et sicut quantum ad intellectum practicum habet huiusmodi naurales affectiones ad similitudinem brutorum, sic similiter habet multas cognitiones et apprehesiones naturales et cogitationes dine [read sine] decursu et deliberatione rationis et sine pleno usu eiusdem."

One must bear in mind that there is a difference between rational operations and cognitive ones. While the internal senses, according to Bacon, perform certain cognitive functions, these are not rational functions as such. Even some of the intellect's functions are cognitive but not rational.[74] The dividing line between nature and reason is no longer found between the sensitive soul and the rational one, but is now drawn within the rational soul itself.

5. Innate Ideas and Divine Illumination Incidental to Knowledge

One of Spruit's arguments against the reception of *species* within the intellect is a passage from the questions on *Physics V*, where Bacon quotes Augustine's view on innate knowledge:

> Many authorities are against this. First is the saying of Augustine "the soul is brought into being in the likeness of all wisdom, carrying within itself the species of all things," and in the third part of *De Anima* [Aristotle writes] that "in a certain measure the soul is everything"; and so, according to these authorities it would seem that the soul has an innate cognition of all things.[75]

In this text, Bacon cites Augustine's hypothesis that the soul is created from the outset having the *species* of all things within it. These *species* are therefore not acquired but innate. Bacon indeed cites Augustine as well as Boethius on this point, but he immediately qualifies that "it is possible to call that intelligence the natural appetite or desire to know, and it is in this way that knowledge is innate."[76] Bacon believed that Augustine and Boethius were right in asserting that the rational soul is created with the *species* of all things in it; and as Wood notes, this was the standard early

---

[74] This distinction is spoken of in Tachau, *Vision and Certitude*, 10.

[75] Bacon, *Quatuor physicorum*, *OHI* 8, 2: "Contra sunt multe auctoritates, prima Augustini dicentis 'anima est ad (similitudinem) totius sapientie procreata species omnium rerum in se gerens' et in tertio *De Anima* 'quodammodo anima est omnia'; et ita per has auctoritates videtur quod anime sit rerum omnium cognitio innata."

[76] Bacon, *Quatuor physicorum*, *OHI* 8, 3: "… potest nominare scientia naturalem appetitum vel desiderium ad sciendum, et hoc modo scientia est innata; et de hoc procedunt ultime auctoritates."

thirteenth-century opinion.[77] Yet he considered that once the soul is united to the body, our access to these innate *species* is hindered. In his first set of questions on the *Metaphysics* he stated that

> .... knowledge can be discussed in two ways: in one way, by the very fact that it is knowledge, it arises as a disposition or affection of the soul, inclining to the affection of the good and the investigation of truth; and this is an imperfect and confused knowledge, which is innate ....[78]

Although in two of his earliest works Bacon mentions functioning innate exemplars, as his thought develops he begins to describe innate knowledge as vague and indistinct.[79] In his mature writings innate knowledge as specific or even general information is out of the question. In the works sent to the Pope innate knowledge is not even discussed, and it is replaced—if at all—by innate capacities. One rare example is found in the *Opus tertium*, where Bacon regards the human ability to articulate arguments and recognize logical fallacies as innate:

> Therefore we have another course of argument besides the one given by the method of Aristotle; yet it is none other than innate. What remains then, is that we know how to provide proofs from nature, and likewise how to refute arguments by destroying false premises and by dividing and distinguishing bad conclusions.[80]

---

77) Wood, "Imagination," 47.

78) Bacon, *Primum metaphisice, OHI* 11, 9: "... de scientia dupliciter est loqui; uno modo ut per hoc quod est scientia nascatur anime dispositio vel affectio inclinans ad boni affectionem et veri speculationem, et hec est scientia imperfecta et confusa, et hec est innata ...."

79) For the details of this development see Wood, "Imagination," 52-53 . The unclarity of innate knowledge and the need for acquired *species* or phantasms is discussed, among other passages, in the *Quaestiones supra libros octo physicorum, OHI* 13, 11; *De causis* 10, *OHI* 12, 74. In the *De causis* 10, *OHI* 12, 75 Bacon states that innate *species* would only be useful for a child as a starting point in his quest for knowledge or after the soul was separated from the body. In this life, however, we are bound to rely on acquired *species.*

80) Bacon, *Opus tertium* 28, ed. J.S. Brewer, in *Opera quaedam hactenus inedita* (London, 1859), 104-105, "Ergo aliud regimen arguendi habemus quam per artem Aristotelis datum; sed non est aliud quam innatum. Relinquitur igitur quod a natura scimus arguere, et similiter dissolvere argumenta per interemptionem propositionum falsarum, et per divisionem et distinctionem malarum consequentiarum." In the *Opus*

To sum up: a soul is born with the innate *species* of all things. Unfortunately, however, it cannot access the information stored in those *species* in this life, for this information gets blurred once it is united to the flesh. Innate *species* only provide a confused starting point to our acquaintance with the world and a general guidance to distinguish true from false reasoning. In no way does innate content furnish distinct and reliable information.

But perhaps divine illumination does. In his early writings Bacon referred to divine illumination as the source of the Platonic exemplars,[81] and as providing information to the possible intellect by a direct infusion of *species*.[82] Yet in much the same way as in his treatment of innate knowledge, he came to discount of divine illumination as a source of knowledge:

> ... the species of the agent intellect are distinct according to themselves, yet because of the weakness of light and their distance from the first cause, there is darkness and obscurity there.... The possible intellect is informed by the agent intellect by gazing upon it, but receives only confused cognition in this way.[83]

Owing to their distance from God, humans cannot make full use of the information supplied by illumination, and it appears to them

---

majus 4.3 (Burke, 121) Bacon treats the ability to comprehend mathematical truths as an innate capacity.

81) Bacon, *Questiones supra undecimum prime philosophie Aristotelis* (*Metaphysica XII*), *primae et secundae*, *OHI* 7, 15-16 and 110; *Quatuor physicorum*, *OHI* 8, 2-3.

82) Bacon, "*undecimum prime philosophie,*" *OHI* 7, 110: "... et hec vocatur intellectus agens, et hec non intelligit rem per administrationem sensuum, set per exempla sibi innata..."; *Octo physicorum, OHI* 13, 9: "Preterea, ex presentia agentis proportionalis cum materia derelinquitur effectus; ergo cum intellectus sit hujusmodi, ergo derelinquit effectum intelligendi in possibili; ergo fit per species agentis. Quod concedo: et fit per species alienas, scilicet per species agentis...." And in the next page (10): "... alia est cognitio precedens istam et hoc per species agentis, et hec non cum fantasmate.... Et est duplex modus ponendi hoc: aut quod intellectus agens informat intellectum possibilem et ponat ibi suas species, aut quia intellectus possibilis intueatur ipsum agentem."

83) Bacon, Octo *physicorum, OHI* 13, 9-10: "... species agentis sunt distincte secundum se, propter tamen defectum luminis et elongationem cause prime est ibi obscuritas tenebrositas... intellectus possibilis informatur quantum ad cognitionem confusam per intuitionem in agente."

as indistinct. In Bacon's mature writings there is no mention of infused *species*, and the illumination coming from the agent intellect serves to activate the possible intellect, thereby enabling it to handle the incoming *species* and grasp the information they bring up. In the *Communium naturalium* Bacon assigns the agent intellect the function of recognizing truthful arguments.[84] He then explains that knowledge through hearing of words is gained:

> ... in part by the teacher showing and explaining, in part by the experience of the senses and in part by the influence of the agent [intellect]. For the thing can be shown and exemplified to vision by the instructor, and the species of the thing arrives in the intellect through sense, and the agent intellect illumines, and in this way stable knowledge (*habitus cognitivus*) is born in the soul ....[85]

The agent intellect helps the possible to stabilize and assimilate the acquired *species*, thus creating cognitive habits.[86] Not a word appears here to link the illumination coming from the agent intellect with abstraction, universality, or any specific content. Divine illumination in Bacon serves to stabilize knowledge, not to construct it. In the *Opus tertium* he writes:

---

84) Bacon, *Communium naturalium* 1.3, *OHI* 3, 289.

85) Ibid., 289-290: "... partim per magistrum ostendentem et exemplificantem, et partim per sensus experienciam et partim per influenciam agentis. Nam res per doctorem potest visui ostendi et exemplificari, et species rei venit ad intellectum per sensum, et intellectus agens illustrat, et sic nascitur in anima habitus cognitivus...."

86) The terms Bacon normally uses when discussing intelligibles are *species* and cognitive habits, by which he means *species* that have settled and stabilized. When discussing the *species* in the intellect as basis for acts of naming by imposition (such a discussion appears in the *De signis* 162 and 164, in Fredborg et al., "An Unedited Part," 75-136, 132-133), he also uses the terms *conceptus mentis*, *cognito* or *cogitatio*, *affectio*, *intellectus rei* and *notitia*. He does not define or differentiate between the various terms, and it seems as though he views them as synonyms. In the *Communium naturalium* 1.2.4.1, *OHI* 2, 110 he defines *conceptus* as an apprehension by which the soul grasps a thing and understands it, and ascribes the ability to form such concepts to animals too. According to Bacon therefore, concepts can be found not only in the intellectual faculties, but in the sensitive ones too. See Tachau's discussion in *Vision and Certitude*, 16.

> … and that which has the power to [acquire] knowledge (*scientia*) is called the possible intellect, and does not have [knowledge] from itself; but when it receives the species of things, and the agent flows into and illumines it, then knowledge (*scientia*) is born in it ….[87]

Acquired *species* remain the main source for *scientia*. Bacon accepts divine illumination as essential to knowledge, but this does not render sensible *species* superfluous, or exempt him from explaining how they are received into the intellect. Moreover, Bacon also fails to explain exactly how divine illumination contributes to *scientia*. The attention he devotes to the study of external experience, as compared with his scant and fuzzy remarks about illumination, is striking. Even if one chooses to disregard the various comments of his Aristotelian lectures and to concentrate on his mature writings alone, one cannot fail to notice that his discussion of spiritual experience is remarkably brief, whereas his treatment of natural, external experience is detailed, systematic and thorough.[88]

It is interesting to note that in his mature writings most of the remarks related to divine illumination appear in the *Opus majus* and are concentrated in the part on philosophy and in the first chapter of the part on experimental science.[89] By contrast, in the *Communium naturalium*, which was Bacon's *scriptum principale* in which he actually carried out the principles laid down in the *Opus majus*,[90] illumination is mentioned only in two places: the first is quoted above, and in another place he states that since our intellect belongs to the realm of time and succession, we cannot perceive spiritual and eternal things in this life without spiritual illumination.[91] Divine illumination is altogether absent from the *DMS*.

---

87) Bacon, *Opus tertium* 23, in Brewer, 74: "… et intellectus possibilis vocatur qui est in potentia ad scientiam, et non habet eam de se; sed quando recipit species rerum, et agens influit et illuminat ipsum, tum nascitur scientia in eo …"

88) Rega Wood, "Imagination," 56, has also drawn attention to this point.

89) Bacon, *Opus majus* 6.1, in Burke, 585-587.

90) See Andrew G. Little, "On Roger Bacon's Life and Works," in idem, ed., *Roger Bacon Essays—Contributed by Various Writers on the Occasion of the Commemoration of the Seventh Centenary of his Birth* (Oxford, 1914), 1-32, 11; Sharp, *Franciscan Philosophy*, 116.

91) Bacon, *Communium naturalium* 1.3.1.8, in *OHI* 2, 175.

In his introduction to this work, Lindberg notes that, "Bacon's work exhibits little in the way of a theology of light .... Bacon was certainly influenced by Augustine's theory of divine illumination, but nothing of this appears in the *DMS* or *DSC*."[92]

One must bear in mind the purpose for which the *Opus majus* was written: it was a research proposal, a plea intended to obtain support and sponsorship from the papal court, including a release from the censorship imposed by the Chapter of Narbonne.[93] As such it was designed to appeal to the potential financer's perceived interests. Accordingly, Bacon assures the Pope in those works that his entire research plan is not opposed to the divine wisdom but is in fact comprised within it, for all wisdom is from God and cannot contradict the scriptures.[94] His identification of the agent intellect with God[95] is made within that context; and while he presents it along with a set of philosophical justifications, it cannot be taken in isolation or explained purely in philosophical terms.

6. Reliable Representation Requires Tight Linkage between Nature and Mind

I am not in a position to judge whether Bacon was making a conscious attempt to "naturalize" the mind or whether this was an unintended outcome from the principles he wished to employ. Nevertheless it seems evident to me that he did call for a strong linkage and a real connection between extra-mental reality and concepts, and that it was this requirement which led him to treat the intellect in naturalistic terms. For as Tachau puts it, once concepts are not seen as created by the mind and are therefore not arbitrary associations or divisions, and once the rules for the reliability of judgments have been laid down, then the *species* of an object becomes

92) Lindberg, *Roger Bacon's Philosophy*, xlii, n. 30.

93) See Lindberg, *Roger Bacon's Philosophy*, xxiii; Brian Clegg, *The First Scientist—A Life of Roger Bacon* (New York, 2003), 80.

94) Bacon, *Opus majus*, 2.3; 2.5; 2.7; 2.9, in Burke, 39, 48, 49, 52 respectively.

95) Bacon, *Opus majus*, 2.5, in Burke, 43-48.

ipso facto veridical.[96] Yet my hunch is that there is more to it than that. After all, Bacon did posit some inferential operations at the sensory level, including comparison, discrimination and correlation.[97] I suggest that what might be at stake here is rather the nature of representation that Bacon requires. Perhaps the best way to understand Bacon's concept of representation would be to contrast it with that of Aquinas. Aquinas distinguished between likeness as agreement in nature and likeness as regards representation. He stipulated that for the purpose of cognition, only likeness as regards representation is required. Aquinas obviously believed that the absence of a natural likeness was no barrier to representational likeness.[98] Bacon, however, disagreed. He required natural likeness all the way through; and this was, in my view, the principal motivation for his naturalistic understanding of the intellect.

Conclusions

Since in his mature writings Bacon detaches the agent intellect from the human soul and identifies it with God, the human intellect is left with a material intellective soul, which is not even wholly rational. The human intellect is deemed material, yet it is so in a special sense. It is not the kind of matter perceived by the senses, nor the kind subject to physical laws. Since Bacon also defines the material *species* as insensible, and since they do not, properly speaking, occupy a well-determined space and are not, in fact, "things" (*res*),[99] then although they are considered natural and an inseparable part

---

96) Tachau, *Vision and Certitude*, 17.

97) For example, the twenty visible intentions are deduced by a proto-syllogistic reasoning from the *species* of light and color. In general, Bacon tends to transfer cognitive operations to the sensitive soul, up to the point of concept formation. Due to considerations of length, this matter is not discussed here.

98) See the discussion in Pasnau, *Theories of Cognition*, 106-107.

99) In the *De causis* 10, *OHI* 12, 41, Bacon emphasizes the intentional and false (*intentionale et vanum*) existence of *species* and contrasts it with real existence (*rem veram*). He seems to hold that the very essence of *species* lies in its being incomplete. Once a *species* fulfills its mission and renders the patient—that is, the one in which it is received—similar to the agent, it ceases to be a *species* and becomes a *res*.

of physical processes, their reception within the intellect's "spiritual matter" does not seem to be problematic. The possible intellect thus becomes the matter into which *species* can be incorporated and express their intelligibility. The intellect is no longer separate from the natural world in any essential way, and is thus better able to access and know it. While the agent intellect becomes a part of God, the possible intellect, at least as far as perception is involved, becomes a part of nature. Spruit's barrier between senses and intellect has been removed.

Magic and the Physical World in Thirteenth-Century Scholasticism

Steven P. Marrone*
Tufts University

Abstract
The turn to modern science in the Scientific Revolution of the seventeenth century is typically characterized as dependent on the novel adoption of a mechanical hypothesis for operations in nature. In fact, the Middle Ages saw a partial anticipation of this phenomenon in the scholastic physics of the thirteenth century. More precisely, it was just the two factors, denial of action at a distance and an emphasis on the primary materiality of causation, that constituted this early mechanism—or "protomechanism." The latter's emergence can be seen most clearly where scholastic thinkers—here, William of Auvergne, Thomas Aquinas and Giles of Rome—confronted the theoretical limits of natural cause and effect in their efforts to determine the reality of magic and locate its place in the natural world.

Keywords
action at a distance, cruentation, magical arts, materiality of causation, mechanical hypothesis, natural magic, occult force or power, protomechanism, separate substances, science of images, Augustine, Avicenna (Ibn Sina), Daniel of Morley, Giles of Rome, Hugh of St. Victor, Robert Grosseteste, Thomas Aquinas, William of Auvergne

For many in the contemporary world, the epitome of reason is science. In the United States in particular, science is further defined quite narrowly to mean natural science. In either case, one can make an argument that the resulting cultural priorities have worked to the benefit or, conversely, to the detriment of the conditions of human existence. Political debates over the consequences of "mod-

* Department of History, Tufts University, Medford, MA 02155, U.S.A. (steven.marrone@tufts.edu).

ernization," "westernization" and "development" provide abundant evidence of the ease with which opposing positions on the issue can be sustained. Leaving aside such normative concerns, let us for the moment simply take as given the common association of "science" and "modernity" as well as the equally commonplace assumption that "science" in the modern world constitutes a special form of the general category of intellectual artifacts known throughout history by that name, a form of it that is especially rational or reasonably worthy of our estimation. With that as ideological background, this article will focus on just one piece of an influential theoretical model according to which modern science, in particular modern natural science, has been conceived.

Historians commonly take stock of what we call modern science by locating its beginnings in the so-called Scientific Revolution of the seventeenth century. Standard for a long time has been the presumption that the ideological momentum propelling this revolution, and in a formal sense summing it up, was a set of ideas and attitudes called, already in the seventeenth century itself, the "Mechanical Philosophy" or, in Robert Boyle's more cautious words, the "mechanical hypothesis."[1] In his fascinating recent book, *Atoms and Alchemy. Chymistry and the Experimental Origins of the Scientific Revolution*, William Newman offers a useful synopsis of the current historical understanding of "mechanism"—as we might label this philosophy or hypothesis—a complex analytical instrument consisting of at least three parts (or alternatively, three different modulations) in differing degrees of emphasis.[2] The first is "structural reductionism," by which he means the ultimate explanation of all natural phenomena as due to the spatial distribution and interaction of tiny corpuscular components, whether they be thought of as absolutely simple and indivisible or not. Second is "denial of action at a distance," which needs no clarification. The third amounts

[1] See, for instance, the title of one of Boyle's papers: "About the Excellency and Grounds of the Mechanical Hypothesis," included in *Selected Philosophical Papers of Robert Boyle*, ed. M.A. Stewart (Manchester, 1979), 138.

[2] William R. Newman, *Atoms and Alchemy. Chymistry and the Experimental Origins of the Scientific Revolution* (Chicago, 2006), 177.

to "rejection of final causes," which again for the purposes of the present essay can pass without comment.

Of these three, I shall be interested in what follows in the rejection of action at a distance, precisely as that characteristic could be considered a necessary attribute of all phenomena occurring in the realm of nature—that is, all those that should be designated "natural." Lurking behind or alongside this was another explicatory demand: insistence on the absolute materiality (or corporeality) of natural causation. Though not explicitly included among Newman's three, this fourth theoretical criterion is imbedded within his first, "structural reductionism," and may be immanent in or inevitably implicated in his second, exclusion of action at a distance. Because the limitation of natural causation to material or corporeal action was still a part of the game in seventeenth-century scientific debate, because in figures like Descartes it was definitely an element of the mechanical understanding of the natural world, and because its ideological presence has so long been taken as an indicator of the mentality of modern science, I have included it as part of a pair with denial of action at a distance. Together, these two will serve for me as identifying indicators of a movement towards at least a part of "mechanism" already in Latin scientific thinking of the high Middle Ages. For brevity's sake, I would like to refer to this medieval and still only partial "mechanism," marked by my two characteristics, as a "protomechanistic" approach or point of view.

Let me begin by suggesting that at the opposite end of the spectrum from those aspects of the mechanical philosophy denying action at a distance and insisting on the materiality of causation lies a vision of action in the natural world where local contiguity is not required nor is materiality a necessary factor in causality, or what we might equally well denominate as the power to induce or generate a natural effect. Such a perspective was, I believe, pervasive in the natural philosophy of the Latin west in the twelfth century—is, I think, invariably taken for granted by historians in what they habitually refer to as twelfth-century Platonism. To my eyes, it was also a property of the eleventh-century Persian Ibn Sina's grand accounting of action in the natural world, so influential in the Latin west from mid-twelfth century on. Though no attempt will be made in the present essay to justify these two claims, the latter will receive at least oblique confirmation in references to Ibn Sina—cited, of course, according to the Latinized version of his name, Avicenna—made by one of the scholastics I shall deal with soon.[3]

To be quick about it, I want to propose that we do not have to wait until the seventeenth century to witness a significant challenge to the intellectual orientation of such Neoplatonizing natural philosophy among the intellectual elite, for already underway in the Latin universities of the thirteenth century was a shift towards emphasizing contiguous activity and materiality of causation as necessary attributes of natural phenomena. In other words, the "protomechanism" I have in mind was already emerging in the high Middle Ages. In the theoretical disputations of the time—among "artists," or what we would call natural philosophers, but more especially also among theologians—it was where magic came under consideration that we find one of the most prominent instances of recourse to this new and protomechanistic perspective. Magic was, in the thirteenth century, well on the way to becoming a vital concern of university-trained intellectuals not only in their moral and legal considerations of legitimate or dangerous activity in the real social world, but also in their ideological joustings over the content, subject and acceptable explanatory models of science, especially natural science. It is hardly surprising this was so, since the very notion of magic evoked the question of the limitations of activity in and by means of nature.

In the centuries after Augustine, "magic" in the Latin west, on the infrequent occasions when the term was even brought to mind, had been subsumed under the broader rubric of "superstition." By "superstition," moreover, Augustine and those who followed him for much of the Middle Ages intended to designate any activity directed towards the veneration or even gratification of demons or demonic powers, requiring their assistance, or sometimes simply inspired by them and done at their behest. It included idol worship as well as just consulting with demons for whatever knowledge or aid they could provide, but also all efforts relying upon

[3] See below, pp. 174-76.

what Augustine, following the standard usage of his time, called expressly the "magical arts."[4] Among the latter could be listed the innumerable divinatory practices of his time, all the incantatory or amuletic procedures for curing illness, astrology in every guise, as well as the many everyday habits and behaviors that Augustine considered otiose and inane and which most of us today mean when we talk of "superstition"—taking care not to walk under a ladder, for example.[5] What tied these "magical" activities together was not just that they were for the most part unreliable or misleading but also that they either depended upon demonic intervention—as in the case of divination, where demons told diviners what to say, at least in all those instances where what was said was true—or were prompted by demonic suggestion—as in even the slightest "superstitious" habit, which invariably served to turn the performer's moral compass away from reliance on and devotion to God. "Magic" by this construal constituted a semantic category characterized by irreligiosity more than anything else, including the often still concomitant attributes of vanity, deceptiveness or outright privation of the truth.

With the influx of works of natural philosophy or cosmological speculation, derivative in large part of the antique and especially Greek philosophical tradition and translated into Latin from Arabic, Hebrew and finally Greek itself, that we start to see in the cultural world of the intellectual elite of western Europe from the twelfth century on, the near-exclusive dominance of the Augustinian notion of "magic," with its capacious boundaries and largely religious connotations, began to be challenged. For the "scientific" literature of the medieval Arabic, Hebrew and Greek traditions contained within itself elements explicitly construed in that very same tradition as magical. Already by the early twelfth century, alarmed scholars, often monks such as Hugh of St. Victor, had been forced

[4] Augustine, *De doctrina christiana* 2, 74 [xx 30] (ed. and trans. R.P.H. Green [Oxford, 1995], 90): "Superstitiosum est quidquid institutum est ab hominibus ad facienda et colenda idola pertinens vel ad colendam sicut deum creaturam partemve ullam creaturae vel ad consultationes et pacta quaedam significationum cum daemonibus placita atque foederata, qualia sunt molimina magicarum artium… ."

[5] Ibid., 2, 74-78 [xx 30-xxi 32], ed. R.P.H. Green, 90-92.

to admit that among the disciplines purported to belong to the arts and sciences of the new learning were those one could do no better than to designate under the simple general heading of "magic" (*magica*). According to Hugh, they fell into five different sorts of learned art: divination or soothsaying and judicial astrology but also fortune-telling (*sortilegia*), sorcery (*maleficia*) and the production of illusions.[6] Although he clearly leaned in Augustine's direction, preferring to see "magic" as capsulized primarily in a moral or religious failing, characterized by turning to demons or falling prey to their wiles, the recent appearance of magical disciplines among the arts almost compelled him to bring to the fore the epistemic qualities of magic as its primary attribute. In any case, there was no ambiguity that he believed it had to be condemned. Its obnoxiousness and failing therefore Hugh felt constrained to characterize as its inherent falsehood. In his words: "Magic is not part of philosophy but rather stands outside it, making false claims [for our attention]. It is [indeed] the master of all iniquity and evil, lying about the truth and truly injuring [our] souls."[7] It is not as if the old Augustinian criticism from a religious standpoint had disappeared. But the claims of magic as art or science had to be confronted, if at all possible confounded as a lie.

Others among the educated, perhaps the majority, were not so convinced that magic in its new guise could be dismissed as nothing more than falsehood and deceit. Among wise Arabs and Jews it had been considered productive of real effects and legitimate pieces of information. Even if some of the products of the so-called magical arts were deceptive and illusory, others most decidedly were not. Reaction to the new arts therefore split into two currents. On one side stood those who were willing to believe that some of the effects and products of some of the magical arts were substantial,

---

6) Hugh of St. Victor, *Didascalicon* VI, 15, ed. Charles H. Buttimer (Washington, D.C., 1939), 132: "Haec [i.e. magica] generaliter accepta quinque complectitur genera maleficiorum: manticen, quod sonat divinatio, et mathematicam vanam, sortilegia, maleficia, praestigia."

7) Hugh, *Didascalicon* VI, 15, ed. Buttimer, 132: "Magica in philosophiam non recipitur, sed est extrinsecus falsa professione, omnis iniquitatis et malitiae magistra, de vero mentiens, et veraciter laedens animos... ."

though illegitimate. For them it was not so much the falsehood of magic that characterized it as unworthy of being counted among the sciences welcome in the educational curriculum—though most of these same people believed that a good deal of what came out of the magical arts was illusory and deceptive. Instead, what counted most was the fact that magic relied not on the impersonal workings of elements in the world of nature but rather on spirits, most importantly invisible spirits, that manipulated these elements somehow to unexpected effect or simply passed along a bit of knowledge on the sly to an aspiring magician. These spirits were, of course, evil spirits—demons. The Augustinian view of magic and superstition had maintained as much for centuries, and to that degree this response to magic fell into an Augustinian mold, although unlike in the case of Augustine it was more concerned with magic as a dangerous form of knowledge than with magic as sacrilegious or idolatrous behavior.

A good example of one such stance is Robert Grosseteste's attack on astrology. Grosseteste was an eminent theologian and then prelate of the church in the early thirteenth century. Though he had practiced astrology as a young man, and clearly believed that much could be learned about the course of events here on earth by looking to the circuits of the stars, as a mature scholar he nonetheless attacked the entire art of drawing judgments from the positions of celestial bodies and condemned even the slightest practice of it. His reason was that demons had to be involved, if only to entice people to dabble in a discipline that threatened to undermine the Christian doctrine of free will. After laying out his philosophical and theological arguments against astrology in his late work, the *Hexaëmeron*, Grosseteste concluded with an omnibus prohibition couched fully in Augustinian terms:

> To draw to a close, let us issue the following warning. Practitioners of the art of astrology are seduced and [in turn] seducers, their teachings impious and profane, written at the inspiration of the devil. Their books, therefore, should be delivered to the flames, and not only they, but also all who consult them, are lost.[8]

---

[8] Robert Grosseteste, *Hexaëmeron* V, xi, 1, ed. Richard C. Dales and Servus Gieben (London, 1982), 170: "Hec tamen in calce volumus admonere, quod huiusmodi

On the other side of the divide, heading in another direction, were those even readier to accept the possibility that some of the magical arts could generate substantial effects or often reveal the truth but who now embraced these same arts with open arms. A case in point is the English scholar of the late twelfth century, Daniel of Morley, who in his *Liber de naturis inferiorum et superiorum* sang the praises of the judicial art of the stars, which he designated as the science of *astronomia*. This latter was, moreover, an extremely inclusive branch of learning. Of it, Daniel proclaimed:

> Concerning the dignity of this [science], we find—as the sages of old have told us—that it is divided into eight parts. They are the science of judgments, the science of medicine, the science of necromancy *secundum physicam*, the science of agriculture, the science of illusions, the science of alchemy, which is the science of the transformation of metals into other kinds, the science of images, which has been passed down to us by the great and universal *Book of Venus* composed by Thoz the Greek, and the science of mirrors. This [last] science is more fruitful and comprehensive than the others, as Aristotle makes clear in his *Book of Burning Glasses*.[9]

Obvious to any educated reader would be the fact that several of these disciplines were identical with those that Hugh of St. Victor had anathematized as magical, not just evil but also unproductive of the truth. To Daniel's eyes, however, there was nothing wrong with these arts. He granted their claim to be included among the sciences and even suggested that among the learned disciplines they held a position of eminence. No mention here of demons. Instead, the magic Daniel had in mind—though he prudently avoided employ-

---

iudices seducti sunt et seductores, et eorum doctrina impia est et profana, diabolo dictante conscripta. Ideoque et libri eorum comburendi; et non solum isti, sed eciam qui consulunt tales sunt perditi."

9) Daniel of Morley, "Philosophia" X, 158, ed. Gregor Maurach, in *Mittellateinisches Jahrbuch* 14 (1979), 239: "De dignitate eius invenitur, quod illius partes, secundum quod dixerunt sapienties primi, octo sunt, scil<icet> scientia de iudiciis, scientia de medicina, scientia de nigromantia secundum phisicam, scientia de agricultura, scientia de prestigiis, scientia de alckimia, que est scientia de transformatione metallorum in alias species, scientia de imaginibus, quam tradit Liber Veneris magnus et universalis, quem edidit Thoz Grecus, scientia de speculis, et hec scientia largior est et latior ceteris, prout Aristotiles manifestat in Libro se speculo adurenti."

ing the word itself—held the key to knowledge of some of the most wonderful, perhaps wondrous, workings of nature, shorn of the intervention of willful, hence unpredictable personalities or spirits.

In light of these last two points of view, accepting at least some of the effects and results of magic but debating over their origin and thus the worthiness of the relevant arts to be taught and practiced, it should thus be clear why, as noted above, magic or magical operations would by the thirteenth century have become a crucial theoretical site on which could be fought out differing claims about just how nature worked. Merely defining magic demanded taking a stand on the boundary between natural operations and those either outside of nature or above it. The fact that it was magic out of all other possible limiting cases that commonly provided this opportunity in the high Middle Ages means therefore that we should expect medieval discussions of magic to have a central place in our investigation of the contours of scholastic natural philosophy, including the question of whether any such "protomechanism" of the sort I am interested in actually existed.

In this essay I attempt no more than to make a contribution towards launching the discussion. In fact, I must confess to being a bit behind the curve. Already scholars such as Béatrice Delaurenti have begun to examine questions of causality, materiality and contiguity in scholastic natural philosophy precisely with reference to debates about the character and efficacy of magical operations and their relation to other phenomena observable in the world around us. [10] To give an idea of some of the evidence that I, myself, have begun to accumulate in support of my proposition about the early signs of mechanism in medieval thought, I shall look at the work of three philosopher-theologians, all of whom at some point exercised the magisterium in the Faculty of Theology at Paris in the thirteenth century. The first is the secular canon and later bishop of Paris, William of Auvergne, who died in the episcopal office in 1249. Second is Thomas Aquinas, Dominican friar and master of theology first at Paris but then also in other *studia* and universities

---

[10] Béatrice Delaurenti, *La puissance des mots. "Virtus verborum"* (Paris, 2007).

at the behest of his Order, whose death occurred in Italy on his way to the Council of Lyon in 1274. Last comes a friar of the Order of Austin Hermits, Giles of Rome, who, having returned to Paris in 1285 as master of theology after suffering the indignity of seeing his original inception into the magisterium impeded in 1277, ended his days as an honored prelate of the church, dying in Avignon in 1316. I hope to demonstrate a progression among the three, from William of Auvergne, still much immersed in the Platonic, Avicennian naturalism dominant in the Latin twelfth century but open to the conception of natural action as confined to the cause and effect of contiguous, material objects, through Thomas, committed to the clarification and firmer establishment of the new perspective, to Giles of Rome, most "mechanistic" of the three, for whom almost no room remained in nature for action at a distance and arising from an immaterial source. I limit myself to the comments of all three on magical effects and their location in the natural order.

William of Auvergne

For William of Auvergne, I rely on two of his treatises, both constituents of a sprawling survey of all reality, from the heights of divinity to the depths of the earth, as it might prove of interest or utility to human beings. The entire collection, which he entitled *Magisterium divinale ac sapientale*, William started compiling in the early 1220s, stopping short of the projected finish in the mid-1240s, just before his death. Of the two pieces of this work examined here, the more important is the treatise *De universo*, the foundations of which might well have been laid in writings composed in the 1220s, all of them, however, greatly reworked in the 1230s, with some parts not receiving their finishing touches until around 1240. Less significant for my purposes, but useful all the same, is *De legibus*, intended to be paired in the Magisterium with a further treatise, *De fide*, and written sometime after 1228, most likely before 1231.[11]

[11] On dating William's writings, see primarily Guglielmo Corti, "Le sette parti del *Magisterium divinale ac sapientiale* di Guglielmo di Auvergne," in *Studi e ricerche di*

Long stretches of *De universo* and passages in *De legibus* are devoted to a frontal attack on many of the magical arts, some of which William claims to have been familiar with from the days of his youth, when he dared to peruse what he later would disavow as the detestable books of necromancers, such as the *De deo deorum* from the Hermetic corpus and the *Liber sacratus*, or *Sworn Book of Honorius*.[12] Whether of necromantic origin or not, however, all the arts William had in mind fell under his rubric, *opera magica*, which by his maturity he considered illicit and evil from beginning to end. For clarity's sake he divided its contents into three subcategories: first, the arts underpinning what we would designate as sleight of hand; second, those concerned with the evocation of false appearances by more complicated manipulation of special substances, natural confections and odd apparatuses; and third those relying on the invocation of demons to work even more startling effects.[13]

Especially the object of William's disdain among the practices associated with the *opera magica* were astrology, which he took to be completely inefficacious insofar as it presumed to predict spe-

---

scienze religiose (Rome, 1968), 289-307, revising but still drawing heavily on the earlier work of Josef Kramp in *Gregorianum* 1 (1920) and 2 (1921).

[12] See William, *De universo* II, 3, 20 (from his *Opera omnia* [Orleans, 1674; reprinted Frankfurt am Main, 1963], I: 1056aH-bE); and *De legibus* 24 (*Opera omnia*, I: 70aF) and 25 (*Opera omnia*, I: 77bD). On the identity of the second of the two works William mentions, see first of all Robert Mathiesen, "A Thirteenth-Century Ritual to Attain the Beatific Vision from the *Sworn Book* of Honorius of Thebes," in *Conjuring Spirits. Texts and Traditions of Medieval Ritual Magic*, ed. Claire Fanger (University Park, Penn., 1998), 143-62, at 145-47; and Richard Kieckhefer, "The Devil's Contemplatives: The *Liber iuratus*, the *Liber visionum* and Christian appropriation of Jewish Occultism," also in *Conjuring Spirits*, 250-65, at 253-54.

[13] William, *De universo* II, 3, 22 (*Opera omnia*, I: 1059aA-bA): "Post haec veniam ad opera hujusmodi, quae opera magica, et ludificationes vel hominum vel daemonum, nuncupantur. Dico igitur in primis, quia horum tria sunt genera: alia namque fiunt agilitate habilitateque manuum … et vocantur vulgariter tractationes vel trajectationes, et sunt magnae admirationis hominibus, donec innotescant modi quibus fiunt. Secundum genus est eorum, quae non habent nisi apparentiam et nihil omnino veritatis, fiunt tamen subtractione vel adhibitione quarumdam rerum… . Praestigium quoque, quo apparet aqua vel fluvius, ubi revera aqua non est, nisi intus vel extra aliquid adhibeatur spectantibus, quod hujusmodi fallaciam visus efficiat, non videtur possibile… . Quapropter hic scilicet in hoc praestigio solum malignorum spirituum ministerium operari videtur."

cific occurrences or to bind humans' will ("[*judicia*] *extend*[*ere*] *in res particulares et voluntarias*"), and what he called the "doctrine (or discipline) of images" (*magisterium imaginum*).[14] Much of his commentary on magic in the two treatises was invested in showing how these two were illegitimate—either false or diabolical or both. But he also took aim at a host of other wondrous but impious acts, such as casting spells or performing obscure, irreligious rites, by which people were brought to believe in the extraordinary powers of magicians and kindred wonder-working seducers into idolatry. These acts, as indeed the manipulation of images in those cases where by appearances it seemed to meet with success, were entirely due to the machinations of demons working invisibly, behind the scene. They consequently fell into the third of William's subcategories of *opera magica*. And as in all such instances where invisible spirits intervened to trick people with their marvelous operations, they worked only to the extent that these spirits called upon the commonplace modes of operation evident throughout the rest of nature. Using natural objects and natural processes of cause and effect, they thus managed—while acting surreptitiously because unseen—to produce results that could not but appear astounding to the uninformed. To those aware of the demons' activity, on the other hand, the very same results would be neither surprising nor regarded as somehow violating the normal rules of nature.[15]

---

[14] For William's attacks on astrology, denying its efficacity almost completely, see William, *De universo* I, 1, 46 (*Opera omnia*, I: 664aG-bE); I, 3, 20 (*Opera omnia*, I: 785aD-bB); and especially II, 2, 76 (*Opera omnia*, I: 929bA): "Quoniam autem apud astronomicos judices, hoc est doctores judiciorum, qui ea extendunt in res particulares et voluntarias, iste modus operationis neque novus neque inopinatus est, conveniens est ut adjuveris quantum possibile est contra errorem eorum, similiter et contra errorem eorum, qui magisterium imaginum professi sunt… ." On the *magisterium imaginum*, see William, *De universo* II, 2, 76 (*Opera omnia*, I: 929bA)—also quoted just above. He called adepts in this same art or learning "*magistri operum*"—see *De universo* II, 3, 12 (*Opera omnia* I: 1039aC). An extraordinary historical investigation of attitudes towards this latter art, more often referred to in the thirteenth century as the "science of images," is Nicolas Weill-Parot's *Les "images astrologiques" au Moyen Âge et à la Renaissance* (Paris, 2002).

[15] William, *De universo* II, 3, 12 (*Opera omnia*, I: 1039aB): "Et intendo facere te scire in sequentibus causas hujusmodi fantasiarum, viasque et modos quibus naturali-

For William, in fact, the vast majority of operations one might observe in the world—including of course most of those considered so far among his *opera magica*—were not only entirely natural but also explicable, for the perceptive and informed observer, by means of the most mundane laws of natural operation. Here is where William's nascent "mechanism"—"protomechanism" as I have called it in deference to the originality of the seventeenth century—emerges most clearly into view. As he explained in unambiguous terms, all natural activity was brought about by one of two modes: either by contrariety, when an agent overcame or erased a contrary attribute of an object falling under its influence, or by assimilation, as when an agent induced a formal similitude of itself in an object upon which it worked.[16] By way of example for the latter sort of operation, William reminded the reader of how something hot made another object upon which it acted hot as well by impressing on it a similitude of its own activating form—that is, the form of heat. Though he gave no example for the first kind of operation, we can easily supply one for him. A solid entity at rest, such as the grassy ground beneath an apple tree, will overcome the local motion downwards of a ripe fruit falling upon it, causing the once-moving fruit likewise to adopt the form of rest. All this was, to be sure, no more than a summary of the standard Aristotelianizing account of formal agency. More interesting, however, is the recognition that no sooner does one reduce the focus to the business of natural activity in the material world, than an additional restriction on the operative process comes into play. As William explained it, natural operations in the world of corporealities required contact of the agent with the object acted upon, whether it be immediate contact—as

ter fiunt. [N]am et ipsi maligni spiritus per res naturales haec omnia faciunt... ." On *magisterium imaginum*, wherever effective, having to operate by the same means, see *De universo* I, 1, 46 (*Opera omnia*, I: 663a[A-B] and 663bD).

16) William, *De universo* II, 2, 76 (*Opera omnia*, I: 929b[C-D]): "Et quod dico ... intendo secundum modum agendi naturalem, qui est secundum duos modos, videlicet interdum secundum victoriam alterius contrariorum super alterum ... . [S]ecundus modus est per similitudinem sive assimilationem, qua agens assimilat sibi patiens, imprimens ei similitudinem per quam agit, sicut calidum calorem et lucidum lumen et ad hunc modum de aliis."

when the apple hits the ground—or mediated contiguity—as when hot embers result in burnt fingers for those touching the andirons on which the embers rest. [17] So far as concerns most operations in the world that we inhabit, therefore, nature works precisely when one material thing is brought into contact with another.

Not that there were no exceptions. Indeed, for human beings, the exceptions accounted for a significant part of what it meant to live and act in the world. The exception most relevant in this regard—and it was surely the primary exception in day-to-day affairs to the normal rule of natural action by means of material contact—involved the case of an embodied soul, such as a human soul or the soul of a dog or cat. Souls had the power to move the bodies they inhabited without recourse to material agency and consequently without reference to anything like local contact or even precisely determinate physical place. And the fact that this was true indicated that William's "protomechanism" extended only so far. Nonetheless, with regard to the ways that human or most other animate beings intervened in their surroundings, the exception was still only partial and quite limited. Souls attached to bodies had the non-corporeal but nonetheless natural power to move those bodies, yet by grace of this initial corporeal effect alone could they then intervene in the rest of the material world around them, either directly by means of an organ of the body or indirectly by further deploying an artificial instrument or tool.[18] Among humans, that was the quickest and commonest way to act.

There was, however, a further exception, and it demands our closest attention. Some operations involving bodies in nature could not be characterized as a formal change or modification obedient to the

---

17) William, *De universo* I, 1, 46 (*Opera omnia*, I:663aA): "Operationes enim corporales per modum naturae per contactum sunt aut mediatum aut immediatum, quod est dicere, quia corpus virtute sua coporali non agit, nisi vel in corpus quod contingit vel in aliud contingens illud."

18) William, *De universo* I, 1, 43 (*Opera omnia*, I:648a[F-G]). This passage is given great attention by Delaurenti in *Puissance des mots*, pp. 126-27, although she interprets the action of the soul upon its own body as itself in conformity to the law of natural action by contact. As I indicated just above, there is no point of contact—no true local contact at all—between a soul and its body.

rules of similarity or contrariety. In such cases the action was frequently due to a hidden force or power (*vis*, *virtus*, *potentia occulta*), still natural but usually unexpected, difficult to describe in concrete terms, and in any event not mundane.[19] Not only, William explained, were the "books of experiments" and "natural histories" loaded with descriptions of these occult yet natural powers, but also much of the art of medicine depended upon them.[20] In instances where an occult power was at work, results were produced that could not be related back to the cause itself according to a formal resemblance or a formal opposition or contrariety, nor understood in any such terms. It was as if a hot body should automatically induce not heat but rather whiteness in another, or something falling on a surface at rest should continue to move, and even faster.

An example of such an operation, so William said, was the power of a sapphire to cure infirmity, restrain ardent passion and calm fears. With reference to this example, in *De legibus* he specifically remarked that, in contrast to the operations of more typical agents in the world, in the case of a sapphire turned to medicinal purposes, it was neither the agent's matter nor its substantial form (nor presumably any accidental form as well) that accomplished the healing or restraining result. For the action of matter here, perhaps he had in mind a process like the solid ground stopping the progress of a falling apple; for substantial form, surely the natural procedures of generation, as when a mature dog engendered a puppy; and for accidental form—had he bothered to mention it—of course a hot body warming that which it touched. In any case, the sapphire

---

19) William, *De universo* II, 3, 21 (erroneously printed as 20) (*Opera omnia*, I: 1058bH): "Sunt et alia, quae commixtionibus et aliis conjunctionibus naturarum fiunt, de quibus multa innotuerunt, plurima autem adhuc in abscondito sunt. Harum igitur naturarum vires et potentias occultas, qui noverunt, multa mirifica faciunt, et multo mirabiliora efficerent, si rerum eis hujusmodi facultas et copia suppeteret et scientia non deesset." See also *De universo* II, 3, 22 (*Opera omnia*, I: 1060aE): "Hic igitur advertendum est tibi, quia virtutes occultae multae et mirabiles sunt in rebus quas attingere non possumus... ."

20) For *libri experimentorum* and *libri naturalium narrationum*, see William, *De universo* II, 3, 22 (*Opera omnia*, I: 1060bF); on medicine, *De universo* I, 1, 46 (*Opera omnia*, I:663bD-64aE); and II, 3, 22 (*Opera omnia*, I:1060aH).

worked the cure or restraint instead "by means of its whole nature" ("*secundum totam naturam*").[21] These latter words surely reflect acquaintance with the Galenic idea of medicinal effect, whenever not by the evident elemental properties of the curative agent, then by the "property of the whole substance."[22] Operations like this might work even at a distance, so William believed.[23] More noteworthy still, he thought of them as belonging to a category of actions he was prepared to designate by—of all words—the adjective "magical." Indeed, in all instances of this sort what was at work, said William, was a "natural magic," resident right there in nature to be made use of in a "natural art of magic."[24] And in sharp contrast to those occasions where he spoke of *opera magica,* this time the magic, so long as it was not applied to harm, was anything but evil, indeed highly to be praised.[25] "Magical" in this positive sense thus pointed for William to a part of nature where operations, even operations among objects fully embedded in the material world, escaped, in a most spectacular and perhaps old-fashioned or Neoplatonizing way, the protomechanism he had elsewhere revealed himself ready to embrace.

---

21) William, *De legibus* 27 (*Opera omnia*, I:86bH-87aA).

22) For the Galenic idea, see Newman, *Atoms and Alchemy*, 141; and Brian P. Copenhaver, "Scholastic Philosophy and Renaissance Magic in the *De vita* of Marsilio Ficino," *Renaissance Quarterly* 37 (1984), 523-54, at 525.

23) Begin with the not precisely pertinent case—because the agent was a soul and not a material form—of the basilisk as presented in *De universo* I, 1, 43 (*Opera omnia*, I: 648aF): "[Q]uia forsitan anima basilisci, et animae quorundam aliorum animalium et quaedam animae humanae, multa operantur et mira valde extra corpora sua... ."

24) For *magia naturalis*, see William, *De legibus* 24 (*Opera omnia*, I: 69bC-D): "Et de operibus hujusmodi est magia naturalis, quam necromantiam sive philosophicam philosophi vocant, licet multum improprie, et est totius <sci>entiae naturalis pars undecima." For *magica naturalis*, see *De universo* I, 1, 43 (*Opera omnia*, I:648a[F-G])—where William continues the passage quoted above, n. 23, with the words: "et illa nominanda sunt et numeranda in ea parte naturalis scientiae, quae vocatur magica naturalis"—and I, 1, 46 (*Opera omnia*, I:663bD); for *ars magica naturalis*, *De universo* II, 3, 22 (*Opera omnia*, I: 1060bF).

25) William, *De universo* I, 1, 46 (*Opera omnia*, I: 663bF); and *De legibus* 24 (*Opera omnia*, I: 69bD).

Thomas Aquinas

Let me proceed to Thomas Aquinas. Two of the sources I shall use are well known: the *Summa contra gentiles*, begun in the late 1250s and finished early the following decade, and the *Summa theologiae*, the composition of which spanned the years from 1266 to 1273. A third is the interesting but little cited letter from sometime between 1269 and 1272, *De occultis operibus naturae*.[26] In the first of these works, the *Summa contra gentiles*, Thomas introduces his most sustained examination of operations associated with the magical arts. He does so by drawing a bead on Ibn Sina—named, to be sure, Avicenna in the Latin text. The critique of Avicenna then serves as an occasion for Thomas to make plain exactly where on the scale of all phenomena acts of magic sit with regard to the operations of nature.

To be precise about it, Thomas's initial target in the section from *Contra gentiles* was Avicenna's understanding of the influence of separate substances—those spiritual entities ranking above human souls on the scale of being but below God—on objects and actions in the sublunar world. As Thomas explained it, Avicenna held that separate substances, whether resident in the heavens or present below the lunar sphere, could, by the power of their own immaterial form alone, immediately bring about actions in the material world in which we live.[27] Because such actions would come about without the intervention of any material object serving as cause or agent, they would surely appear to most of us as practically miraculous. In fact, of course, since technically speaking they would not violate Avicenna's idea of natural action, they would not constitute authentic miracles but rather just wondrous events—wondrous, that is, from our subjective, limited point of view.[28] A possible example

[26] See Joseph B. McAllister, *The Letter of Saint Thomas Aquinas De occultis operibus naturae ad quemdam militem ultramontanum* (Washington, D.C., 1939), 16.

[27] Thomas, *Summa contra gentiles* III, 103 ([Marietti edition] Turin, 1935), 344a: "Fuit autem positio Avicennae quod substantiis separatis multo magis obedit materia ad productionem alicujus effectus quam contrariis agentibus in natura."

[28] Thomas, *Contra gentiles* III, 103, 345a: "[L]icet tales effectus simpliciter miracula dici non possint, quia ex naturalibus causis proveniunt, mirabiles tamen redduntur nobis... ."

was when a spirit—in extraordinary cases it might, thought Avicenna, even be a human soul, if the latter were sufficiently purified and separated from the influences of the body in which it dwelled—brought about a cure in a sick person merely by means of its apperception of the person to be cured.[29] To Avicenna's mind, a similar process came into play in the more mundane phenomenon of "bewitchment," when a soul supercharged by an emergent feeling of hatred worked harm on the object of its evil intent merely by means of its gaze.[30]

The spectrum of events suggested by Thomas's words was admittedly extraordinarily wide, and perhaps uncomfortably vague. Included would seem to be everything from exceptional interventions into this lower world by higher spirits to some of the more marvelous and mysterious actions traditionally attributed to especially powerful human souls. In any case, there can be no doubt that Thomas thought Avicenna had in mind here a class of operations and events overlapping those we have seen referred to by William of Auvergne as "*opera magica*." This is plain, moreover, despite the fact that Thomas had not yet employed the word "magic" or any of its cognates. More relevant, for the moment at least, was Thomas's claim that the active process Avicenna intended to suggest with each of his examples was simply corollary to his more celebrated notion of how the separate substances—in strict terms, celestial souls—that animated and moved the planets and stars provided, by means of a strictly Neoplatonic effluence or pouring out, the substantial forms of all objects in the material, natural world.[31]

---

29) Thomas, *Contra gentiles* III, 103, 344a-b: "[U]nde ponit quod ad apprehensionem praedictarum substantiarum sequitur interdum effectus aliquis in istis inferioribus, vel pluviarum, vel sanitatis alicujus infirmi, absque aliquo corporeo agente medio… . Et per hunc modum dicit quod, si anima sit pura, non subjecta corporalibus passionibus et fortis in sua apprehensione … ad ejus apprehensionem sanetur aliquis infirmus… ."

30) *Ibid.*, 344b: "Et hoc ponit esse causam fascinationis, quia anima alicujus, vehementer affecta in malevolentia, habet impressionem nocumenti in aliquem… ."

31) Thomas, *Contra gentiles* III, 103, 344b: "Haec autem positio satis consona est aliis suis positionibus. Ponit enim quod omnes formae substantiales effluant in haec inferiora a substantia separata, et quod corporalia agentia non sint nisi disponentia materiam ad suscipiendam impressionem agentis separati… ." Just before these words,

Aquinas proceeded to reject Avicenna on both principal counts—that is to say, with regard to his theory of the generation of substantial forms as well as his explanation of the origin of many wonderful and even magical operations. He did so expressly by wielding what he characterized as an Aristotelian doctrine of natural action among material bodies exclusively by means of contact and local motion. According to Thomas, Aristotle had proven beyond a doubt that the forms of material things did not derive by infusion from separate substances either operating from the heavens or present in the sublunar world but instead from other material forms themselves resident in other material objects. The way material form induced a new material form in a second object was by acting upon the object to produce a similitude of itself.[32] Required in the process was an act of local motion, either that by which the agent object was brought into contact with the object it acted upon to produce a formal alteration or the more complicated movement by which an object, often a living being, generated another entity similar to itself by inducing a form just like its own in matter to which it had at least a mediated access.[33] Of course, all of this—except, to be sure, the technical concept of a process of generation—we have seen before in William of Auvergne, from the demand that natural activity in the world of material things result only upon contact of the agent with the object acted upon to the reference to working by assimilation. Though Thomas did not mention the

---

Thomas had reminded his readers of Avicenna's view that among separate substances were the "animae vel motores orbium." For the early thirteenth-century understanding of this view as underlying Avicenna's doctrine of the "dator formarum" or "agent intellect," see Steven P. Marrone, "The Philosophy of Nature in the Early Thirteenth Century," in *Albertus Magnus und die Anfänge der Aristoteles-Rezeption im lateinischen Mittelalter*, ed. Ludger Honnefelder *et al.* (Münster, 2005), 115-57, at 135-41.

32) Thomas continued the quotation with which n. 31 above begins as follows: "[Q]uod quidem non est verum, secundum Aristotelis doctrinam, qui probat quod formae quae sunt in materia non sunt a formis separatis, sed a formis quae sunt in materia; sic enim invenitur similitudo inter faciens et factum."

33) Thomas had already introduced this principle in *Contra gentiles* III, 102, 343b-44a: "Omnis autem effectus qui in his inferioribus producitur, per aliquam generationem vel alterationem necesse est ut producatur. Oportet igitur quod, per aliquid localiter motum, hoc proveniat... ."

second of William's two modes of natural operation, working by contrariety, it is likely that he saw it as falling under his category of action by similarity productive of formal alteration. He did, on the other hand, take the time to explain how embodied spiritual substances such as human souls produced such alterations on objects in the material world in which they lived. They did so by activating the bodies in which they resided—a non-material capability they possessed, purely naturally as William also held, by virtue of their being souls in body—which then in turn either produced further changes in other parts of themselves or impinged on other bodies around them along the standard lines of natural material cause and effect.[34]

As for those wondrous and pseudo-miraculous operations, which we have compared to many of what William had called *opera magica* and which Avicenna had wanted to see as the direct effect of separate celestial substances, Thomas again followed William's path by attributing them to spirits circulating here below. Whether acting on their own or invoked by human wonder-workers, these spirits—describable in more Christian language as angels or demons—manipulated one material object in their power the same way the human body lay in the power of the soul in order consequently to act upon other material objects to render an effect.[35] It was only because the common observer did not see the spirits or understand the natural

[34] The explanation is offered as an answer for how the soul's apprehension of something—for instance, a frightening object—can cause a corporeal effect—in the relevant case, coldness and shivering. See Thomas, *Contra gentiles* III, 103, 344a, and in response, 344b: "[H]ujusmodi autem passiones accidunt cum aliquo determinato motu cordis, ex quo sequitur ulterius immutatio totius corporis, vel secundum motum localem, vel secundum alterationem aliquam; unde adhuc remanet quod apprehensio substantiae spiritualis non alterat corpus, nisi mediante motu locali."

[35] Thomas, *Summa contra gentiles* III, 103, 344b-45a: "Substantia igitur spiritualis creata propria virtute nullam formam inducere potest in materiam corporalem, quasi materia ad hoc sibi obediente ut exeat in actum alicujus formae, nisi per motum localem alicujus corporis. Est enim hoc in virtute substantiae spiritualis creatae ut corpus obediat sibi ad motum localem; movendo autem localiter aliquod corpus, adhibere potest aliqua naturaliter activa ad effectus aliquos producendos… . [Q]uum res aliquas naturales vel angeli vel daemones adhibent ad aliquos determinatos effectus, utuntur eis quasi instrumentis quibusdam… . Conveniens est igitur quod ex ipsis

process by which the bodily action occurred that such operations were thought of as miraculous. In fact they remained, as William, too, had insisted, fully natural and largely explicable in terms of local motion and body-to-body contact.[36] At this point, moreover, Thomas finally made explicit reference to magic. Having explained the more stereotypical wonders wrought by spirits and separate substances working largely on their own, Aquinas continued by considering an especially arcane type of exercise that he chose to identify as the works produced by the "magical arts."[37] From what follows it is evident that with this latter term Thomas meant precisely what William had referred to as the *magisterium imaginum*. Not surprisingly, once more he had recourse to the same explanatory narrative as his predecessor. What occurred in such phenomena depended upon the invisible intervention of demonic spirits invoked by the master of the images, which spirits of course performed their wonders in accordance with the natural procedures of material action.[38] Thomas merely added that the images in such instances should be thought of as intentional signs of communication to the demons, much like words.[39]

Where Thomas seems to have differed from William was in the degree of attention he gave to alternative ways of explaining more worldly phenomena of the wonder-producing sort, whether purportedly natural or magical. Like William, Aquinas conceded that there was sometimes strange business in nature, whereby inexplicable results were produced by what would otherwise seem to be ordinary means. The magnet was a case in point, whose elementary attributes as an identifiable material object would not suffice to

---

rebus naturalibus proveniant aliqui altiores effectus, ex hoc quod spirituales substantiae eis utuntur quasi instrumentis quibusdam."

36) The fact is made clear in the lines quoted above, n. 28, which immediately follow the passage just given in n. 35.

37) Thomas, *Contra gentiles* III, 104, 345a: "... opera ... quae per artes magicas fiunt."

38) The complete argument in this regard is long and sometimes wordy, involving *Summa contra gentiles* III, 104-106, 345b-49b.

39) In fact, William had already briefly anticipated Thomas on this point as well. See William, *De legibus* 27 (*Opera omnia*, I: 89aD).

explain its attractive powers.[40] In such instances, one had to realize that there were hidden or occult powers and forces at work, entirely natural for all their hiddenness but not reducible to the normal mode of material cause and effect by similarity.[41] Thomas was in accord with William that demons themselves sometimes resorted to using such forces to fool people into thinking they were working miracles. The truth was, of course, that they were merely taking advantage of common ignorance about the extent of natural operation.[42] Yet Thomas seems to have regarded the domain of such operations as relatively limited in comparison to the rest of natural activity. At least he spoke about them much less frequently than did William. And it is interesting that on at least one occasion where he touched upon an instance of the workings of a hidden force that William likewise had commented upon—the occult power of a sapphire—he accounted for it as following from the agent's "specific form," in contrast to William's apparently Galenic "whole nature."[43] Although there is no room in this article to offer an explanation, I think of this as a consequence of Thomas's deeper understanding and acceptance of Aristotle's theory of natural cause and effect among material objects by way of the elemental properties of things.[44]

Surely related to this same intensification of Aristotelianizing theory at the expense of recourse to "occult forces"—whether or not

---

40) Thomas, *De occultis operibus naturae*, n. 3 (in McAllister, *The Letter of Saint Thomas*, 191).

41) Thomas, *De occultis operibus*, n. 6 (McAllister, 193).

42) See Thomas, *Summa theologiae* II-II, 178, 2, resp. ([Biblioteca de Autores Cristianos] Madrid, 1955), 3, 1101b, which refers back to *Summa theologiae* I, 114, 4, resp. (Biblioteca de Autores Cristianos), 1, 817a.

43) Thomas, *De occultis operibus*, n. 14 (McAllister, 195). For a start in understanding how "specific form" might be considered a replacement for Galen's "whole nature," see Brian P. Copenhaver, "Natural Magic, Hermetism and Early Modern Science: The 'Yates Thesis,'" in *Reappraisals of the Scientific Revolution*, ed. David C. Lindberg and Robert S. Westman (Cambridge, 1990), 261-301, at 272-73; and "Scholastic Philosophy and Renaissance Magic," 540-41.

44) Suggestive in this regard is Newman's rendition of Thomas on Aristotle's view of the relation between the substantial form of a material object and its constituent elements, in *Atoms and Alchemy*, 36-37.

I am right about the sapphire—was Thomas's appropriation into his attack on presumed magic, especially the magic of signs, of a complicated theoretical model for how the celestial bodies, both planets and stars, intervened by means of their motions in the production of a multitude of natural forms—as opposed to wondrous formalities purportedly evoked by images in the *artes magicae*—within the material world.[45] Thomas's ability to explain all such interventions was dependent on the higher order integration into his analytical armory of a model for the generation of substantial forms derivative of Aristotle's theory of natural generation in the material world.[46] The presence of such powerful theoretical tools for analyzing natural operations did not, to be sure, necessitate that Thomas would leave less room in his natural philosophy for something like William's natural magic—a phrase that Thomas would never have been willing to employ—but it did make that result more plausible and more likely. With Thomas, our "protomechanism," a convoluted and greatly Aristotelianizing sort of "mechanism" to be sure, could be seen to entrench itself ever more deeply than in William's day into at least a part of the scholastic milieu.

Giles of Rome

I arrive at last at Giles of Rome. Here we enter more cramped theoretical quarters than were afforded by the grand philosophical considerations of magic in William and Thomas. In fact, we shall focus on a single and quite peculiar scholastic disputed question about how, in a very specific circumstance, a reputedly magical operation might occur. The question has to do with the bleeding—cruentation—of a corpse in the presence of the murderer responsible for the demise of its former animating inhabitant. Belonging to a series of similar questions raised in scholastic dispute over several years, it forms part of an episode in academic politics that Alain Boureau has introduced us to in his *Théologie, science et censure au XIIIe siècle*, revolving around Dominican allegations that John Pecham,

---

45) Thomas, *Summa contra gentiles* III, 104, 346a and 347a.

46) For a start, see Thomas, *Summa theologiae* I, 65, 4; I, 91, 2; and I, 115, 4.

Franciscan, student of Bonaventure's and eventually Archbishop of Canterbury, was responsible for the death of his fellow English bishop, Thomas of Cantiloupe.[47] It was a query posed to Giles, probably in order to sow discomfort among members of the Faculty of Theology, in a quodlibetal disputation held at Paris in 1290. The question, number 25 in Giles's *Quodlibet V*, asks whether the wounds of a victim killed in an act of violence would bleed in the presence of the killer.[48] The very fact that such an inquiry into the reality of magic could be handled in so expert and technical a way by a theologian of the late thirteenth century, such as Giles, gives an indication of how well the efforts of those like William and Thomas had prepared the theoretical ground and advanced the technical terms of the debate.

In question 25, Giles accepted, as did most of his university colleagues, that such cruentation could and did occur. The issue was how to explain it. As might be expected, the matter drew its urgency—its theoretical as opposed to political urgency, that is—from the fact that the phenomenon under scrutiny amounted to almost a type of what could be categorized in the scholastic ambience of the day as a magical operation. Under this category would have stood, as we have seen of course already exemplified in the works of William and Thomas, all those wondrous occurrences not readily explicable with reference to the mundane and unremarkable laws of ordinary natural operation that educated minds would by expectation be familiar with. By way of response, Giles ran down a list of potential explanatory accounts for the occurrence from among a stock of standards familiar in the Latin West at least since the twelfth-century collections of Salernitan questions.[49] What interests us is exactly which of these explanations he took as possible, in reality, and which he did not. I shall rapidly summarize.

---

47) See Alain Boureau, *Théologie, science et censure au XIIIe siècle. Le cas de Jean Peckham* (Paris, 1999), 237-87.

48) Giles of Rome, *Quodlibet* V, 25 (from his *Quodlibeta* [Bologna, 1481]), 108vb: "[Q]uerebatur de occiso per uiolentiam, utrum ad presentiam occidentis plage emanent sanguine."

49) See *The Prose Salernitan Questions*, ed. Brian Lawn (London, 1979), 130 (q. 269).

First Giles laid before the reader an explication that a group of authorities put forth, proposing that, because a trace of the victim's blood had been left on a knife the returning murderer was still holding or on his clothes, the blood remaining in the corpse was drawn out of the wound by attraction when the killer approached sufficiently closely.[50] The theoretical basis for such an account, added Giles, would have been the general principle of natural operation whereby like attracted like. In this case, since the likeness did not involve a quality, such as heat, but rather the substantial nature of the blood, the general principle had to be reinforced by drawing upon the further notion of a particular kind of attraction, the idea for which he traced back to Avicenna, who had designated it as a mode of operation "by the whole species" (*a tota specie*).[51] From what follows in the text it would appear that without committing himself definitively to the validity of such an Avicennian mode, Giles thought that in any case the rule of attraction would have to be understood as subject to an additional condition. An attraction between two bodies of similar nature, whether qualitative or substantial, was conceivable only in the case where the two possessed the similarity according to differing degrees of perfection.[52]

Theoretical niceties aside, the explanation would not stand no matter how construed. As Giles read the underlying principle, with the requisite condition attached, it was the blood in the victim's corpse that should have attracted the traces of blood present on the

---

[50] This paragraph and the next depend upon Giles, *Quodlibet* V, q. 25 (Bologna edition, ff. 109ra-b). The first potential explanation is introduced on f. 109ra: "Dicunt enim aliqui quod ex sanguine remanente in cultello uel in uestibus occidentis, quia a simili sit attractio, ideo non potest contingere quod adueniente occisore <non> fiat attractio sanguinis existentis in corpore occisi."

[51] Ibid., 109ra: "Unde et quidam talem tractum a tota specie uocant secundum quod distinguit Avicenna triplicem tractum: uidelicet a calido, qui sit a natura elementari, a uacuo, qui uidetur fieri a celesti natura, <et> ibi a tota specie, qui uidetur reduci ad naturam mixti."

[52] Ibid., 109ra-b: "... si sit attractio a simili, hoc est quia imperfectum trahitur ad perfectum.... [Q]uare si simile agat in simile et si attrahat simile, non erit omnino simile, cum unum se habebit ut actus, aliud ut potentia, unum secundum quod huiusmodi ut imperfectum, aliud ut perfectum."

person of the killer.[53] Important for us at present, however, is that the whole account drew on the tradition of natural but occult forces operative in the world that we have already seen emergent in both William of Auvergne and Thomas Aquinas. Indeed, Giles's Avicennian phrase "a tota specie" would seem to be kindred to both William's denomination of an occult but natural action "by the whole nature" and Thomas's description of an analogue as by the agent's "specific form." To this degree, Giles's understanding of natural operation was continuous with that of his two predecessors. More significant still is the fact that he found a way to exclude that hermeneutical line from his own approved response. Indeed, he made fun of it, saying that nothing was "stupider" (*stolidius*) than to search for a way to make the rule of similarity, even if refined by linking it to operation "from the whole species," work to prove that blood would attract blood.[54]

With the first proposed explanation neatly disposed of, Giles turned to a second that other authorities advanced, relying not on an occult natural cause but instead on a perfectly routine operation involving material bodies, local motion and contact between agent and acted-upon. Such a mechanistic—or "protomechanistic"—account was one Giles evidently felt quite comfortable with. This was how one would hope actions could be accounted for in the natural world. In the particular case at hand, the explanation ran that in the violence of the act certain material spirits or vapors were generated in both the murderer and the victim. Parts of these spirits or vapors were exchanged through the eyes on the occasion of a mutual glance at the instant of the murder, so that when murderer and corpse were brought together again, the parts sought to return to the source whence they came, following a presumed rule that a part seeks to go back to the whole. The resultant commotion in the matter of

---

53) Ibid., 109rb: "[Q]uia perfecta ratio sanguinis magis reseruabitur in eo … qui est in corpore quam in illo qui est in cultello uel in uestibus, per talem attractionem <illa pars> sanguinis qui est in corpore non proced<e>t ad exteriora sed magis <ista pars> sanguinis qui est in cultello uel in uestibus redibit ad corpus."

54) Ibid.: "Quamuis totum hoc dictum sit fantasticum et <nihil> sit stolidius quam ex hoc in potentia consimilare." The text in the Bologna edition is obviously corrupted here, but the general point remains clear enough.

the dead body then led to the emanation of a certain quantity of blood.[55] Despite its down-to-earth attractiveness, however, this was again an answer Giles was forced to reject. Simply put, he proclaimed, we do not observe in nature a regular process of part returning to whole or whole returning to part, except, in the former case, where the original whole is located in the material's natural place.[56] The explanation was thus believable, because expressed in unexceptional terms of mechanistic material act, but there was unfortunately no evidence to support the ostensible rule of nature to which it appealed.

Having twice come up empty-handed, Giles retreated to explanatory ground neither he nor any other scholastic, protomechanistic or not, could find objectionable.[57] When the phenomenon of cruentation did occur—and general opinion confirmed that it sometimes did—the reason would have to lie with one of three causal accounts.[58] First, it could be the result of a direct act of God, by his providence intervening supernaturally to ensure that so horrific a crime did not go unpunished.[59] Second, it could be the work of demons, acting the way William and Thomas had explained with regard to reprehensible works of magic by secretly manipulating material objects so as to procure, by the natural procedures of material

55) Ibid., 109rb.

56) Ibid., 109rb-va: "Sed cum isti intendant in hoc rationem reddere naturalem, quia nihil tale uidemus in naturalibus, imo magis uidemus oppositum quam propositum … quia non est de natura totius quod tendat ubi est sua pars, nec etiam est de ratione partis quod uadat ad totum nisi quatenus esset totum in loco proprio et pars esset extra locum." On the natural motion of any body towards its proper place, which finds its most famous instantiation in the natural fall of heavy bodies—that is, the phenomenon of *gravitas* or gravity—it is worth looking at the discussion by James A. Weisheipl, "The Principle *Omne quod movetur ab alio movetur* in Medieval Physics," *Isis* 56 (1965), 26-54, at 30-32 and 35-40.

57) This paragraph relies on Giles, *Quodlibet* V, q. 25, 109va.

58) Already ibid., 108vb (*ratio in contrarium*), it was announced that *publica fama* held cruentation to be an authentic occurrence.

59) See Giles's initial listing of the three (ibid., 109va): "His itaque praelibatis, uolumus reddere rationes et causas quomodo hoc possit contingere. Dicemus enim triplici de causa hoc posse esse, uidelicet ex prouidentia Dei, ex falla<c>ia demonum, et ex contingentia casuali."

action and local motion, an apparently wondrous effect. They would of course do so in the hope of inciting some sort of superstition among gullible humans. Alternatively, an instance of cruentation might be simply a matter of chance, a case of mere fortuity.

In all this from Giles, there is not much to enlighten us in detail about theories of natural operation or grand visions of magic as either compatible with or counter to nature. He simply answers the question, drawing upon arguments made many times before and thus needing little in the way of elucidation or justification. But in light of what has already been said about his two predecessors, William and Thomas, even Giles's terseness reveals a lot about how he approached the operations of nature in general. To his eyes, the possibilities of action in the sublunar world were effectively limited to either material causality by contact or divine, and thereby supernatural, intervention. There might be occult natural causes, working even at a modest distance, but they were few and far between and in any case hard to apply to particular instances. Where wondrous things happened, if God was absent, then demons stealthily applying the rules of nature should immediately be suspected as involved. Otherwise, one should simply resist making connections and drawing conclusions, realizing that circumstances often catch us by surprise. A medieval "mechanism," remarkably anticipatory of that of the seventeenth century, was definitely setting in.

The Debate over the Nature of Motion: John Buridan, Nicole Oresme and Albert of Saxony. With an Edition of John Buridan's *Quaestiones super libros Physicorum, secundum ultimam lecturam*, Book III, q. 7

Johannes M.M.H. Thijssen*
Radboud University Nijmegen

Abstract
The so-called 'Buridan school' at the University of Paris has obtained a considerable fame in the history of science. Pierre Duhem had made some bold claims about the achievements by John Buridan and his 'pupils' Nicole Oresme and Albert of Saxony in the field of medieval dynamics. Although generally, Duhem's views are no longer accepted, the idea of a 'Buridan school' has survived. This idea is, however, misleading. John Buridan, Nicole Oresme and Albert of Saxony should rather be viewed as members of an intellectual network. While interested in similar philosophical themes and understanding each other's conceptual language, they also disagreed about numerous topics. One case in point is the nature of motion, as discussed in their respective *Questions on the Physics*. Despite the common features of the language in which they discuss motion, the three thinkers defend different positions. This article compares the three sets of *Questions on the Physics* and presents a critical edition of Buridan's "ultima lectura", Book III, q. 7.

Keywords
John Buridan, Nicole Oresme, Albert of Saxony, *fluxus formae*, *forma fluens*, local motion, motion as *aliter et aliter se habere*, *modus rei*, Buridan school, Pierre Duhem, John Murdoch, Edith Sylla, Bishop Etienne Tempier, Averroes, Aristotle, *Quaestiones super libros Physicorum secundum ultimam lecturam*

* Faculty of Philosophy, Radboud University Nijmegen, P.O. Box 9103, NL-6500 HD Nijmegen, The Netherlands (hthijssen@phil.ru.nl).

There was no such thing as the Buridan school, and this is a paper about it.[1] At the beginning of the twentieth century, Pierre Duhem (1861-1916) announced the existence of the Buridan school and attributed a pivotal role to it in making science modern. He pushed the origin of modern science back to the fourteenth century and claimed that the University of Paris was its location: "La Faculté des Arts de Paris allait enfanter la Science moderne."[2] One of his most celebratory statements of the role of Paris and of Buridan can be found in the preface to his *Études sur Léonard de Vinci*. He there went even further and claimed that Christian faith had vanquished both pagan Aristotelianism and Neoplatonism:

> Cette substitution de la Physique moderne à la Physique d'Aristote a résulté d'un effort de longue durée et d'extra-ordinaire puissance. Cet effort, il a pris appui sur la plus ancienne et la plus resplendissante des universités médiévales, sur l'Université de Paris. Comment un parisien n'en serait-il pas fier? Ses promoteurs les plus éminents ont été le picard Jean Buridan et le normand Nicole Oresme. Comment un français n'en éprouverait-il pas un légitime orgueil? Il a resulté de la lutte opiniâtre que l'Université de Paris, véritable gardienne, en ce temps-là, de l'orthodoxie catholique, mena contre le paganisme péripaticien et néoplatonicien. Comment un chrétien n'en rendrait-il pas grâce à Dieu?[3]

Elsewhere, Duhem created a link between John Buridan and his 'disciples' Nicole Oresme and Albert of Saxony, at times including Themon Judaeus and Marsilius of Inghen as well:

> Le philosophe de Béthune [i.e., John Buridan] n'est pas seul à professer cette Dynamique; ses disciples les plus brillants, les Albert de Saxe, et les Nicole Oresme, l'adoptent et l'enseignent; les écrits français d'Oresme la font connaître même à ceux qui ne sont pas clercs.[4]

---

1) The opening sentence is inspired by Steven Shapin, *The Scientific Revolution* (Chicago, 1996), 1.

2) Pierre Duhem, *Le système du monde. Histoire des doctrines cosmologiques de Platon à Copernic*, 10 vols. (Paris, 1913-1959), IX:697.

3) Pierre Duhem, *Études sur Léonard de Vinci. Ceux qu'il a lus et ceux qui l'ont lu*, 3 vols. (Paris, 1906-1913), 3:xiii-xiv. This passage is also quoted in English translation in H. Floris Cohen, *The Scientific Revolution. A Historiographical Inquiry* (Chicago, 1994), 53.

4) Duhem, *Études sur Léonard de Vinci*, 3:viii.

In his view, these thinkers together constituted the Buridan school, which advocated a brand of French positivism.[5]

In an important article, published in 1991, John Murdoch examined the crucial role of Duhem's work in shaping the historiography of medieval science, either by extension or by criticism. He demonstrated how Duhem's claims and the wealth of sources which he made available in French translation set the research agenda for historians of the next four decades or so.[6] In addition, John Murdoch's analysis illustrated that, as a matter of fact, Duhem's thesis contained at least four different ingredients. Only one of these concerned the claim that Buridan's mechanics was the larva which had been turned into adult form by Galileo Galilei. Or, in other words, that the essence of seventeenth-century science was already there in the fourteenth century.[7]

Many historians of science and philosophy have felt uneasy with Duhem's interpretation of fourteenth-century natural philosophy as an anticipation of seventeenth-century mechanics. Early critics such as Anneliese Maier and Marshall Clagett believed that Duhem's views were not justified by the medieval sources.[8] However, the 'Buridan school' had been born, for even Maier and Clagett did not question its existence and considered it the most prominent center of medieval science, together with the school of Bradwardine at Merton College.

In a previous article I have argued that there are neither institutional nor doctrinal reasons to conclude that there ever existed a Buridan school. Nicole Oresme and Albert of Saxony were not Buridan's students in any formal sense of the word. They belonged to different nations, whereas students usually took their degrees with masters of their own nation. In the case of Albert of Saxony, we

---

5) See further, for instance, Duhem, *Système du monde,* VI:729.

6) John E. Murdoch, "Pierre Duhem and the History of Late Medieval Science and Philosophy in the Latin West," in *Gli studi di filosofia medievale fra otto e novecento,* eds. Ruedi Imbach and Alfonso Maierù (Rome, 1991), 253-302.

7) Duhem, *Système du monde,* VIII: 200: "La Mécanique de Galilée, c'est, peut-on dire, la forme adulte d'une science vivante dont la Mécanique de Buridan était la larve." See also Murdoch, "Pierre Duhem," 256-257.

8) Murdoch, "Pierre Duhem," 274-279.

even know that he took his degree in 1351 under master Albert of Bohemia.[9] Moreover, the works by John Buridan, Nicole Oresme, and Albert of Saxony have, so far, not shown the type of unified doctrinal position which is usually taken as a characteristic of a school orthodoxy. In this respect, the so-called Buridan school is in stark contrast with, for instance, the Thomist or Scotist schools. For these reasons, it is time to depart from the traditional way of thinking about a Buridan school, and to try instead to picture it as an intellectual network or community. Its organizational base was the arts faculty at Paris. The university and the church supplied the social and material conditions for the existence of this network. At a micro-level, the contemporaneous intellectuals of this network engaged in specific social activities, such as lecturing on Aristotle's texts and discussing. They were focused on their own arguments, philosophical disagreements and conceptual language.[10] Their intellectual rivalry and opposition were focused on certain topics, such as, for instance, the ontological status of quantity.[11] At the time, however, there was a fair amount of doctrinal agreement and of shared interest in certain philosophical issues and texts, in particular Aristotle's *Physics*.[12] Precisely this aspect of intellectual networks may have led previous historians to believe that there was a school of followers, clustered around a leader, John Buridan. Another factor may have been that previous scholars did not have the benefit

[9] Henri Denifle and Émile Chatelain, eds., *Auctarium Chartularii Universitatis Parisiensis*, vol. I (Paris 1894), 149, 150 and 152 and further Johannes M.M.H. Thijssen, "The Buridan School Reassessed, John Buridan and Albert of Saxony," *Vivarium* 42 (2004), 18-42, esp. 21, and 23-25.

[10] Interesting sociological observations about intellectual networks can be found in Randall Collins, *The Sociology of Philosophies. A Global Theory of Intellectual Change* (Cambridge, Mass., 1998), esp. 1-15 and 51-53. Collins' analysis of the "Paris Network" on 490-492 is outdated.

[11] Thijssen, "Buridan School Reassessed," esp. 29-38 discusses the rival positions which John Buridan and Albert of Saxony took in the debate about quantity.

[12] This aspect is nicely brought out in Edward Grant's study of fourteenth-century commentaries on Aristotle's *De caelo*. From his inventory and analysis of question-titles it is clear that some masters at the arts faculty had shared philosophical interests, though not necessarily adhering to the same views. Edward Grant, *Planets, Stars and Orbs. The Medieval Cosmos, 1200-1687* (Cambridge, 1994).

of the critical editions of some of the crucial texts, which can now be studied from this new perspective.

In this paper I will attempt to give some more substance to the idea that John Buridan, Nicole Oresme, and Albert of Saxony constituted an intellectual community or network, rather than a school. One of the topics that attracted the attention of this community was the nature of motion. In their question-commentaries on Aristotle's *Physics*, all three thinkers are engaged in a debate about the question what motion really is. The dating of their commentaries is elusive. Elsewhere, I have suggested that Albert of Saxony's *Quaestiones* on the *Physics*, has to be placed inbetween John Buridan's so-called *tertia lecture* and the *lectura ultima* of his *Quaestiones* on the *Physics*. Albert knew Buridan's *tertia lectura*. Buridan, in his turn, responded to Albert's text in his own *ultima lectura*, written after the *tertia lectura*.[13] This relative chronology, based on doctrinal analysis and text comparisons, finds support in independent evidence about the tentative dates of origin of these works. Buridan's *tertia lectura* is dated around 1350, whereas the *ultima lectura* was composed sometime between 1352 and 1357.[14] Albert of Saxony's *Quaestiones super libros Physicorum* are to be dated after 1351. This date is suggested by one of its copies, whose introductory remarks tie the text to Albert's opening lecture (*principium*) on Aristotle's *Physics*.[15] Since Albert of Saxony incepted in the summer of 1351,

13) Thijssen, "Buridan School Reassessed, 40-42. This relationship was previously suggested by Jürgen Sarnowsky, *Die aristotelisch-scholastische Theorie der Bewegung. Studien zum Kommentar Alberts von Sachsen zur Physik des Aristoteles* (Aschendorff, 1989), 50-51, and 58-60.

14) The evidence concerning the dates of composition of Buridan's two question-commentaries on the *Physics* was already known to Edmond Faral and Anneliese Maier. See further Johannes M.M.H. Thijssen, *John Buridan's Tractatus de infinito, Quaestiones super libros Physicorum secundum ultimam lecturam, liber III, quaestiones 14–19. An Edition with an Introduction and Indexes* (Nijmegen, 1991), xx-xxi and Bernd Michael, *Johannes Buridan: Studien zu seinem Leben, seinen Werken und zur Rezeption seiner Theorien im Europa des späten Mittelalters,* 2 vols., (Ph.D. dissertation, Freie Universität Berlin, 1985), 2:606-608.

15) It is the manuscript Erfurt, Wissenschaftliche Allgemeinbibliothek, F. 345, which was written in 1360 in Cologne. Its *incipit* reads as follows: "Quoniam dicit Philosophus 'a communibus et prioribus prius est inchoandum', ideo propter informationem

it seems that the *Quaestiones* on the *Physics* was the first work that he read at Paris after the completion of his studies there. Nicole Oresme's *Quaestiones* on the *Physics* are dated around 1347. This date is based on an argument *ex silentio*: after John of Mirecourt's condemnation in 1347 which appeared to include the very characteristic theory of *modi rerum* (about which more below), Oresme seems to have abandoned it.[16] It is likely that Albert of Saxony used Oresme's *Quaestiones* on the *Physics* in his own commentary on that text.[17]

What Kind of Entity is Local Motion?

In a seminal paper of 1978, John Murdoch and Edith Sylla pointed out that in the Middle Ages there was no "science of motion" as a separate discipline, in contradistinction to, say, optics or alchemy.[18] Motion and change play such a crucial role in Aristotle's definition of nature that the medieval "science of motion" would actually

---

intuicionemque iuvenum scolarium in isto primo libro *tanquam pro principio meo* volo istam questionem disputare: Utrum scientia...." See Sarnowsky, *Die aristotelisch-scholastische Theorie der Bewegung*, 18-19, and 49. The text may even imply that the first *quaestio* harks back to Albert's *principium*.

16) See Stefan Kirschner, *Nicolaus Oresmes Kommentar zur Physik des Aristoteles* (Stuttgart, 1997), 29. An important qualification has been added in Stefano Caroti, "'Les modi rerum'... encore une fois. Une source possible de Nicole Oresme: le commentaire sur le livre 1er des *Sentences* de Jean de Mirecourt," in *Quia inter doctores est magna dissensio. Les débats de philosophie naturelle à Paris au XIVe siècle*, eds. Stefano Caroti and Jean Celeyrette (Florence, 2004), 195-223. There, Caroti demonstrates that Mirecourt's 1347 condemnation was actually not aimed against the theory of *modi rerum*, but that it was merely implied, because Mirecourt mentioned it in his commentary on the *Sentences*. Thus, the theory came to be included in the condemned excerpts (*articuli*).

17) Jürgen Sarnowsky, "Nicole Oresme and Albert of Saxony's Commentary on the Physics. The Problems of Vacuum and Motion in a Void," in *Quia inter doctores est magna dissensio*, eds. Caroti and Celeyrette, 161-175.

18) John E. Murdoch and Edith Sylla, "The Science of Motion," in *Science in the Middle Ages*, ed. David C. Lindberg (Chicago, 1978), 206-265. Note that chapter 3 ("The Philosophical Setting of Medieval Science") by William A. Wallace in the same volume, refers to "Buridan's pupils Albert of Saxony and Nicole Oresme" (113).

coincide with Aristotelian natural philosophy.[19] Hence, rather than discussing the whole of medieval natural philosophy, the authors confined themselves to certain areas regarding motion and change. One of the topics singled out was the problem of the nature of motion, which was of major concern to late medieval thinkers.[20]

As is well known, Aristotle had made contradictory statements concerning the ontological status of motion. In *Physics*, book III (200b32-201a10) he maintained that motion is not something over and above the things in motion. In other words, motion does not constitute a separate category, but belongs to the same category as what is gained by motion, i.e. the category of Place in the case of local motion.[21] In the *Categories* (11b1-8), however, Aristotle had claimed that motion fell into the category of *passio* or Affection.

Averroes tried to reconcile these incompatible statements by pointing out that in the *Physics*, Aristotle had set forth the more correct view, whereas in the *Categories*, he had maintained the more common view. Averroes's explanation of Aristotle's view hinges on an analysis of motion from two different perspectives. Motion, if considered from the terminus toward which it tends, only differs from it in its degree of "more or less," i.e. in its degree of perfection. If, however, motion is considered as a process (*via*) towards perfection or actuality, and, as a consequence, is different from the perfection it attains, it belongs to a category of its own. When seen as a road towards actuality, motion cannot coincide with that actuality.[22] The same twofold analysis of motion recurs in Averroes's commentary

---

19) Aristotle, *Physics* 192b9-193a30.

20) In addition to the section in Murdoch and Sylla, "Science of Motion," 213-218, see especially Anneliese Maier, *Zwischen Philosophie und Mechanik* (Rome, 1958), esp. 61-143; Cecilia Trifogli, *Oxford Physics in the Thirteenth Century (ca. 1250-1270). Motion, Infinity, Place and Time* (Leiden, 2000), esp. 37-86, and further Johannes M.M.H. Thijssen, "The Nature of Change," in *The Cambridge History of Medieval Philosophy*, ed. Robert Pasnau, forthcoming.

21) In this paper, I will restrict myself to the medieval discussion of local motion (*motus*). Aristotle at times uses *kinêsis* (translated as *motus* in the Middle Ages) to cover change in general, and sometimes in the more restricted sense of motion, i.e., local motion.

22) Averroes, *Physica* in *Aristotelis Opera cum Averrois Commentariis*, IV (Venice, 1562-1574; reprinted Frankfurt, 1962), bk. III text 4, f. 87ra-b.

on *Physics* V. There it is couched in the terminology of change "according to matter" and "according to form." According to matter, change and its terminus belong to the same category; according to its form, one must view change as a transmutation that takes place in time and constitutes a category of its own, namely that of Affection (*passio*).[23]

In the thirteenth and fourteenth centuries, these alternative analyses of motion came to be captured under the formulas *forma fluens* and *fluxus formae*, a distinction that medieval authors usually attributed to Albert the Great.[24] According to the *forma fluens* theory, change is nothing but the form successively gained by the changeable body. In the case of local motion, the *forma fluens* is the place successively attained by the mobile body. In other words, motion is the same as the perfection or form it acquires, but it represents that form in a state of flux. It is important to note that the flowing character of the form is not posited in the form itself, but results from the degree of actualization of the form in the subject. Thus, the view of motion as *forma fluens* did not contradict the common medieval view that forms are unchangeable.[25] The *fluxus formae* theory, on the other hand, maintained that change is not the form acquired but is "the flux" of that form– that is, the flow, the process, or the road towards an actuality or perfection. These distinctions lay, at least implicitly, in the background of fourteenth-century discussions of Aristotle's statement that there is no change over and above real things.

Albert of Saxony provides a nice starting-point for mapping this debate. In book III, of his *Questions* on Aristotle *Physics*, he discusses whether local motion involves the existence of a flux (*fluxus*)

---

23) Averroes, *Physica*, Bk.V, text 9, f. 215ra.

24) *Alberti Magni Opera Omnia, Physica. Pars I, Libri 1-4*, ed. P. Hossfeld (Münster, 1987), bk. III, tr.1 c.3, 151. See further Maier, *Zwischen Philosophie und Mechanik*, 73-77 and E.J. McCullough, "St. Albert on Motion as Forma Fluens and Fluxus Formae," in *Albertus Magnus and the Sciences, Commemorative Essays 1980*, ed. J.A. Weisheipl (Toronto, 1980), 129-153, for a discussion of Albert's views and of Maier's interpretation thereof.

25) See Maier, *Zwischen Philosophie und Mechanik*, 78-83.

which is distinct from the mobile object and the place.[26] Interestingly, Albert divides his discussion over two questions, one dealing with the views of Aristotle and Averroes (III, q. 6), the other (q. 7) dealing with the view according to faith and the truth (*secundum nostram fidem et secundum veritatem*). Both questions are intertwined in that certain arguments from q. 6 are only resolved at the end of q. 7.[27]

Albert's discussion of the views of Aristotle and Averroes is embedded in a systematic survey of the several different interpretations that were circulating. Apparently, some thinkers believed that Aristotle and Averroes adhered to a *fluxus* theory; others believed that they did not hold a *fluxus* theory: *motus non est fluxus*. Albert himself is an advocate of the latter opinion. He too believes that according to Aristotle and Averroes nothing more than mobile body and place are required to account for local motion. However, if motion is not a flux additional to the mobile thing and the place, what, then, is it? Albert reviews two different descriptions of local motion, before offering his own. According to one description, being locally moved (*localiter moveri*) involves "continuously being in another way than before with respect to another body" (*continue se habere aliter et aliter ad unum aliud corpus*). This description of local motion, however, is inadequate, because it fails to distinguish between bodies that are at rest and bodies that are in motion. For if a mobile body moves in relation to a body which is at rest, the body at rest too is in another way than before with respect to another body, namely precisely with respect to the body that is really moving.[28]

Others claim that being locally moved implies "being in another place than before" (*fiat in alio loco quam prius*). However, this description too is inadequate, according to Albert, for God could cause a body to be in one place, and then in another place without any

26) Albert of Saxony, *Expositio et Quaestiones in Aristotelis Physicam ad Albertum de Saxonia attributae*, ed. Benoît Patar, 3 vols. (Louvain-La-Neuve, 1999), 2:501-520.

27) The section *ad rationes* at the end of q. 7; Albert of Saxony, *Expositio et Quaestiones*, 2:519, ll. 20-52 provides the resolution of the six arguments *in oppositum* put foward in q. 6, on 506 l. 78—507, l.12.

28) Ibid., 508, ll. 24-31.

local motion taking place (for instance, by creating a body in one place and then putting it in another place).[29]

According to Albert, the correct interpretation of Aristotle's and Averroes's view on local motion entails that a mobile body "continuously and successively is in another and another place" (*continue et successive esse in alio et alio loco*). This description is derived from a more general description of being moved (*moveri*) as "being continuously in another way than before" (*continue aliter et aliter se habere quam prius*). Note that Albert's own interpretation of Aristotle and Averroes is a subtle modification of the two rejected descriptions of local motion. Positing an additional flux is not necessary. Local motion is not essentialy different from alteration, a type of change which only involves the mobile body and the forms successively acquired or lost.[30]

An important objection against Albert's interpretation is that it does not seem to apply to the motion of the last sphere of the heaven. Since the last sphere is not surrounded by another body, and hence is not in a place, it could not be in motion according to the description advocated by Albert. He responds to this instance by pointing out that the last sphere, indeed, does not move locally in its totality (*secundum se totam collective sumptam*). It does move locally, however, with respect to its parts (*ratione suarum partium*), because they are continuously in another and another place.[31]

The local movement of the last sphere, and even of the entire universe, recurs in the next question (book III, q. 7). As a matter of fact, this question is entirely devoted to a consideration of the divine case that God would turn the universe into one continuous body and rotate it from East to West, or in some other way. In that case, local motion would, indeed, lack an external point of reference. So, if the universe moves, i.e. continuously is in another way than before, it can only be so because of something that is intrinsic to the mobile body and yet distinct from it. Albert concludes

29) Ibid., 508 l. 33: "Sed breviter ambo isti modi dicendi deficiunt." The refutation follows on 508 ll. 33-509, l. 44.

30) Ibid., 508, ll. 18-23 and 509, ll. 45-510, l. 66.

31) Ibid., 510, ll. 67-511, l. 80.

that this element which inheres in the mobile body is a flux or the motion itself, which the mobile body successively acquires.[32] Precisely as a flux, the universe's relation to it can be described as continuously being in another way than before (*aliter et aliter se habere*).

The same position was defended by John Buridan. He too believes that local motion is a flux intrinsic to the mobile body. Buridan develops his position in both question-commentaries on the *Physics*, in the *tertia lectura* (book III, q. 7) and in the *ultima lectura* (book III, q. 7). The differences between the two versions are minimal. The same arguments of the *tertia lectura* recur, in the same order, if not in the same wording, in the *ultima lectura*.[33] In other questions, the differences between the two versions can be substantial.[34]

An interesting contrast with Albert's discussion is that Buridan is totally unconcerned with the correct interpretation of Aristotle and Averroes on the nature of motion. At the opening section of his own solution, he merely observes that the "philosophers of old" (*antiqui*) held no doubts about this question (i.e. the nature of motion). They unanimously agreed that local motion was different from the mobile body and the space traversed.[35] If Buridan includes

32) Ibid., 517, ll. 66-68: "Septima conclusio: in omni mobili quod movetur localiter, volentes admittere casus divinos oportet ponere fluxum seu motum inhaerentem mobili qui successive illi mobili acquiritur."

33) I have checked the *tertia lectura* in the ms. Vat. Chig. E VI 199, fols. 43rb-44va against the edition of the *ultima lectura*, given below. One exception is the mention of two probabilities by which Buridan supports his main thesis: "Et pro confimatione praedictorum possunt adduci aliquae probabilitates..." This passage was dropped in the *ultima lectura*. Another difference is the brief introductory phrase to Buridan's own solution (*determinatio*): "Ista quaestio reputatur valde difficilis et tenent moderni quod motus non sit res vel dispositio alia a mobili et loco vel aliis a duobus simul. Tamen contrarium videtur mihi." These differences were already noted by Maier, *Zwischen Philosophie und Mechanik*, 127-128.

34) Thijssen, "Buridan School Reassessed," 29-42 discusses the differences between the two versions with respect to quantity and impetus theory.

35) See the edition below, p. 205: "Antiquiores non dubitaverunt de ista quaestione, sed concorditer concesserunt motum localem esse aliam rem a mobili et loco, sed iam posteriores moderni propter rationes praedictas posuerunt quod motus non sit alia res a mobili." Note the contrast with Buridan's *tertia lectura* cited in the previous note,

Aristotle and Averroes among these *antiqui*, he is contradicting Albert's assessment of their views. In any case, he too believes that local motion is a separate entity and thus disagrees with some *posteriores moderni*, some more recent contemporaries.

In order to bring out the true nature of local motion, Buridan invokes God's power to move the universe. He couches his appeal to God's omnipotence in a reference to Tempier's condemnation, in particular to article 49.[36] If it is conceded that God could indeed move the universe in rectilinear or circular motion, how should one imagine this movement? Buridan's answer is clear: by admitting that motion is a quality inhering in the mobile body.[37] In both his question commentaries, John Buridan defends that local motion is a flux intrinsic to the mobile body. More explicitly than Albert, Buridan characterises this flux as a quality or property of a mobile being, but of such a nature that it is purely successive. He believes that motion is a property, such as whiteness, and thus can maintain that supernaturally speaking there can be motion without anything moving, just as there can be whiteness without a white object.[38] In other words, accidents can exists independently, if God so wishes.

But can there be local motion without the existence of space (*locus*)? At first sight, it seems that the local motion of the universe is something of a contradiction. The universe is not *located* in a space (*locus*), and yet can move *locally*. Buridan explains that the phenomenon of local motion is grounded in perception. In the common course of nature we perceive that a thing moves, because

---

which contains no reference to the views of *antiqui*. Was this reference induced by Albert of Saxony's discussion of Aristotle and Averroes?

36) See Roland Hissette, *Ênquete sur les 219 articles condemnés à Paris le 7 mars 1277* (Louvain, 1977), 118.

37) See Buridan's text which is edited below. Another passage in which Buridan indicates that 'being in motion' is an accident of the mobile body, comparable to 'whiteness' is quoted by Maier, *Zwischen Philosophie und Mechanik*, 129: "...Nos autem dicimus quod omnis motus est subiective in mobili, scilicet in eo quod movetur, per realem inhaerentiam, sicut albedo esset in pariete."

38) See the edition, p. 210: "<2> Ad aliam dico quod non plus reputarem impossibile quod esset motus et nihil moveretur vel mutaretur, quam quod esset albedo et nihil esset album. Neutrum est possibile naturaliter, et utrumque est possibile supernaturaliter."

there is a reference object with respect to which the motion takes place. Thus, we do not perceive the local motion of the universe, nor the movement of a boat next to another boat of the same velocity.[39]

As mentioned above, Albert of Saxony was familiar with Buridan's question-commentary. It is plausible that he derived the consideration about God's omnipotence to move the universe from Buridan's *tertia lectura* and found the argument compelling. However, Albert provides this argument with a new context. He was quite satisfied with Aristotle's and Averroes's view that motion is *not* a flux, if motion is described in the right way, i.e. as continuously and successively being in another and another place. Even the local motion of the last heavenly sphere (which is not situated in a place), can be accounted for and does not require the assumption of an additional flux which inheres in the mobile body. When seen from the perspective of Aristotle and Averroes, local motion is not different from alteration. Both types of change can be described as *forma fluens*: in the case of alteration as the successive impression of a quality (for instance, heat) upon the changeable body, and in the case of local motion as the successive aquisition of various places by the mobile body. If, however, one introduces Buridan's consideration of God's motion of the entire universe into the debate, Albert apparently is convinced to accept the flux theory. Alteration

---

39) See the text below, p. 210: "Sed de hoc quod dicitur, quod implicat contradictionem esse motum localem et non esse locum, ego dico quod motus ultimae sphaerae vel navis in fluvio non dicitur localis quia necesse sit quod secundum illum mutetur locus, sed quia secundum communem cursum naturae omne quod movetur illo loco variat de facto habitudinem localem vel situalem ad aliquod aliud. Et omnino ille motus quem vocamus localem potest non esse localis, quia nullus mutaretur locus nec situs ad aliquam aliam rem, sed tunc non possem illum percipere. Non ergo vocatur motus localis, quia ad ipsum sit locus neccessarius, sed quia percipi non potest nisi appareret mutatio loci vel situs rei ad aliam rem; unde existentes in navibus in mari velociter et simul motis non percipiunt quod illae moveantur." The point about the perception of motion in two moving boats, also occurs in Oresme. See Stefano Caroti, "La position de Nicole Oresme sur la nature du mouvement (Quaestiones super Physicam III, 1-8): Problèmes gnoséologiques, ontologiques et sémantiques," *Archives d'histoire doctrinale et littéraire du moyen âge* 61 (1994), 303-385, esp. 317 n. 51, and 314-320 for a discussion of the perception of motion according to Oresme.

and local motion can no longer be considered the same type of change, because the possible local motion of the universe requires the acceptance of a distinct flux. In other words, for theological reasons, Albert shifts to a flux theory of motion, and this flux is explained in the same way as Buridan, i.e. as a inherent quality of the mobile object.

Where now does Oresme fit into this debate? In his question commentary on Aristotle's *Physics*, Oresme devotes book III, qq. 2-7 to a discussion of the nature of motion.[40] He examines five different views, before presenting his own theory. In his overview, Oresme presents himself as an advocate of the *fluxus* theory. Yet, his view of the nature of *fluxus* is entirely different from that of Buridan and Albert of Saxony. He rejects the idea that the *fluxus* is an inherent quality, such as a form. He disqualifies this interpretation of *fluxus* as "the worst possible view" (*omnium pessima*). How then should the fluxus be understood? Oresme introduces a new ontological entity, the *modus rei* or way of being, to explain the phenomenon of motion. Motion is nothing but the mode or condition of the mobile object, i.e., its condition of traversing spaces in succession. The successiveness of the mobile body, however, should not be taken in the sense of a successive thing (*res successiva*) that is distinct from it.[41]

---

[40] I have used the edition in Kirschner, *Nicolaus Oresmes Kommentar.* In the apparatus, this edition indicates when it diverges from the edition in Caroti, "Position de Nicole Oresme," 303-385.

[41] The crucial texts are Kirschner, *Nicolaus Oresmes Kommentar*, 228, ll. 90-94: "Ex predictis potest elici quinta opinio, scilicet quod motus est res successiva distincta simpliciter a permanentibus. Et potest dupliciter intelligi: primo quod sit una res inherens significabilis incomplexe, sicut una forma, et sic non est verum; secundo quod sit condicio vel modus ipsius mobilis, et sic est verum," and further 234, ll. 147-156: "Alia, que ponit quod est fluxus ad modum unius forme distincte, sicut esset albedo vel anima vel aliquod tale, est omnium pessima; tamen, si intelligatur quod non sit talis forma vel talis res, sed modus vel condicio ipsius mobilis, tunc est verissima et probabilior et facilior inter omnes et concordat dictis Aristotelis et philosophorum." The refuted view resembles that of Buridan. See also Caroti, "La position de Nicole Oresme," 315. Note that Buridan compared 'being in motion' to accidents such as 'whiteness,' a view explicitly rejected by Oresme. See the text cited in note 37.

Oresme determines being moved (*moveri*) as continuously being in another way than the mobile object was before. Like Buridan and Albert, he insists that the mobile body is in another way with respect to *itself*, not in relation to an external point of reference. The considerations which lead Oresme to this description of motion sound familiar and rely on an analysis of the circular motion of the heaven and on alteration. If there were only one body, so Oresme argues, it could not move in a circular motion, unless one assumes that it is in another way than before with respect to itself. The same holds true for alteration, such as, for instance, the heating of water. The body in which the change takes place, is in another way than before with respect to itself.[42]

Conclusion: An Intellectual Community in Fourteenth-Century Paris

Ideas do not beget ideas. Rather, they come into existence as a results of interaction between thinkers who are connected in intellectual communities. One of the characteristics of such an intellectual community is the gradual evolution of its own conceptual framework and the perspective it provides on its own problems and arguments.

The debate about the nature of motion suggests that John Buridan, Nicole Oresme and Albert of Saxony were members of such an intellectual community. Their discussion of the nature of motion

---

[42] Oresme reviews five different descriptions of "being in motion" (*moveri*). Only the fifth and last one has his approval: Kirschner, *Nicolaus Oresmes Kommentar*, 231, ll. 37-43: "Quinta est descriptio melior et vera quod 'moveri' est 'aliter se habere continue quam ipsum mobile prius se habebat respectu sui et non respectu cuiuscumque extrinseci,' et illa probatur, quia, si esset unum solum corpus, non videtur quin posset moveri circulariter, et sic non se haberet aliter quam prius nisi respectu sui ipsius, et ita de alteratione quod, si non esset nisi unum corpus, adhuc posset alterari et corrumpi, sicut aqua calefacta, et ad nihil aliud aliter se haberet." From what follows, it is clear he was thinking of the heaven (*caelum*) and the cosmos (*orbis*). See ll. 52-54: "Ad secundum concedo quod partes talis mobilis, sicut partes celi et unius orbis, continue una se haberet ad aliam uno modo, non tamen ad se ipsam, nec similiter totum, et hoc satis patet de celo."

was cast in a shared conceptual framework. Being in motion was described as *aliter et aliter se habere quam prius*, and as *fluxus*. The crucial issue in the debate came to be whether the local motion of the universe is comparable to other types of change, such as alteration. Yet within this general framework, these thinkers took rival positions. Buridan and Albert of Saxony agreed that the *fluxus* character of motion should be interpreted as an inherent quality or disposition (*dispositio*) of the mobile object. Although Oresme describes 'being in motion' with a term that seems to approach *dispositio*, namely *condicio* or *modus*, his interpretation is completely different. In his view, 'being in motion' is a state of being of the mobile thing, but adds nothing to it. As a matter of fact, in Oresme's interpretation the *fluxus* is almost turned into a *forma fluens* theory. This theory also receives much sympathy from Albert of Saxony. According to him, the *forma fluens* theory was the theory genuinely advocated by Aristotle and Averroes. Albert only gave it up for the *fluxus* theory because he found Buridan's arguments concerning God's omnipotence compelling. With more evidence culled from the commentaries on the *Physics* by John Buridan, Albert of Saxony and Nicole Oresme, and from other texts as well, we should eventually be able to map the inner structure of this intellectual community in fourteenth-century Paris.

The Edition

In Book III, q. 7 of his *Quaestiones super libros Physicorum, secundum ultimam lecturam*, John Buridan discusses the nature of motion. The text has not been edited before. Excerpts of the text were transcribed and included by Anneliese Maier in her discussion of John Buridan's views.[43] Maier used the old printing of Paris 1509 making corrections on the basis of the manuscripts Vat. Lat 2163 and 2164.

The present edition of Buridan's *quaestio* is based upon the manuscript København, Det Kongelige Bibliotek, Ny kongelig Samling, cod. 1801 fol. (C). The Copenhagen manuscript has been fully

---

43) Maier, *Zwischen Philosophie und Mechanik*, 121-130.

collated with the manuscript Cracow, Biblioteka Jagiellońska, cod. 1771 (G). The text transmitted in C has been retained throughout. The use of the subjunctive in C is careless, or in any case, not according to classical usage. Yet it has not been corrected, except when readings of C are manifestly corrupt, in which case they have been corrected by G. The choice of C and G as base manuscripts for this edition is based on the analysis of the textual tradition and circulation of Buridan's *Quaestiones*.[44]

Punctuation and division into paragraphs are according to current usage. The spelling has been homogenized according to classical standard orthography (as, for instance, codified in the *Oxford Latin Dictionary*). The medieval *e* for *ae* has not been retained, and *u* and *v* have always been distinguished. Variant readings of the type igitur/ergo, ille/iste, item/iterum have not been reported in the critical apparatus.

There is repeated evidence to the effect that either the scribe of C had access to readings of other manuscripts, or that the manuscript was corrected. Occasionally, C notes that another witness gives an alternative reading. These references are introduced in the margin by the letters 'al' [*alia lectio*]. All these instances are included in the critical apparatus. Quite frequently, the scribe or corrector of C made corrections by way of deletions, interlinear, and marginal corrections. In those instances, the corrected reading of C has been taken into consideration for the edition, without indicating that this reading has been subject to correction.

[44] The results of the examination of the textual tradition of John Buridan's *Quaestiones super libros Physicorum, secundum ultimam lecturam* (transmitted in 32 manuscripts), have so far been presented in Johannes M.M.H. Thijssen, *John Buridan's Tractatus de infinito* (Nijmegen, 1991) and Dirk-Jan Dekker, *De tijdfilosofie van Johannes Buridanus* (*d. ca. 1360*). *Een historisch-wijsgerige studie met editie van Buridanus' Quaestiones super octo libros Physicorum Aristotelis* (*secundum ultimam lecturam*), *IV, 12-16* (Ph.D, dissertation, Radboud University Nijmegen, 2003). The textual transmission will also be discussed in the Introduction to the forthcoming edition of books III and IV of John Buridan's *Quaestiones super libros Physicorum, secundum ultimam lecturam*.

<Quaestio 7. Utrum motus localis est res distincta a loco et ab eo quod localiter movetur>

Quaeritur septimo[1] utrum motus localis est res distincta a loco et ab eo quod localiter movetur.

<Arguitur quod non>

<1> Arguitur quod non, quia si totum potest salvari sine re addita mobili et loco, frustra illa poneretur,[2] et hoc est inconveniens; sed sine hoc possunt salvari. Probatio: quia esset motus localis si continue a mobile esset supra aliam et aliam partem spatii *b*, licet non poneretur aliud esse, et salvaretur successio et prioritas et posterioritas per diversas partes spatii quae secundum situm habent ordinem et positionem circumscripto alio addito.

<2> Item. Sequeretur quod Deus potest separare et separatim conservare motum sine mobili et loco, immo ipsis annihilatis, quod videtur inconveniens, quia tunc esset motus et nihil moveretur.

<3> Item. Quomodo intenderetur velocitas motus, cum non remaneat pars prior cum posteriore? Et tamen sic debet fieri intensio, scilicet per additionem partis ad partem remanentem in eodem subiecto.

<4> Item. Commentator et alii ponunt quod motus est de essentia termini ad quem;[3] et in motu locali terminus ad quem non est nisi locus qui acquiritur; ergo motus localis est de essentia loci.

<5> Item. Potest argui quod motus non esset. Sed hoc dictum est in alia quaestione,[4] ideo dimitto.

<Oppositum>

<1> Oppositum arguitur, quia nec esse loci nec esse mobilis consistit in fieri, immo utrumque est perfecte factum, nisi sit aeternum; sed esse motus localis vel temporis consistit in fieri aliud post aliud; ergo esse[5] motus non |# est esse loci, vel eius quod movetur;

---

1 quaeritur septimo] septimo quaeritur consequenter G. 2 poneretur] ponerentur C. 3 Averroes, *In Physicam*, VI, comm. 48, F. 4 Johannes Buridanus, *Quastiones super libros Physicorum*, III, q. 6. 5 esse] ille C. # 72vb.

ergo non est de essentia alicuius eorum, nec per consequens est aliquid eorum. Consequentia patet, quia idem est esse hominis, essentia hominis et homo.[6]

<2> Item. Tam mobile quam locus est naturae permanentis; et motus non, sed naturae successivae.

<3> Item. Cum tempus sit motus, prout debet videri in quarto libro,[7] si tempus non est mobile nec locus, sequitur etiam quod neque motus. Sed ego probo[8] quod tempus non est mobile nec locus, quia dicit Aristoteles quarto huius[9] quod nullae partes diversae temporis simul sunt, nisi una includit aliam; sed mobilis et loci omnes partes simul sunt; igitur etc.

<4> Item. In quolibet instanti temporis[10] sempiternum est sphaera lunae et locus eius,[11] cum numquam desinant; et in nullo instanti est motus, prout[12] habetur in[13] sexto[14] huius;[15] igitur etc.

<5> Item. Arguitur quod motus non sit locus, quia motus est subiective in eo quod movetur, ut dicitur tertio huius;[16] et non est locus in eo quod movetur, sed in locante forte quiescente. Etiam ultimae sphaerae est per se motus, immo primo, et eius non est per se locus. Immo vel ultimae sphaerae non est locus, vel ille secundum Commentatorem[17] est locus centri, et ille non est motus ultimae sphaerae.

<6> Item. Arguitur quod motus non sit ipsum mobile, scilicet quod movetur. Motus enim[18] est actus ipsius mobilis, ut dicitur tertio huius;[19] et idem non est actus sui ipsius; igitur etc.[20]

<7> Item. Motus est in mobili subiective, ut dicitur tertio huius,[21] et idem non est subiective in se.

<8> Item. Cum locus qui acquiritur sit terminus ad quem et locatum sit mobile, manifestum est quod mobile non est de essen-

---

6 homo *add.* prout manifestari debet quarto Metaphysicae G. 7 Aristoteles, *Physica*, IV, 10, 218a33-218b1. 8 probo] dico G. 9 Aristoteles, *Physica*, IV, 10, 218a11-13. 10 temporis] tempus C. 11 eius] est G. 12 prout] ut G. 13 in *om.* G. 14 sexto] quarto C. 15 Aristoteles, *Physica*, VI, 3, 234a31. 16 Aristoteles, *Physica*, III, 2, 202a13-14. 17 Averroes, *In Physicam*, IV, comm. 43, fol. 142v G-H; IV, comm. 45, fol. 144r B; and VIII, comm. 84, fol. 432r E-F. 18 enim *om.* G. 19 Aristoteles, *Physica*, III, 2, 202a7-8. 20 igitur etc. *om.* G. 21 Aristoteles, *Physica*, III, 2, 202a13-14.

tia termini ad quem; et tamen motus est de essentia termini ad quem[22] secundum Commentatorem.[23-24]

<9> Item. Si terra moveretur sursum, non est verum quod ille motus sit illa terra, quia ille motus est innaturalis et violentus ipsi terrae; terra autem non est sibi ipsi[25] innaturalis et violenta.

<10> Item. Sequeretur quod materia esset motus, si Deus eam solitarie moveret; et sic ipsa esset actus, quod est inconveniens.

<Quid nominis motus>

Antiquiores non dubitaverunt de ista quaestione, sed concorditer concesserunt motum localem esse aliam rem a mobili et loco, sed iam posteriores moderni propter rationes praedictas posuerunt quod motus non sit alia res a mobili. Sed[26] ad videndum[27] oportet supponere quid nominis, quia sine hoc non potest esse disputatio, ut patet[28] quarto *Metaphysicae*, et in libro *Posteriorum*, et etiam in libro *De sensu*, ubi dicitur quod quid nominis est principium doctrinae.[29] Omnes ergo concedunt quod motus est mutatio quaedam, et moveri mutari. Et in quinto huius[30] dicit Aristoteles: "et est per se notum quod mutari est aliter et aliter se habere prius et posterius, vel saltem est prius aliqualiter se habere, et posterius taliter non se habere, aut e converso." Unde Aristoteles dicit sic: "quoniam autem omnis mutatio est a quodam in |# quoddam (manifestat utique nomen. Post aliud enim aliquid et aliud[31] quidem prius aliud autem monstrat posterius)."[32] Et Commentator dicit sic: "hoc est per se manifestum, quoniam dum res fuerit in eadem dispositione, tunc[33] illic non erit transmutatio."[34]

22 et…quem *om.* (*hom.*) G. 23 Commentatorem *add.* ergo etc. G. 24 Averroes, *In Physicam*, VI, comm. 48, F. 25 ipsi *om.* G. 26 sed] etiam G. 27 videndum *add.* de hoc G. 28 patet] apparet G. 29 *non inveni.* 30 *non inveni.* # 73ra. 31 et aliud om. CG. 32 Aristoteles, *Physica*, V, 1, 224b35. 33 Averroes, In Physicam, V comm. 7, f. 211ra. 34 tunc] termini C.

<Conclusiones>

Tunc ergo ego· pono conclusiones.

<1> Prima[35] est quod possibile est[36] ultimam sphaeram moveri motu quo movetur sine loco. Probatur sic: quia, si ultima sphaera et alia fierent unum continuum per potentiam divinam, ita quod[37] totus mundus esset unum corpus continuum, tunc nullus esset locus secundum Aristotelem,[38] quia nulla esset superficies corporis continentis divisi et tangentis. Unde Aristoteles ponit totum mundum non habere locum, nisi ratione partium, quarum una locat aliam, quia continet eam et est divisa ab ea et tangens ipsam;[39] hoc enim requiritur ad hoc quod sit locus. Unde, si Deus omnia corpora annihilaret praeter istum lapidem, ipse[40] lapis non amplius esset in loco. Et tamen illo casu posito adhuc esset possibile quod Deus moveret simul circulariter totum mundum. Hoc probo per quendam articulum Parisius condemnatum,[41] in quo dicitur quod Deus non potest[42] movere simul totum mundum motu recto (error). Et non est ratio quare magis posset movere ipsum motu recto, quam motu circulari.

Item. Sicut motu divino movet omnes sphaeras caelestes simul cum ultima sphaera, ita potest[43] omnia alia, scilicet inferiora, volvere simul. Et si ipse possit[44] omnia volvere simul, cum modo[45] sint ad invicem discontinuata, non minus hoc potest,[46] si essent facta unum continuum. Ergo potest[47] totum mundum movere, licet non esset locus.

Item. Oporteret[48] concedere, si totus mundus esset unum continuum, quod Deus extra posset formare unum granum milii tangens illum mundum, et quod illo grano milii formato quiescente Deus potest[49] sic volvere illum mundum quod continue alia et alia pars eius tangeret illud granum milii;[50] et tamen illo posito ille

35 prima *add.* conclusio G. 36 est] esset G. 37 quod] ut G. 38 secundum Aristotelem *om.* G. 39 Aristoteles, *Physica*, IV, 5, 212b8-11. 40 ipse] iste G. 41 See Roland Hissette, *Ênquete sur les 219 articles condemnés à Paris le 7 mars 1277* (Louvain, 1977), 118. 42 potest] posset G. 43 potest] posset G. 44 possit] potest G. 45 modo] non G. 46 potest] posset G. 47 potest] posset G. 48 oporteret] oportet G. 49 potest] posset G. 50 milii *om.* G.

mundus non haberet aliquem locum. Et etiam, si Deus illo grano formato posset sic volvere illum totum mundum,[51] ita posset sine illo.[52]

<2> Secunda conclusio est[53] quod ultima sphaera non solum ex eo movetur quod se habet continue aliter et aliter ad ipsam terram,[54] vel ad aliquod aliud corpus. Probatio: quia non minus moveretur, si omnia alia volverentur cum illa[55] sine alio motu eorum;[56] et tamen tunc non se haberet per talem motum aliter et aliter |# ad[57] aliquod aliud corpus.

Item. Non minus oportet se habere aliter et aliter quod movetur motu recto, quam quod movetur circulariter. Et tamen ad moveri recte non oportet se habere aliter et aliter ad aliud corpus, quia si a Deo totus mundus moveretur simul motu recto, non propter hoc se haberet aliter et aliter ad terram, licet quiesceret sicut nunc habet, scilicet[58] si terra volveretur. Ergo ex aliter se habere ad terram non sequitur quod moveatur. Ideo non solum ex eo movetur quod aliter se habet ad terram vel ad aliud corpus, quoniam ita diceremus de alio corpore, sicut diximus de terra.

Item. Sequeretur quod ex motu deformi terrae vel partium eius ultima sphaera deformiter et irregulariter moveretur, quod est falsum. Consequentia patet, quia ex tali motu irregulari terrae vel partium eius ultima sphaera deformiter et irregulariter se haberet ad terram aliter et aliter.

<3> Tertia conclusio est[59] quod ultimam sphaeram moveri est eam intrinsece aliter et aliter se habere prius et posterius. Probo, quia per quid nominis 'moveri' est aliter et aliter se habere prius et posterius;[60] et tamen moveretur, licet non se haberet aliter et aliter prius et posterius ad aliquod extrinsecum, ut apparuit per conclusiones praecedentes;[61] ergo etc.

Sed aliqui respondent quod moveri est aliter et aliter se habere ad aliquod quiescens, aut simpliciter, si aliquid quiescit, aut sub

51 illum totum mundum] totum mundum istum G. 52 illo *add.* grano G. 53 est *om.* G. 54 terram *om.* C. 55 illa] ea G. 56 eorum] earum C. # 73rb. 57 ad] et G. 58 scilicet] sicut C. 59 est *om.* G. 60 probo…posterius] probatur ipsam moveri est eam se aliter et aliter habere ut patet ex quid nominis *in marg.* G. 61 Johannes Buridanus, *Quaestiones super libros Physicorum*, III, q.7, concl. 1-2.

condicione quia, si aliquid quiesceret, se haberet ad illud aliter et aliter. Sed ista evasio nihil valet, quia possibile est quod ultima sphaera moveretur de facto, licet nihil de facto quiesceret; ergo ipsa nullo modo se haberet de facto aliter et aliter ad aliquod quiescens nec ad aliquod extrinsecum. Ergo si non se haberet aliter et aliter intrinsece, ipsa nullo modo se haberet aliter et aliter de facto, ideo nullo modo mutaretur de facto. Nam ad mutari requiritur aliter et aliter se habere simpliciter de facto, et non solum sub condicione.

Item. Numquam habitus debet describi per privationem sibi oppositam, immo oportet quod fiat e converso. Sed iste terminus 'quiescere' est privatio opposita huic termino 'moveri'; ergo mala est descriptio dicens et declarans quid nominis 'moveri' quod moveri sit aliter et aliter[62] se habere ad aliquod[63] quiescens. Nam loco huius termini 'quiescens' pono eius descriptionem, tunc idem omnino describeretur per se ipsum, scilicet sic: "moveri est aliter et aliter se habere ad illud quod est aptum natum moveri et[64] |# non movetur," et hoc est manifeste[65] inconveniens.

<4> Quarta conclusio est[66] quod motus ultimae sphaerae non est sphaera illa, nec locus eius. Primo manifestum est quod non est locus eius, quia possibile est quod moveretur, licet non haberet locum, ut dictum est;[67] et quia, si habet locum, tamen ille est divisus ab ea, motus autem eius non est divisus ab ea, cum dictum sit quod ipsa intrinsece aliter et aliter se habet.[68] Etiam[69] nec ille motus est illa sphaera, quia ut[70] dictum fuit in quaestione de distinctione figurae a figurato,[71-72] non est imaginabile vel possibile quod aliquid se habeat aliter quam se habebat ante, nisi hoc sit ad[73] aliquod extrinsecum, vel nisi hoc sit propter aliquod[74] esse quod ante non erat, aut non esse quod ante erat; sed duo primi modi non habent[75] locum in motu ultimae sphaerae, ut patet ex dictis; ergo oportet concedere tertium modum, et tamen quantum ad substantiam

62 et aliter *om.* (*hom.*) G. 63 aliquod *om.* G. 64 et *rep.* C. #73va. 65 manifeste] manifestum C. 66 est *om.* G. 67 Johannes Buridanus, *Quastiones super libros Physicorum*, III, q. 7, concl. 1. 68 Johannes Buridanus, *Quaestiones super libros Physicorum*, III, q. 7, concl. 3. 69 etiam] sed etiam G. 70 ut] sicut G. 71 a figurato] ad figuratum G. 72 Johannes Buridanus, *Quaestiones super libros Physicorum*, II, q. 3. 73 ad *om.* G. 74 aliquod] aliquid G. 75 habent] habet G.

ultimae sphaerae nihil est quod non esset ante, et nihil erat ante quod non sit modo; ergo aliud a sphaera est quod ante non[76] erat, vel e converso, et hoc non est nisi motus vel partes eius, igitur etc.

Item. Motus ultimae sphaerae non est mutatio eius substantialis, nec in ordine ad aliquod extrinsecum, nisi hoc acciderit[77] ad aliud aliter se habere intrinsece, ut praedictum est. Ergo est mutatio secundum dispositionem aliam a substantia sphaerae et sibi inhaerentem.

Item. Aliter et aliter se habet intrinsece. Ergo est alteritas alicuius ab aliquo intrinsece; et non substantiae sphaerae ad se ipsam circumscripto omni alio; ergo est alia dispositio, et illa est motus eius.

<5> Quinta conclusio est quod motus ultimae sphaerae est distinctus ab ultima sphaera et a loco eius, si habeat locus: quia est, et non est hoc nec illud; ergo etc.

<6> Sexta conclusio est quod motus ultimae sphaerae est res pure successiva, cuius scilicet est pars prior et pars posterior non manentes simul, quia si esset res permanentis naturae, tunc secundum illam ultima sphaera non se haberet aliter et aliter prius et posterius plus, quam secundum eius magnitudinem, vel figuram, vel alia eius accidentia permanentia, quod est falsum.

<Ad rationes >

Tunc igitur respondendum est ad rationes.

<1> Ad primam manifestum est quod sine dispositione superaddita[78] non potest[79] salvari quod ultima sphaera se habeat aliter et aliter intrinsece, sicut se habet.

<2> Ad aliam[80] dico quod non plus reputarem impossibile[81] quod esset motus et nihil moveretur vel mutaretur, quam quod esset albedo et nihil esset album. Neutrum est possibile naturaliter, et utrumque est possibile supernaturaliter. Sed de hoc quod dicitur, quod impli-

76 non] nec G. 77 acciderit] exciderit G. 78 superaddita] habita G. 79 potest] possit G. 80 aliam] secundam G. 81 impossibile *om.* G. # 73vb.

cat contradictionem esse motum localem et non esse |# locum, ego dico quod motus[82] ultimae sphaerae vel navis in fluvio non dicitur localis quia necesse sit quod secundum illum mutetur locus, sed quia secundum communem cursum naturae omne quod movetur illo loco variat de facto habitudinem localem vel situalem ad aliquod[83] aliud. Et omnino ille motus quem vocamus 'localem' potest[84] non esse localis, quia nullus mutaretur locus nec situs ad aliquam aliam rem; sed tunc non possem illum percipere. Non ergo vocatur motus 'localis' quia ad ipsum sit locus neccessarius, sed quia percipi non potest[85] nisi appareret mutatio loci vel situs rei ad aliam rem; unde existentes in navibus in mari velociter et simul motis non percipiunt quod illae moveantur.

<3> Ad aliam dictum fuit prius quod forma ita bene[86] redderetur intensa, si plures gradus vel maiores gradus generarentur simul, sicut si generarentur unus prius et alter posterius cum permanentia eorum. Modo a fortiori lucido et[87] fortiore motore ceteris paribus generarentur plures gradus simul luminis vel velocitatis et maiores, quam a debilibus; ideo esset intensius lumen et intensior velocitas.

<4> Ad aliam dicitur quod termini intrinseci, qui sunt de necesitate motuum quos vocamus 'locales', non sunt loca, sed sunt partes extremae illorum motuum, sicut partes extremae lineae sunt termini lineae.

<5> Ad aliam[88] rationem responsum fuit in alia quaestione.[89] Haec de quaestione.

---

82 motus *add.* alias locus *sup. lin.* (*a.l.*) C. 83 aliquod] aliquid G. 84 potest] posset G. 85 potest] posset G. 86 forma ita bene] ita bene forma G. 87 et *add.* a G. 88 aliam] ultimam G. 89 Johannes Buridanus, *Quastiones super libros Physicorum*, III, q. 6.

John Buridan and Critical Realism

Edith Dudley Sylla*
North Carolina State University

Abstract
In this paper I examine what John Buridan has to say in his *Quaestiones in Analytica Posteriora* relevant to the subalternate mathematical sciences, particularly astronomy. Much previous work on the scholastic background to the Scientific Revolution relies on texts that were written in the late sixteenth or seventeenth centuries. Here I am interested in texts that might reflect the context of Copernicus, and, in particular those before 1500. John Buridan and Albert of Saxony were fourteenth century authors influential in Cracow in the fifteenth century, whose conception of science may be characterized as "critical realism." Their view would support the autonomy of astronomy, as well as the idea that sciences may progress over time.

Keywords
John Buridan, Albert of Saxony, Nicholas Copernicus, astronomy, subalternate sciences, *moderni*, critical realism, "*ideo quasi mendicare oportet intellectum humanum.*"

Introduction

In this paper I describe the distinguishing characteristics of the image of subalternate mathematical sciences as found in the *Questions on the Two Books of Aristotle's Posterior Analytics* of John Buridan, and to a lesser extent in the *Very Subtle Questions on the Books*

* Department of History, North Carolina State University, Raleigh, NC 27695-8108, U.S.A. (edsssl@ncsu.edu). I would like to dedicate this paper to John Murdoch and thank him for questioning my translation in a previous paper (unfortunately after the paper was already published) of "*ideo quasi mendicare oportet intellectum humanum,*" which has led me to reconsider the phrase in this paper. I would also like to thank Bill Newman, stern editor, for refusing to accept earlier versions of this paper until I had worked out my ideas more fully.

of the Posterior [*Analytics*] of Albert of Saxony. The main source for my discussion will be the transcription of Buridan's *Questions on the Two Books of Aristotle's Posterior Analytics* [hereafter *QuestPostAn*], made by Hubert Hubien from MS Liège, Bibliothèque de l'Université, cod. 44 C (= Grandjean 648), which is available electronically.[1] This work, however, has much in common with the *Questiones subtilissime Alberti de Saxonia super libros posteriorum* [hereafter *Questiones subtilissime*] published at Venice in 1497.[2] Earlier scholars have tried to untangle the *Quaestiones* of Buridan from those of Albert of Saxony, but much work remains to be done.[3] There are also manuscripts of two versions of an *Expositio* of the *Posterior Analytics* ascribed to Buridan. The 1985 dissertation of Bernd Michael still represents the *status quaestionis*.[4] I suggest that the nature of Albert of Saxony's *Questiones subtilissime* as published at Venice in 1497 could be explained if Albert, in his writing on the *Posterior Analytics*, began from Buridan's *QuestPostAn* rather than from the bare text of the *Posterior Analytics* itself. Albert's work is shorter than Buridan's and includes large subtle or sophistical digressions that might

---

1) The transcription by Hubert Hubien, *Iohannis Buridani Quaestiones in duos libros Aristotelis Posteriorum Analyticorum*, was brought to my attention by the bibliography of Jack Zupko, *John Buridan. Portrait of a Fourteenth-Century Arts Master* (Notre Dame, 2003). It is available electronically on the website of Peter King: <<http://individual.utoronto.ca/pking/resources/buridan/QQ_in_Post_An.txt>> I thank Hans Thijssen for providing me with a link to the electronic text.

2) This was the only version of Buridan's or Albert's questions to receive an early printed edition. It has been reprinted (Hildesheim, 1986) with Albert's *Questiones subtilissime in libros Aristotelis de celo et mundo* (Venice, 1492).

3) See, for example, J.M.M.H. Thijssen, "Buridan, Albert of Saxony and Oresme, and a Fourteenth-century Collection of Quaestiones on the Physics and on De Generatione et Corruptione, *Vivarium* 24 (1986), 70-80.

4) Bernd Michael, *Johannes Buridan: Studien zu seinem Leben, seinen Werken und zur Rezeption seiner Theorien im Europa des späten Mittelalters*. Inaugural-Dissertation der Freien Universität Berlin (Berlin 1985). For works on the *Posterior Analytics* see items 11 (480-494; works of Buridan) and 30 (898-908; questions now ascribed to Albert of Saxony). Hubien's transcription omits what Michael lists, 487, as question 11 in the Buridan *Quaestiones* manuscript, no doubt because it appears out of place at the end of the work. I have consequently not been able to check the question, "utrum in omni demonstratione propter quid praemissae sint causae conclusionis et non in demonstratione quia."

be supposed to appeal to students, while following the overall structure of the *QuestPostAn* as found in the Liège manuscript. I will here give some samples from the *Questiones subtilissime* to suggest how it lives up to its title.

The research for this paper was done as part of a project sponsored by the Max Planck Institute for the History of Science in Berlin under the title "Before Copernicus." One of my tasks as a member of that project is to examine theories of science in the fifteenth century that might have been relevant to Copernicus. In a previous paper examining the *Quaestiones* on the *Posterior Analytics* of Walter Burley, I argued that Burley's conception of science kept the door open to scientific change and new knowledge derived from experience.[5] There is no evidence that I know of, however, that Burley's *Questions on the Posterior Analytics* were available in Cracow or in other centers where Copernicus worked (the questions are extant in only one manuscript, currently in Gonville and Caius College, Cambridge). In this paper, therefore, I have chosen to examine the same issues in the works of John Buridan and Albert of Saxony, because they *are* known to have been influential in the fifteenth century in Cracow.[6] In future work, I will examine other commentaries

5) Edith Sylla, "The Status of Astronomy between Experience and Demonstration in the Commentaries on Aristotle's *Posterior Analytics* of Robert Grosseteste and Walter Burley," in *Erfahrung und Beweis: Die Wissenschaften von der Natur im 13. und 14. Jahrhundert*, ed. Alexander Fidora and Matthias Lutz-Bachmann (Berlin, 2007), 265–91.

6) See e.g. Paul W. Knoll, "The Arts Faculty at the University of Cracow at the End of the Fifteenth Century," in *The Copernican Achievement*, ed. Robert S. Westman (Berkeley, Los Angeles, London, 1975), 137-156, and, in the same volume, Nicholas H. Steneck, "Commentary in Defense of Context," 157-164. For logic in particular, see André Goddu, "The Logic of Copernicus's Arguments and His Education in Logic at Cracow," *Early Science and Medicine* 1 (1996), 28-68 at 38, "the following authors were often cited in logical works: Giles of Rome… Radulphus Anglicus (evidently Ralph Strode), John Buridan, Marsilius of Inghen (a follower of Buridan), Paul of Venice (an Averroist), Nicholas Orbellis and Peter Rosellus (both Scotists), and John Versoris (influenced by both Albert the Great and Thomas Aquinas). See also Mieczyslaw Markowski, "Dialektische und rhetorische Argumentation an der Krakauer Universität im 15. Jahrhundert," in *Argumentationstheorie. Scholastische Forschungen zu den logischen und semantischen Regeln korrekten Folgerns*, ed. Klaus Jacobi (Leiden,

on the *Posterior Analytics*, for instance those of Albertus Magnus, Thomas Aquinas, Marsilius of Inghen, Paul of Venice, John Versor, and of other scholastics whose views were known or influential before 1500.

To examine the ideas of Buridan and Albert of Saxony on the *Posterior Analytics* is all the more called for—beyond the fact that they may have been influential at the time of Copernicus—because previous work on the Aristotelian background to the Scientific Revolution has privileged versions of Aristotelianism that differ to a greater or lesser degree from the version represented by Buridan and Albert of Saxony.[7] It used to be assumed that "Renaissance Aristotelians" could be treated as a single group. As James Lennox so humorously put it, the label "Galileo's Aristotelian opponent" is "a uniformly colored blanket hiding a vast variety of philosophical and scientific shades and colors."[8] More recently, finer distinctions have been made between various flavors or strands of Aristotelianism, but the emphasis on later versions and on versions unlike that of Buridan and Albert of Saxony remains.

A good example of research that uses very late Aristotelian sources to represent what scholastic Aristotelians held is Zvi Biener's 2008 dissertation from the University of Pittsburgh, which attempts to show that the new science of Galileo and Descartes *more geometrico* draws on themes from the medieval Aristotelian conception of subalternate sciences. Biener argues that the Aristotelian position can be represented by the work of Eustachius a Sancto Paulo and Francesco

---

1993), 577-587. In an email of 22 October 2008, André Goddu kindly summarized for me Markowski's more detailed findings as published in his articles in Polish: the views of the *moderni* were dominant at Cracow in the first half of the fifteenth century; the period 1461-1464 was dominated by Johannes Versor; the period 1464-1474 by Thomas Aquinas; the period 1475-1483 by what Markowski calls a 'historical Aristotelianism' with attention to Giles of Rome; 1484-1489 by Albert the Great; and 1490 to 1524 by doctrinal pluralism. Thus the situation at Cracow when Copernicus was a student was highly eclectic.

7) This would hold, for instance, for Jacopo Zabarella. See Heikki Mikkeli, "Giacomo Zabarella," in *Stanford Encyclopedia of Philosophy* <http://plato.stanford.edu/entries/zabarella> last accessed 8 October 2008.

8) James G. Lennox, "Aristotle, Galileo, and 'Mixed Sciences,'" in *Reinterpreting Galileo*, ed., William A. Wallace (Washington, D.C., 1986), 49.

Suarez, said to represent "the liberal Jesuit position."[9] Given Descartes' attendance at La Flèche, Jesuit Aristotelianism may indeed be the relevant stripe of Aristotelianism to compare to the ideas of Descartes, but the choice is less obvious in the case of Galileo.[10] One of the positions of Eustachius a Sancto Paolo that Biener takes to have had a possible influence on Descartes holds that all of the sciences are at least in a loose sense (*non omnino proprie*) subalternated to metaphysics.[11] Moreover, Biener quotes Descartes' statement in his *Principia* that, "The only principles which I accept, or require, in physics [*physica*] are those of geometry and abstract mathematics; these principles explain all natural phenomena" (*Principles* II §64).[12] If the foundations of science are *a priori* and if science is supposed to be certain, then this would quite naturally lead to Descartes' program to set aside what he had been taught in order to re-establish science on his own new and more secure *a priori* foundations.

Biener's purpose in his dissertation is to study the scholastic mixed-mathematical, middle, or subalternate sciences as pointing the way to the much broader early modern mathematization of science and to a revised classification of the sciences.[13] What I am suggesting here is that the picture Biener paints of the scholastic subalternate sciences as background to Descartes and Galileo should not, without

---

[9] Zvi Biener, "The Unity of Science in Early-Modern Philosophy: Subalternation, Metaphysics and the Geometrical Manner in Scholasticism, Galileo and Descartes," Ph.D. dissertation, University of Pittsburgh, 2008 < http://etd.library.pitt.edu/ETD/available/etd-03092008-215336/unrestricted/Biener_2008.pdf> retrieved 28 September 2008. Eustachius a Sancto Paolo's textbook, *Summa philosophiae quadripartita*, was first published in Paris in 1609 (Biener, 49). Francisco Suárez, *Disputationes metaphysicae* was published beginning in 1597 (Biener, 50). According to Biener, "liberal Jesuit scholasticism" (Stephen Menn's term) is "mostly devoid of dedicated Ockhamists, Averroists, Scotists and Thomists" (Biener, 51).

[10] Cf. William A. Wallace, *Galileo's Logic of Discovery and Proof* (Boston Studies in the Philosophy of Science 137) (Dordrecht, 1992), as well as the many other publications of Wallace, and Rivka Feldhai, *Galileo and the Church. Political Inquisition or Critical Dialogue?* (Cambridge, 1995), ch. 10-12.

[11] Biener, "The Unity of Science," 60 and 147.

[12] Ibid., 9.

[13] Ibid., 11.

further study, be assumed to apply to the background of Copernicus.

Buridan, Albert of Saxony, and the *moderni* as Critical Realists

How one might best describe the trends in the history of the theory of science, and particularly of the subalternate sciences such as astronomy, in one or another location during the fourteenth and fifteenth centuries depends to a great extent on the issues that one is most interested in, and a definitive picture is not currently available. For the theory of science available in Cracow in the fiftheenth century, however, one alternative approach was that of John Buridan and those whose views were like his.[14] After considering other choices, I have decided, *faute de mieux*, to refer to Buridan, Albert of Saxony, Marsilius of Inghen, and others like them as *moderni* and to their way of approaching problems as the *via moderna*, although they might also have been labeled Ockhamist, nominalist, or terminist, depending upon the historian's sensitivities as to the meanings of these terms.[15] At the very least, Buridan and Albert were the sort of thinkers who, in the fifteenth century, came to be labeled *moderni* or followers of the *via moderna* even if this label was not attached to them by their contemporaries.[16] In the fourteenth century, the people later labeled *moderni*, including John Buridan and Albert of Saxony, but also William of Ockham and even Walter Burley, tended also to be Averroists, voluntarists, and

---

14) Cf. Markowski, "Dialektische und rhetorische Argumentation."

15) Since Hans Thijssen begins his paper in this volume by saying "There was no such thing as the Buridan school," I have thought it best not to use the term, "the Buridan school."

16) See Maarten J.F.M. Hoenen, "*Via Antiqua* and *Via Moderna* in the Fifteenth Century: Doctrinal, Institutional, and Church Political Factors in the *Wegestreit*," in *The Medieval Heritage in Early Modern Metaphysics and Modal Theory, 1400–1700*, ed. Russell L. Friedman and Lauge O. Nielsen (Dordrecht, 2003), 9–36, and the works he cites by William J. Courtenay and others. See also Pepijn Rutten, "*Contra Occanicam discoliam modernorum*: The so-called *De universali reali* and the Dissemination of Albertist Polemics against the *via moderna*," *Bulletin de Philosophie Médiévale* 45 (2003), 132-165.

empiricists.[17] Whereas the *moderni* I am concerned with had their heyday in the fourteenth and early fifteenth centuries, there was another grouping of people, later labeled *antiqui*, who were more concerned with the certainty of science and therefore with *a priori* or rational foundations and with the unity of science (at least at first examination, the views Biener describes seem to be more like the *antiqui*). Although the central positions of the *antiqui* were laid down in the thirteenth century (hence the label *antiqui*), similar positions became popular again the sixteenth and seventeenth centuries. The *antiqui* included those who favored Avicenna over Averroes, and many Thomists such as Aegidius Romanus.[18] The reemergence of the *antiqui* position in the *Wegestreit* of the fifteenth and sixteenth centuries may have been connected, as far as commentaries on the *Posterior Analytics* are concerned, with new editions and translations of the Greek commentators on Aristotle and possibly with the new availability of Averroes's large commentary on the *Posterior Analytics*.[19]

---

17) See Edith Sylla, "The A Posteriori Foundations of Natural Science; Some Medieval Commentaries on Aristotle's *Physics*, Book I, chapters 1 and 2," *Synthèse* 40 (1979), 147–87, and eadem, "Averroism and the Assertiveness of the Separate Sciences," in *Knowledge and the Sciences in Medieval Philosophy*. ed. Reijo Työrinoja, Anja Inkeri Lehtinen, and Dagfinn Føllesdal (Proceedings of the Eighth International Congress of Medieval Philosophy 3) (Helsinki, 1990), 71-80. Note that I say "tended," by which I mean only something like "more often than not."

18) In the commentary on the *Posterior Analytics* found among Galileo's early papers, one question ascribes alternate positions on the types of demonstration to Avicenna and to Averroes. See Galileo Galilei, *Tractatio de Praecognitionibus et Praecognitis and Tractatio de Demonstratione,* trans. William F. Edwards; intro. and notes William A. Wallace (Padua, 1988), 98–101. Avicenna is said to hold that there is only one species of demonstration, the *propter quid* demonstration that gives the cause of an effect. On the other hand, Averroes is supposed to hold that there are three species of demonstration: *propter quid, quia,* which proves existence but not cause, and a third type, the *potissima,* which proves both the cause and the existence of the effect. Galileo, or the author from whom he is excerpting, holds a third position, ascribed to Themistius, Philoponus, Algazali, and Thomas, saying that there are only two species of demonstration, the *propter quid,* and the *quia.*

19) Walter Roy Laird, "The *Scientiae Mediae* in Medieval Commentaries on Aristotle's *Posterior Analytics*" (Ph.D. dissertation, University of Toronto, 1983), ch. VIII, on Jacopo Zabarella, 218-222.

The tendencies of the *antiqui* were reinforced as a result of the decisions of the Council of Trent, which was first convened in 1545, two years after Copernicus's death. In an earlier paper, I argued that the climate created by the decisions of the Council of Trent helps to explain why Galileo was condemned in part for holding that the Copernican theory was "probable."[20] In that same climate, there was a tendency to consider the *moderni* to be "skeptics."[21]

Rather than skeptics, Buridan and the other *moderni* are better called, in Alan Musgrave's terms, "critical realists."[22] Musgrave's argument, against Pierre Duhem and those following him, was that the great astronomers of antiquity were not instrumentalists.[23] A key tool in his analysis was to distinguish "instrumentalism" from "critical

---

20) Edith Sylla, "Galileo and Probable Arguments," in *Nature and Scientific Method*, ed. Daniel Dahlstrom, Studies in Philosophy and the History of Philosophy, vol. 22, (Washington, 1991), 211-234.

21) Although the intellectual strife of the time was mainly the result of theological controversies, the resulting passions spilled over to what seemed to be philosophical issues, such as nominalism versus realism. Indeed, during the fifteenth-century *Wegestreit*, as Hoenen describes it, each side accused the other of heresy, often presenting a distorted picture of what the other side held. Hoenen, "*Via Antiqua* and *Via Moderna*," 16, 22 ff. From the point of view of the *antiqui*, the *moderni* were tearing down the harmonious synthesis of knowledge put together by Thomas Aquinas and the *antiqui*. This has meant that some historians have looked primarily to the *antiqui* as potential carriers of the torch toward modern science (I do not have space here to determine whether Aquinas himself or the early modern Thomists should simply be considered *antiqui* in this matter or if a more complicated taxonomy is in order; the followers of Albertus Magnus are another group whose role must be taken into account). On the other hand, some historians of science have let Thomas Aquinas stand in for the scholastic opponents of the new science for the simple reason that English translations of Thomas Aquinas's work are more readily available than translations of the *moderni*. These and similar factors have contributed to the lack of attention to the *moderni* as potentially laying out a framework congenial to progress in mathematical astronomy. (I should note that, based on a reading of Hans Thijssen's paper in this volume, I should perhaps recognize that Pierre Duhem previously promoted the role of the *moderni* in the origins of modern science.)

22) Cf. Jack Zupko, "Buridan and Skepticism," *Journal of the History of Philosophy* 41 (1993), 191-221.

23) Alan Musgrave, "The Myth of Astronomical Instrumentalism," in *Beyond Reason: Essays on the Philosophy of Paul Feyerabend*, ed. Gonzalo Munévar (Boston Studies in the Philosophy of Science 132) (Dordrecht, 1991), 243-280. The article previously

realism." Whereas instrumentalists hold that "even a theory that is wholly correct does not describe anything but serves as an instrument for the prediction of facts that constitute its empirical content," critical realists may "concede the conjectural or hypothetical character of some scientific theory," but nevertheless hold that "a hypothesis is still a true or false description of reality, rather than a mere 'mathematical hypothesis.'"[24] The mistake of historians claiming that ancient and medieval astronomers were instrumentalists is to assume that the only alternative to instrumentalism is what Musgrave calls "dogmatic realism," holding not only that scientific theories are true or false descriptions of reality, but also that we can *know for sure* which theories are true and which are false. While aiming for theories that are true descriptions of reality, critical realists may admit in certain situations, for instance when there are competing or even conflicting viable theories, that science has not yet achieved certainty, while not giving up the search. A similar argument applies to Buridan, Albert of Saxony, and arguably to the *moderni* as a group—in questioning scientific dogmatism, they were not skeptics but critical realists.[25]

To Recommence the Inquiry into Principles

To see how a critical realist might operate in practice it is useful to look at Ibn al-Haytham. Near the beginning of his *Optics*, Ibn al-Haytham expresses clearly the attitude that a critical realist might take in the mathematical science of optics. He begins by explaining that previous scholars have been divided into two camps, with the more physical thinkers holding that rays come from objects into the eye (intromission theories of vision) and the more mathematical

---

appeared in German in *Versuchungen Aufsätze zur Philosophie Paul Feyerabends* (Frankfurt, 1980).

24) Ibid., 243-244.

25) Cf. Amos Funkenstein, "The Dialectical Preparation for Scientific Revolutions," in *The Copernican Achievement*, ed. Robert S. Westman (Berkeley, 1975), 165-203, at 192, "The key methodological term in both Buridan's analysis and its revision by Oresme is *persuasio*. Both of them do not seek absolute demonstrations, but rather sufficient certitude."

thinkers holding that some sort of visual rays go out from the eye (extramission theories). When there are two conflicting theories like this, he says, there are several possibilities:

> Now, for any two different doctrines, it is either the case that one of them is true and the other false; or they are both false, the truth being other than either of them; or they both lead to one thing which is the truth. [In the latter case] each of the groups holding those two doctrines would have failed to complete its inquiry and, unable to reach the end, has stopped short of it. Alternatively, one of them may have reached the end but the other has stopped short of it, thus giving rise to the apparent difference between the two doctrines, although the end would have been the same had the investigation been pushed further. Disagreement may also arise in regard to the subject of any inquiry as a result of a difference in methods of research, but when the inquiry is rightly conducted and the investigation intensified, agreement will emerge and the difference will be settled.[26]

In view of the disagreement in optics, Ibn al-Haytham proposed to "recommence the inquiry into the principles and premises, beginning our investigation with an inspection of the things that exist and a survey of the conditions of visible objects."[27] From observations and induction, al-Haytham said he would ascend gradually and in orderly fashion. He admitted that he was "not free from that human turbidity which is in the nature of man," but says, "we must do our best with what we possess of human power."[28] In his *Reportata Parisiensia*, John Duns Scotus said that the subalternate sciences may obtain the evidence for their principles from experience, as had been proved by Ibn al-Haytham in his *Perspective*. And the trend I see in the ideas of Buridan may have roots in the work of Duns Scotus, who, among other things, may have transmitted to his European contemporaries Ibn al-Haytham's conception of sciences like optics combining mathematics and empiricism.[29]

---

26) A.I. Sabra, trans. and intro., *The Optics of Ibn al-Haytham. Books I—III. On Direct Vision* (London, 1989), bk. I, ch. 1, sec. [5], 5. Sabra already drew attention to this passage as embodying a concept of experiment as a method of proof.

27) Ibid., sec. [6], 5.

28) Ibid., 6.

29) Alexander Fidora, "Subalternation und Erfahrung: Thomas, Heinrich und Duns Scotus," in *Erfahrung und Beweis. Die Wissenschaften von der Natur im 13. und 14.*

I propose that an early modern astronomer like Copernicus might have seen himself as following a path in astronomy like the path followed by Ibn al-Haytham in optics, recommencing the experiential inquiry into the principles and premises within an established field where there were conflicting theories. The image of mathematical sciences presented in Buridan's *QuestPostAn* and Albert of Saxony's *Questiones subtilissime* supports this conception of the role of an astronomer. As critical realists, the *moderni* endorsed an image of science that they found in Aristotle's *Posterior Analytics*. They were confident in the possibility of human knowledge of the natural world, and yet they did not dogmatically assert that current scientific theories were entirely certain. Instead of the certainty of science, they emphasized its empirical origins, and they treated each scientific discipline as separate, rather than supposing that all knowledge could be synthesized into a consistent whole under the umbrella of metaphysics or philosophy in general.

Subalternate Sciences and the "Prohibition of *Metabasis*"

In doing my research I have used the dissertations (and later articles) of Steven Livesey and Roy Laird.[30] Steven Livesey has made a special investigation into what he calls Aristotle's "prohibition of *metabasis*" or the attempt to combine material from disparate sciences within a single demonstration. He is particularly interested whether medieval

---

Jahrhundert. Experience and Demonstration. The Sciences of Nature in the 13th and 14th Centuries, eds. Alexander Fidora and Matthias Lutz-Bachmann (Berlin, 2007), 195-206, at 205. In his "Induction according to Duns Scotus," in the same volume, 207-222, at 221 Steven Marrone expresses surprise at finding such an attitude in Scotus: "Was Duns then on the verge of proposing a natural science of probability or at least of a sceptical and almost early-modern less-than-absolute certitude? His words would surely point in that direction, though it is hard to believe that anyone in his day was truly prepared for so radical a departure from the epistemic standards of the still-regnant orthodoxy of an Aristotelianiazing theory of knowledge." My suggestion here is that it was only later that Aristotelianism became so orthodox, rather than empirical.

30) Steven Livesey, "*Metabasis*: The Interrelationship of the Sciences in Antiquity and the Middle Ages," (Ph.D. dissertation, University of California, Los Angeles, 1982); Laird, "The *Scientiae Mediae*."

thinkers overcame the prohibition of metabasis in order to use mathematical arguments within natural philosophy. While benefiting from the sources and analysis that Livesey provides, my interest here is a different one, because I am especially interested in attitudes toward astronomy, which was already a mathematical science in antiquity. Thus I am more interested in the status of astronomy as an autonomous mathematical science, and alert to a possible lack of independence of astronomy vis-à-vis natural philosophy.

In studying what Aquinas called the "middle sciences" between mathematics and natural philosophy, including astronomy, perspective (another name for optics), and harmonics, Roy Laird has studied which authors make the middle sciences subalternate to natural philosophy as well as to mathematics and which do not. According to Laird, whereas a few philosophers such as Aegidius Romanus and Paul of Venice suggested that perspective is subalternate both to geometry and to natural philosophy, Aristotle, and following him Robert Grosseteste, Thomas Aquinas, Walter Burley, and others supposed that perspective is subalternate only to geometry.[31] John Buridan and Albert of Saxony not only do not make perspective subalternate to natural philosophy, but they follow earlier thinkers in supposing that, if anything, natural philosophy, at least the relevant parts of it, may be under perspective. Thus experience of the rainbow includes facts that are explained by perspective. In like manner, one could treat the data of astronomy or the experience of astronomical phenomena as evidence leading to principles of mathematical astronomy from below or *a posteriori*.

Thus in Aristotle's terms, astronomy was said to be "subalternated" to geometry, and harmonics to arithmetic. In some cases, the higher subalternating science was said to be able to demonstrate *propter quid*, or on account of what, while the subalternated science could only demonstrate the same conclusion *quia*, or that.[32] This has sometimes been understood to mean something like Descartes' later claim

---

31) Laird, "The *Scientiae Mediae*," 140 and 194: "That the intermediate sciences are subalternated to both mathematics and to *physica*—a notion first suggested by Aegidius ...—turns out to be for Paul their distinguishing feature."

32) See ibid., 4-5; Alistair Crombie, *Robert Grosseteste and the Origins of Experimental Science, 1100-1700* (Oxford, 1953). Sometimes it is mistakenly thought that geometry

(see above) that the only principles he needed for physics were those of geometry and abstract mathematics, since mathematical principles could be used to prove physical conclusions. In fact, for the *moderni*, the vast majority of the principles of physics and even of astronomy come from experience and not from a higher science.

Geometry and *dignitates* in the Subalternate Sciences

This is the situation as envisioned by the *moderni*. If, in sciences like perspective and astronomy subalternated to geometry, geometry is to be used by the perspectivist or astronomer, it is only on the condition that geometrical principles or conclusions are "contracted" to match the subject matter of perspective. Thus where geometry speaks of lines, the geometrical theorems used in perspective are rephrased to speak of visual lines or rays.[33] The same outcome can be described by saying that the geometrical material appearing in perspective or astronomy does not appear explicitly at all. In Buridan's terms, the principles of any science include the subject, passions, and *dignitates*, which are principles for demonstrating passions of their subjects. Geometry enters perspective or astronomy not in the role of subject terms, but through the *dignitates* in the sense of warrants for inferences. In the syllogisms of the subalternate sciences such as optics and astronomy, all the explicit propositions are optical or astronomical, whereas the geometrical *dignitates*, when they are used, are most commonly tacit inference rules.

Many scholastic authors make this clear. In his commentary on the *Posterior Analytics*, bk. I, tract. III, ch. 3, Albertus Magnus wrote that *dignitates* seldom appear explicitly in ostensive demonstrations, or, if they do appear, they are "contracted" to the subject of the special science concerned.[34] He went on to explain how this would

---

can prove *proper quid* the conclusions of optics, when in fact geometry can only prove these propositions "in their universal," not as optical.

33) See Sylla, "Status of Astronomy," 278-282.

34) Albertus Magnus, *Posteriorum Analyticorum Liber Primus* in *Opera Omnia*, vol. 2, ed. August Borgnet (Paris, 1890), 74, 75: "Est autem quoddam principium commune quod est de quodlibet affirmare vel negare, quod aliquando demonstrationem

play out in geometry and in special sciences subalternated to geometry.[35] Likewise, in his *Summa Logicae*, part III, ch. 4, William of Ockham states clearly that of the propositions required for demonstration, some are not explicitly part of the demonstration. Among the propositions that do not appear explicitly in the demonstration are such laws of logic as the law of non-contradiction. In the mathematical sciences including astronomy as well as arithmetic or geometry, a proposition such as "when equals are taken from equals, equals remain" would be presupposed without appearing explicitly, unless with the terms "contracted" to fit within the subject matter of the subalternate science.[36] Geometry might have a demonstration *propter quid* of a proposition analogous to a proposition in optics, but it would concern only triangles or other geometrical figures and not light rays: it would not prove *propter quid* an optical conclusion. On the other hand, a perspectivist might demonstrate a conclusion *propter quid* where, in his demonstration, the stated premises and conclusion would be optical, but the unstated "maxim" by which the conclusion followed from the premises would be geometrical, rather than simply logical.

ostensivam ingreditur, sed raro.... Hoc autem principium quod commune est valde, quod dicit quod omne quod est vel non est, est affirmare vel negare, requiritur in demonstratione quae est ad impossibile; quia illa per illud confirmatur. Hoc autem non semper accipit quaelibet scientia particularis demonstrativa secundum suam communitatem universaliter, sed contractum ad suum genus in quantum illi generi sufficiens est...."

35) Ibid., 77-79.

36) William of Ockham, *Summa Logicae*, = *Opera Philosophica*, vol. I (St. Bonaventure, 1974), 509-510: "Est autem primo sciendum quod propositionum requisitarum ad demonstrationem quaedam sunt partes demonstrationis, sicut duae praemissae et una conclusio, et quaedam non sunt partes demonstrationis. Et vocantur dignitates et maximae vel suppositiones, quae sub propria forma non ingrediuntur demonstrationem, virtute tamen illarum propositionum aliquo modo scientur praemissae demonstrationis.... Propositio autem requisita ad demonstrationem, non tamquam pars, subdividitur, quia quaedam est talis quod necesse est quemlibet docendum habere eam, cuiusmodi sunt tales 'quidlibet est vel non est', 'de quolibet affirmatio vel negatio' et huiusmodi. Quaedam sunt tales quas non est necesse quemlibet docendum habere, sed necesse est aliquos artifices speciales eas habere, sicut est de istis 'aliquid est mobile', 'si ab aequalibus aequalia demas' etc."

In his *Summulae de dialectica*, Buridan talks about the principles of a discipline that ground the consequences rather than appearing as premises:

> But I say that in the demonstrations of these conclusions, not only are two first principles required, namely, the two premises, but also several others, for a demonstration requires not only the evidentness of the premises but also the evidentness of the consequence. But that consequence is a proposition, albeit a hypothetical one. And so, if the consequence is evident in itself, then it is an indemonstrable principle....[37]

Here the warrants for the inferences that others called *dignitates* Buridan labels "consequences." Thus what happens in the case of the subalternated sciences like astronomy is that the rules of reasoning in mathematics are taken to be as certain as the rules of reasoning in logic. They allow consequences (i.e. inferences) to be drawn. With all this in mind, it is time to look at the texts of Buridan's and Albert of Saxony's questions on the *Posterior Analytics* for support of the characterizations of the position of the *moderni* that I have given.

Questions and Answers of Buridan and Albert of Saxony on the *Posterior Analytics*

Three questions from Buridan's *QuestPostAn* and the related questions in Albert of Saxony's *Questiones subtilissime* will be used to characterize the aspects of their conception of science relevant here: 1) that science or human knowledge of the world is possible; 2) that sciences like astronomy use mathematics together with principles based on experience; and 3) that, although natural scientists should have confidence in their principles, the existing state of knowledge is not perfect and at some point an improved set of principles might be called for.

---

[37] John Buridan, *Summulae de dialectica*, trans. Gyula Klima (New Haven, 2001), 714.

Book I, Q. 2. Whether it is Possible for us to Know Something.[38]

Buridan's conclusion is that just as fire is naturally inclined to heat, and heavy bodies to descend, so is the human intellect naturally inclined to understand objects sufficiently presented to it and also naturally inclined to comprehend the truth of first complex principles.[39] For science certitude of truth, certitude of assent, and evidence (or evidentness) are required. The greatest evidentness belongs to propositions, such as logical first principles, to which it is impossible to dissent. But there is also natural evidentness. In the natural case, we can be deceived by supernatural causes, but not in the common course of nature. Natural evidentness is sufficient in natural science. Although experience that is scanty and not carefully examined may deceive, when there is a great deal of experience examined in diverse cases, experience does not deceive. Although one cannot give a formal proof that the universal proposition one arrives at is true, one accepts the proposition as a known and naturally evident universal principle.[40]

---

38) Compare Albert of Saxony, *Questiones subtilissime*, qq. 3-4, ff. 3vb-5va: "Utrum, quia aliqui antiqui dixerunt quod impossibile est nos aliquid scire, sicut achademici; alii autem dixerunt quod bene possibile est nos aliquid scire, sed impossibile est nos scire aliquid de novo, sicut Platonici; ergo primo videndum est an possibile sit nos aliquid scire; secundo an possibile sit nos aliquid scire de novo."

39) Buridan, *QuestPostAn*: "Unde debetis notare quod cum ponantur tres operationes intellectus, in earum qualibet sunt aliqua per se manifesta ex natura et inclinatione ipsius intellectus ad veritatem. Unde sicut ignis est naturaliter inclinatus ad calefaciendum et grave ad descendendum, ita intellectus est naturaliter inclinatus ad intelligendum obiecta sibi sufficienter praesentata, et etiam naturaliter inclinatus ad comprehensionem veritatis primorum principiorum complexorum; ideo non oportet quod per aliud primum iudicium declaretur...."

40) Ibid.: "Sed de evidentia debetis notare quod 'evidentia' multipliciter accipitur. Uno modo proprissime, et tunc evidentia propositionis dicitur secundum quam intellectus per suam naturam cogitur propositioni assentire et non potest ei dissentire; et isto modo diceret Aristoteles quod primum principium est nobis evidens. Secundo modo 'evidentia' dicitur quia cuilibet apparet et per nullam rationem humanam nisi sophysticam [*corr. ex* subiectivam ref. Albert of Saxony, *Questiones subtilissime*, f. 4rb] posset oppositum apparere; et isto modo sunt evidentia principia naturalia et conclusiones naturales. Et notandum quod haec evidentia non dicitur proprie 'evidentia' : quia circa tales propositiones evidentes intellectus posset decipi per causam supernaturalem; quia deus posset facere ignem sine caliditate, et posset facere in sensu meo et conservare

In his *Physics*, Buridan essentially makes accepting the truth of principles derived from repeated sense, memory, and experience a norm of the natural philosopher or student of ethics. He says that "We should concede that there cannot naturally be a vacuum as known by that means which is sufficient for positing and conceding principles in natural science," and "One who does not want to concede such declarations in natural and moral science is not worthy of having a large part in them."[41]

---

speciem sensitivam sine obiecto, et ita per istam evidentiam tu iudicares ac si obiectum esse praesens, et iudicares falsum. Tamen illa evidentia naturalis bene dicitur naturalis, quia secundum illam non potest homo decipi stante communi cursu naturae, licet deciperetur per causam supernaturalem; et haec evidentia sufficit ad naturalem scientiam.... Istis visis, facile est solvere rationes.... 7. Ad aliam rationem... saepe per rationem intellectus evidenter corrigit errorem sensus.... 9. Ad aliam dicentem quod experientia saepe fallit, dicitur primo quod multa sunt principia evidentia non credita per experientiam, sed per manifestam terminorum inclusionem ad invicem, vel exclusionem, scito quid nominis. Sic enim est evidens quod homo est animal et asinus est animal. Deinde principia sumpta per experientiam adhuc habent naturalem evidentiam. Quia licet experientia pauca et parum examinata saepe fallat, tamen experientia multa et bene in diversis casibus examinata numquam fallit. Et quando ultra opinatur quod experientia numquam gratia formae concludit universale principium, quia numquam fit in omnibus singularibus, respondet Commentator, secundo *Physicorum*, quod licet inductio, sive experientia inductiva, non concludat gratia formae, tamen intellectus, ex eius naturali inclinatione ad veritatem, percipiens multotiens ita fieri quod non potest nec potuit recipere instantiam, nec videre esse rationem quare in aliis debeat esse aliter, ipse concedit universale principium tamquam notum et evidens evidentia naturali et possibili circa tali."

41) Buridan, *Sutilissime questiones super octo phisicorum libros Aristotelis* IV, 7 (Paris 1509; reprint Frankfurt, 1964), f. 73v, reply to the question, "Utrum possibile est vacuum esse": "Item omnis propositio universalis in scientia naturali debet concedi tanquam principium quae potest probari per experimentalem inductionem sic quod in pluris singularibus ipsius manifeste inveniatur ita esse et in nullo nunquam apparet instantia. Sicut enim bene dicit Aristoteles quod oportet multa principia esse accepta et scita sensu memoria et experientia. Immo aliquando non potuimus scire quod omnis ignis est calidus, sed per talem inductionem experimentalem apparet nobis quod nullus locus est vacuus... . Ideo debemus concedere quod non potest naturaliter esse vacuum tanquam scitum per illum modum qui est sufficiens ad ponendum et concedendum principia in scientia naturali." Also *Super octo physicorum libros* I, 15, ff. 18vb-19ra: "Secunda conclusio est quod necesse est omne quod fit naturaliter fieri ex subiecto presupposito vel in subiecto presupposito. Forma enim non fit integraliter ex subiecto presupposito, sed in eo. Sed tamen non puto quod hec conclusio sit

It should be understood that this is not a "skeptical" position, but one that recommends to natural scientists a justifiable degree of confidence in their results.[42] Albert of Saxony gives essentially the same answer to this question, but he spices it up with typical subtle or sophistic condiments. The point that the normal course of nature and our confidence in the veracity of our experience may be broken into by God's absolute power is phrased to allow the scholastic favorite "you are an ass":

> Whether it is possible for us to know something. It is argued that it is not possible… for it is possible that by some divine power there is produced in me a species representing heat and that heat is destroyed and cold induced and the action of the cold on sense is suspended…. By this argument it could be proved that it is not evident to me that you are a man, but that I may rationally doubt whether you are an ass or a goat. For the visible species which is in the organ of my visual power representing to me that you are a man could be conserved in my soul while you were transubstantiated into an asinine substance by the power of the First Cause.[43]

---

demonstrabilis. Sed est declarabilis per inductionem in qua non inventa est instantia. Sic probat enim eam Aristoteles et tale reputari debet principium in scientia naturali. Aliter enim tu non posses probare quod omnis ignis est calidus, quod omne reubarbarum est purgativum colore, quod omnis magnes vel adamas est attractivus ferri. Et tales inductiones non sunt demonstrationes, quia non concludunt gratia forme, cum non sit possibile inducere in omnibus suppositis, sicut dicitur secundo *Posteriorum* quod multa principia indemonstrabilia fiunt nobis manifesta sensu, memoria, vel experientia. Experientia ex multis significationibus et memoriis deducta non est aliud quam inductio in multis singularibus per quam intellectus, non videns instantiam nec rationem instandi, cogitur ex eius naturali inclinatione ad veritatem concedere propositionem universalem. Et qui non vult tales declarationes concedere in scientia naturali et morali non est dignus habere in eis magnam partem."

42) Thus I think that previous historians who have labeled Buridan a skeptic are mistaken. Cf. Peter King, "Jean Buridan's Philosophy of Science," *Studies in the History and Philosophy of Science* 18 (1987), 109-132, at 109. Although King states that Buridan "attempts to chart a course through these skeptical waters," he says that, "what Buridan gives up in the final analysis is nothing less than the truth itself" (110).

43) Albert of Saxony, *Questiones subtilissime*, f. 3vb: "An possibile sit nos aliquid scire et arguitur quod non, quia non possumus de aliquo habere evidentiam sine formidine ad oppositum. Ergo non est possibile nos aliquid scire. Consequentia tenet, ex eo quod scientia est evidentia sine formidine ad oppositum. … et antecedens probatur, quia non possumus de aliquo habere notitiam nec evidentiam per sensum nec

Of course, Albert's reply to this argument is that although we cannot have the highest evidentness from sense, nevertheless we can have natural evidentness, which is sufficient for us to know.[44] But then Albert goes on to another scholastic favorite, the division of time into small parts. It is not evident, he says, that you were a man for a year, although it appeared to me for a year that you were a man. Because it is possible that you were only a man for half a year. The First Cause might have created you for an insensible part of a year, then annihilated you for another insensible part of the year, then recreated you, and so forth up to the end of the year. For the whole year you would have appeared to me as a man, but you would only have been a man for half a year.[45] This argument is already part of the reply to the principal arguments and so receives

---

per intellectum, ergo nullo modo.... non per sensum. Nam quantumcumque appareat mihi quod ignis per sensum sit calidus, tamen propter hoc non habeo iudicium et evidentiam de hoc quod ignis sit calidus sine formidine. Nam possibile est quod per aliquam potentiam puta divinam producatur in sensu meo species representativa caliditatis, et quod caliditas destruatur et frigiditas inducatur et actio frigiditatis in sensum suspendatur. Isto posito apparet mihi quod ignis esset calidus et in veritate esset frigidus. Et cum casus positus sit possibilis, cum Prima Causa sit agens liberum, sequitur quod quotienscumque video ignem habeo dubitare an caliditas ignis sit corrupta et frigiditas inducta et species caliditatis in sensu sit representata et actio frigiditatis in sensu sit suspensa per Primam Causam.... Per istam rationem probaretur quod non est mihi evidens te esse hominem, sed quod rationabiliter habeam dubitare te esse asinum vel capram. Nam species visibilis que est in organo meo virtutis visive representans mihi te esse hominem posset in animo meo conservari te transubstantiato in substantiam asini per potentiam Primae Causae. Et quia hoc Prima Causa potest facere quando vult, sequitur quod quandocunque apparet mihi quod tu sis [corr. ex. scit] homo, habeo dubitare utrum sis asinus."

44) Ibid., f. 4va: "Ad rationes. Ad primam dico quod bene probatur quod mediante sensu non possumus habere de aliquo evidentiam summam. Nihilominus mediante ipso possumus habere evidentiam naturalem que sufficit nobis ad hoc quod sciamus."

45) Ibid., f. 4va, "Unde evidentia summa non est mihi evidens quod tu per unum annum fuisti homo quamvis per totum annum apparuit mihi quod tu esses homo. Unde possibile est quod tu per integrum annum continue apparuisti mihi homo. Et tamen non nisi per medium annum fuisti homo. Si enim Prima Causa per unam partem insensibilem anni te crearet, et iterum per aliam partem insensibilem te annihilaret, et iterum per sequentem insensibilem te recrearet, et sic de aliis usque ad finem anni. In fine anni verum esset dicere: per integrum annum mihi apparuisti homo, et tamen non nisi per medietatem anni fuisti homo, ex eo quod solum per tempus insensibile

no reply. From Albert's jocularity it should not be inferred, however, that he was not serious about the subject matter at hand, nor that he was skeptical, but only that he wanted to keep the students awake and paying attention. Just before he began to reply to the principal arguments he had continued to talk of asses while stating his clear position:

> For the solution of certain arguments it should be noted that something is said to be evident in two ways. There is the highest evidence [or evidentness] and there is natural evidentness. The highest evidentness of a proposition is when the intellect by its nature is compelled to assent to the proposition and cannot dissent from the proposition. In this way of speaking of evidentness we say that the principle "of anything, it is etc. [i.e. or it is not]" is evident. And speaking similarly of evidentness it is evident to me that I exist. But natural evidentness is when something is evident to us in the sense that by no human argument unless a sophistic one can the opposite be made plausible. And in this way natural principles and natural conclusions are said to be evident. And note that this is not evidentness in the strictest sense, because the intellect concerning things that are evident in this way can be deceived by supernatural power. Thus, as it was argued, a supernatural cause could conserve the species and similitude of a man in an eye, for example Socrates in my eye, and transmute that Socrates into an ass, and then Socrates would appear to me as a man but nevertheless he would be an ass. Therefore speaking of the highest evidentness it is not evident to me that you are a man, nor similarly is it evident to me that fire is hot. And those for whom this natural evidentness would not suffice would not be skillful demonstrators.[46]

---

non fuisti et iterum per tempus insensibile fuisti et iterum non fuisti per tempus insensibile...."

46) Ibid., ff. 4rb-va, "Pro solutione aliquarum rationum notandum quod dupliciter dicitur esse evidentia. Quedam dicitur evidentia summa, quedam naturalis. Evidentia summa alicuius propositionis est secundum quam intellectus per suam naturam cogitur assentire illi proposioni et non potest dissentire illi propositioni. Isto modo loquendo de evidentia dicimus istud principium: de quolibet esse etc. esse evidens. Et similiter loquendo sic de evidentia evidens est mihi quod ego sum. Evidentia autem naturalis est quando aliquid est nobis sic evidens quod per nullam rationem humanam nisi sophysticam oppositum posset apparere. Et isto modo principia naturalia et conclusiones naturales dicuntur esse evidentes. Et nota quod hec non est evidentia proprissime dicta, quia intellectus circa illa que taliter sunt evidentia posset decipi per potentiam supernaturalem. Unde sicut arguebatur: causa supernaturalis posset speciem et similitudinem hominis in oculo puta Sortes in oculo meo conservare, et illum Sortem in asinum transmutare, tunc Sortes appareret mihi homo et tamen esset asinus. Ergo loquendo de evidentia summa non est mihi evidens quod tu es homo, et similiter

This last sentence mirrors the similar sentences quoted from Buridan earlier, indicating that the natural philosopher ought to have confidence in the principles and conclusions of natural philosophy. Indeed, the person who does not accept principles based on carefully examined experience is not worthy of having a large part in natural or moral science.

Q. 23. Whether it is Possible for a Demonstrator to Descend from One Genus to Another.[47]

Buridan admits that it is difficult to say whether it is possible for a demonstrator to descend from genus to genus because the question has hardly been discussed and it involves the issue of the distinction of the sciences. Sciences have unity because of the unity of their subjects, but the subject of one science can contain within itself the subject of another science contracted in some way, as the subjects of perspective and astronomy contain within themselves geometrical magnitude contracted to the specific subject area. The same *dignitates* can appear in diverse sciences, contracted to each science, as the axiom "if you subtract equals from equals, equals remain," becomes in geometry, "if you subtract equal magnitudes from equal magnitudes, equal magnitudes remain," and in arithmetic "if you subtract equal numbers from equal numbers, equal numbers remain." When the conclusions of a subalternating science are accepted as principles in a subalternated science, they are accepted not in their generality, which exceeds the bounds of the subalternated science, but as "contracted" to fit the subalternate science.[48] This not infrequently happens:

est mihi evidens quod ignis est calidus, et quibus non sufficeret illa evidentia naturalis non essent habiles ad probandum."

[47] Cf. Albert of Saxony, *Quaestiones subtilissime*, q. 24, "Utrum possibile sit demonstrantem descendere de genere in genus," ff. 18vb-19rb.

[48] Buridan, *QuestPostAn*: "Ista quaestio, ut mihi videtur, est difficilis, primo quia valde modicum discussa est inter philosophos et doctores, secundo quia tangit ad modum distinctionis scientiarum, et est multum difficile assignare unde et quo modo scientiae accipiant originaliter suam distinctionem..... Deinde, dico, quantum ad principia, quod numquam sic descendetur de una scientia in aliam quod principium vel conclusio unius fiat principium alterius nisi aliquo modo diversificetur per contractionem,

> Now it is certain that natural science often supposes from mathematics, as in the consideration of the rainbow and of the proportions of motions.[49]

Albert of Saxony's answer to the parallel question is similar and without sophistical decoration.

Q. 25. Whether the Mathematical Sciences are the Most Certain of the Other Sciences.[50]

Buridan concludes that mathematical sciences are quite uncertain (*multum incerta*), but that the means of demonstration in mathematical sciences is most certain. So there is a conditional certainty in mathematics: if the principles are true, then the conclusions are true. So, for example, geometry is true if continua are not composed of indivisibles, which is a geometrical principle, not to be determined or demonstrated by the geometer himself.[51]

---

quoniam scientia communior non utitur principiis propriis scientiae specialis, et etiam scientia specialis non utitur principiis vel conclusionibus scientiae superioris in tota eorum communitate, sicut apparet, sed cum contractionibus, ut dictum fuit. Tamen, ultimo, dico quod si per 'descensum de una scientia in aliam' intelliges quod principia vel conclusiones unius scientiae fiant cum quadam contractione principia alterius scientiae, tunc valde multipliciter contingit descendere de una scientia in aliam. Quoniam quandocumque supponit aliqua ab alia, contingit dicto modo descendere de illa a qua supponitur in illam quae supponit."

49) Buridan, *QuestPostAn*: "Modo certum est quod scientia naturalis saepe supponit a mathematica, ut in consideratione de iride et de proportionibus motuum."

50) Cf. Albert of Saxony, *Quaestiones subtilissime*, q. 26, "Utrum scientie mathematice sint certissime," ff. 19vb-20rb.

51) Buridan, *QuestPostAn*: "Credo etiam quod isto modo mathematica scientia est multum incerta. Quia principia mathematicalia indigent bene explanatione per superiores scientias; et etiam multae conclusiones mathematicae indigent expositione eius quod est dubium et a multis non concessum, <ut> scilicet quod linea non sit composita ex punctis; verbi gratia, conclusio in geometria est <quod possibile est> omnem lineam per aequalia dividere, quod esset impossibile de linea punctorum imparum, et etiam ita ista conclusio esset falsa quod diameter est incommensurabilis costae, et sic de multis aliis; tamen geometria non habet determinare nec demonstrare illam dubitationem. Ultimo <modo> certitudo attenditur ex certo modo demonstrandi suppositis principiis. Et tunc esset concedendum quod mathematicae demonstrationes essent in primo gradu certitudinis: quia demonstrationes mathematicae maxime observant

Albert reaches the same answer, that the inferences of mathematics are the most certain, but the principles are not.[52] He uses the occasion of the question to give the students some mathematical practice. For instance, applying the maxim "if equals are subtracted from equals, equals remain" to proportions (ratios in modern terminology), he first gives examples that misapply the maxim to ratios, before explaining the proper sense in which the maxim is true.[53] In other principal arguments, like Buridan, he raises the problem of subtracting or adding an indivisible to a finite quantity.

formam syllogisticam et modum et maxime sunt ex propriis mediis. Et si est dubitatio in huiusmodi demonstrationibus mathematicis, est maxima certitudo ex suppositione principiorum, et est tamquam certitudo condicionalis quod si ita est sicut <principia ponunt, ita est sicut> conclusio ponit."

52) Albert of Saxony, *Questiones subtilissime*, f. 20ra-b: "Tertio modo certitudo scientie attenditur penes hoc quod principia illius scientie sunt notissima et evidentissima non ab alia presupposita et probata. Quarto modo certitudo scientie attenditur penes hoc quod illa scientia presuppositis suis principiis et concessis, firmissime et evidentissime concludit et demonstrat conclusionem ex illis principiis.... Tertia conclusio: loquendo de certitudine tertio modo dicta scientia metaphysica omnium aliarum scientiarum est certissima. Patet quia eius principia sunt certissima, nec presupponit ea ab alia scientia, sicut tamen alie scientie faciunt. Igitur etc. Quarta conclusio: loquendo de certitudine quarto modo scientia mathematica aliis scientiis est certior. Probatur quia scientie mathematice certiorem habent modum demonstrandi suas conclusiones et magis evidenter. Unde principiis mathematice presuppositis et concessis necesse est conclusiones ex illis principiis demonstratas concedere. Et ille consequentie quibus conclusiones mathematice inferuntur ex suis principiis sunt sic certe et evidentes: quod statim concessis principiis oportet conclusionem concedere. Et propter talem certitudinem et evidentiam consequentie conclusionis ex premissis scientie mathematice reputantur certiores aliis scientiis. Et hoc intendebat Aristoteles in littera."

53) Ibid., f. 19vb. He takes the equal ratios 8 to 4 and 6 to 3. He first proposes subtracting 2 from the first term of each of these ratios. When that is rejected, he proposes subtracting 1 from all four terms. Neither of these procedures results in equal ratios. In reply to the principal arguments, f. 20rb, he proposes subtracting "half a double ratio" from each of the ratios, and subtracting a "third part of a double ratio" from each, and finds in each case that the remaining ratios are equal, in the first case "half a double ratio" remaining for each.

Natural Evidentness and the Possibility of Scientific Change

If I am right that Buridan and Albert of Saxony were critical realists, then it is also relevant to consider what action they might have proposed if, in the course of working *within* an already established scientific discipline, the scientist began to sense flaws in the structure, and especially in the principles of the science. Did their image of science, and in particular of a subalternate mathematical science like astronomy, provide for the possibility of scientific change, as I claimed in my earlier article was the case for Walter Burley?

It is widely agreed that Greek and medieval astronomy set itself the problem of saving the phenomena by use of uniform circular motion. This is sometimes called the Platonic axiom, but in Aristotelian astronomy the principle of uniform circular motion of the star sphere would surely be based on experience, just like the principle that fire is hot. On the other hand, the principles that the sun, moon, and other planets move with combinations of uniform circular motions would be an extrapolation from the principle concerning the star sphere, but one that had been shown justified by the successes of Eudoxus, Hipparchus, Apollonius, and the other early astronomers. Thus, I claim that the so-called Platonic axiom was never taken as dogmatically as is often implied.

In the *Almagest*, Ptolemy simply introduced the principle of uniform circular motion of the heavens without justification. For instance, in Book III, ch. 3, on the hypotheses for uniform circular motion, he wrote:

> Our next task is to demonstrate the apparent anomaly of the sun. But first we must make the general point that the rearward displacements of the planets with respect to the heavens are, in every case, just like the motion of the universe in advance, by nature uniform and circular.[54]

In the fifteenth century, Albert of Brudzewo, professor of astronomy at the University of Cracow at about the time when Copernicus

[54] Ptolemy, *The Almagest*, trans. and annotated by G.J. Toomer (New York, 1984), 141.

was a student, wrote in his commentary on Georg Peurbach's *Theorica nova planetarum*:

> Even though it is a first principle of astronomy that the sun moves regularly in its eccentric (and therefore one should not dispute in astronomy with anyone who denies this principle), nevertheless this principle can be demonstrated by the subalternating science, that is mathematics, as follows: the sun describes equal angles in equal times around its center [of motion] and cuts off equal arcs, therefore it is moved uniformly.... [55]

How does mathematics demonstrate the principle that heavenly bodies move in uniform circular motion? If Brudzewo's statement is examined carefully, mathematics enters his demonstration only on the assumption that the positions of the sun in relation to the center at different times are known. Then geometry is used to show that such positions are consistent with a uniform circular motion. This is crucial to understanding what is going on: mathematics is used as a way of inferring from observed positions of the sun at stated times that the sun is undergoing uniform circular rotation (and, since this is the case, that its circular rotation is eccentric to our point of observation). Of course, the difficulty here is that the inference from the measured positions to the theory is not unique—if there is a valid epicycle model, there will also be a valid eccentric model, and vice versa. Nevertheless. the predicted positions, as far as observation is concerned, will be the same.

In the first book of the *Physics* Aristotle had argued that a geometer need not debate those who deny his principles and likewise a physicist need not debate those who deny that things move, since this is a first principle of physics.[56] If, as was the case, from Greek antiquity it was a principle of mathematical astronomy that the

---

55) Albertus de Brudzewo, *Super theoricas novas planetarum*, ed. Ludovicus Antonius Birkenmajer (Cracow, 1900), 30-31: "Etsi Solem in suo ecentrico regulariter moveri sit primum principium in astronomia (ideo cum negante illud, non est amplius in Astronomia disputandum), tamen tale principium potest per scientiam subalternantem, scilicet mathematice demonstrari sic. Sol in temporibus aequalibus aequales super centrum suum describit angulos et aequales resecat arcus, ergo aequaliter movetur...."

56) Sylla, "The A Posteriori Foundations of Natural Science."

heavenly bodies move in uniform circular motion, under what circumstances might an astronomer dispute concerning it?

As a critical realist in Alan Musgrave's terms, John Buridan believed that the principles of any genuine science should be certain, but may in fact be violated if for no other reason than that God, by his absolute power, can break the laws of nature. If long experience leads to a conclusion, no cases are known in which it does not hold, and no arguments against it have been proposed except sophistical ones, then a natural philosopher ought to assert it as true or else give up any claim to be a natural philosopher.[57] On the other hand, if there was evidence against a principle, or arguments that were not sophistical, might a scientist indeed adopt a new or changed principle?

In his questions on the *Physics*, Buridan had faced the problem of what to do, not when experience contradicted a physical theory, but when the injunction of the theologians that the philosopher should not deny that God can do anything that is not a logical contradiction forced the natural philosopher to ask what would happen if God broke into the normal course of nature, for instance, by annihilating everything inside the sphere of the moon.[58] Would there be a vacuum where the earth and its atmosphere had previously been, or would there be nothing, not even empty space, that is, no measurable extension? Seeing that natural philosophy did not provide an unambiguous answer to such a question, and that one was supposed simply to believe it (*hoc simpliciter credendum est*), Buridan concluded that it is necessary for the intellect so-to-speak to beg (*ideo quasi mendicare oportet intellectum humanum*). Having mistranslated this phrase in an earlier article, I now suggest taking the correct translation as a pointer to what the astronomer or natural philosopher might do when his principles turn out to be

---

57) See texts in note 41.

58) Edith Sylla, "'*Ideo quasi mendicare oportet intellectum humanum*': The Role of Theology in John Buridan's Natural Philosophy," in *The Metaphysics and Natural Philosophy of John Buridan*, ed. J.M.M.H. Thijssen and J. Zupko (Leiden, 2001), 221-45. In this paper I mistranslated the Latin phrase in the title to say that one must "beg the intellect" rather than that the intellect must beg.

inadequate or faulty in some respect.[59] It is, after all, the intellect with which we know principles, whereas we have scientific knowledge of demonstrated conclusions. What would it mean for the intellect to beg?

In the question of his *Physics* where Buridan made this comment, he had been discussing what could be said about the duration of three angels created before the creation of the rest of the cosmos, if first God created one alone and then two more together. Then it seemed that there was nothing by which duration could be measured unless by the angels themselves or by God, but none of them are extended in a temporal dimension. Having commented that the human intellect must so-to-speak beg, Buridan went on:

> On this it seems to me plausible (*probabile*) to say that the duration of angel A or of angel B or C is nothing else but angel A or B or C ... and nevertheless it appears to me that angel A was prior to angel B, not because it was in a prior time, but because it existed when B did not exist. And it can be said that that priority was temporal in a conditional sense, because if time coexisted with those angels, as is the case with those that now coexist with time, then angel A would have been prior in time to the succeeding time in which angel B or angel C existed ... from an exterior denomination... even in a certain proportion....[60]

So, if Buridan's intellect is begging here, he is making suggestions of what it might be plausible to say, rather than asserting or demonstrating.

Begging or Borrowing?

Does the phrase "ideo quasi mendicare oportet intellectum humanum" in the context of God's absolute power help to elucidate what Buridan might recommend to the practitioner of any science who begins to doubt the principles of a given science for some reason

---

59) Ibid., 229-230. Buridan, *Physics*, f. 82rb, says: "adhuc est fortis dubitatio," "sequuntur alie magne inconvenientie" and "istud est difficile quia sensus vel imaginatio non cadit super talia sed solum intellectus et adhuc apud intellectum forte non potest convinci per rationes et ex sensibus deductas quod dicti casus sint possibiles."

60) Buridan, *Physics*, f. 82ra: Sylla, "*Ideo quasi mendicare*," 230.

other than God's absolute power? In Buridan's Question 25 of *QuestPostAn*, discussed above, he said that the principles of geometry are not in fact certain, but the geometer assumes them as true and does not debate them. If there were to be any debate about geometrical principles, according to Buridan, it would be the task of the physicist or natural philosopher to engage in it, but not of the geometer.[61] In question 23 of *QuestPostAn*, after saying that natural science supposes principles from mathematics, Buridan goes on immediately to bring up the converse case in which it might be supposed that natural philosophy "descends" into geometry:

> Similarly, and conversely, the mathematician supposes [*supponit*] from natural [philosophers], as that the continuum is always divisible into divisibles, or also that it is not composed of indivisibles. For it is necessary to suppose this, because if a continuum were composed of indivisibles, almost all the demonstrations and conclusions of geometry would be false.[62]

Might "supposing" in this case be the result of the intellect so-to-speak begging? Could it be said that the mathematician "borrows" the principle that the continuum is always divisible into divisibles from natural science?

William of Ockham himself uses forms of the word *mendicare* in his *Dialogus*. Ockham's question is whether theology or canon law is the discipline most appropriate for determining heresies. His answer is that theology is the most appropriate discipline and that,

---

61) See text in note 51. Buridan says the same thing in his *Physics* I, 5, f. 7ra, "Hec ergo declaro, quia magnum dubitabile est et fuit apud antiquos, utrum corpus esset compositum ex punctis indivisibilibus vel non, sed esset divisibile in semper divisibilia. Et illam dubitationem non potest geometer tractare per suam scientiam, sed tractanda est per phisicam vel per metaphisicam, et tamen geometer habet supponere quod continuum non sit compositum ex indivisibilibus, quia si esset compositum ex indivisibilibus, omnes pene conclusiones geometrie essent false."

62) Buridan, *QuestPostAn*: "Similiter, e converso, mathematicus aliqua supponit a naturalibus, ut quod continuum sit divisibile in semper divisibilia, vel etiam quod non sit compositum ex indivisibilibus. Hoc enim oportet quod supponat, quoniam si compositum esset continuum ex indivisibilibus, quasi omnes demonstrationes et conclusiones mathematicae essent falsae. Similiter, omnes scientiae supponunt a metaphysica et a logica universalissima quae sibi contrahunt...."

if canon law pronounces on heresies, it is because it "borrows" principles from theology:

> Therefore such a decision is known to pertain chiefly to theologians; it does not pertain to canonists, however, except in so far as they are known to borrow (*mendicare*) some theological matters from theologians.[63]

Here John Kilcullen and John Scott, who posted the text and translation of the *Dialogus* electronically have translated *mendicare* by "borrow"—does it make a difference if *mendicare* means "borrow" rather than "beg"? "Borrow" does seem to be a reasonable translation, since another text says that the moon borrows (*mendicat*) light from the sun.[64] The twelfth-century grammatical text *Promisimus* writes that the arguer sometimes borrows propositions in the sense of taking them fully formed.[65]

So here we have the suggestion that the geometer would take the principle that the continuum is always divisible into divisibles from natural science not as demonstrated (natural philosophy cannot demonstrate a principle of geometry, because that would cross disciplinary boundaries), but simply as a loan, to be used as an indemonstrable first principle (perhaps in the form of an implicit inference warrant). In this sense, a subalternating discipline does not prove the principles of a subalternate discipline, but may help

---

63) <http://www.britac.ac.uk/pubS/dialogus/t1d1.html>, accessed 28 October 2008, bk. I, ch. 2, "Ergo ad theologos talis diffinitio principaliter noscitur pertinere, ad canonistas autem non spectat nisi in quantum aliqua theologica noscuntur a theologis mendicare." I discovered this and the following uses of *mendicare* by Googling *mendicare* with various other relevant words.

64) Alain de Lille, *Distinctiones dictionum*, as quoted by Virginie Minet-Mahy, "Quelques traces d'une 'théorie du texte' dans l'allégorèse en moyen Français...," *Le Moyen Age* at http://www.cairn.info/revue-le-moyen-age-2004-3-page-595.htm, accessed 30 October 2008.

65) Karin Margareta Fredborg, "The 'Promisimus,'" in *Medieval Analyses in Language and Cognition*, eds. Sten Ebbeson and Russell L Friedman (Copenhagen, 1999), 191-205 at 201 (found by Google search 28 October 2008): "Quid<am> dicunt quod disputator disponit argumentationem in animo suo antequam eam proferat, et sic in uno intellectu illa argumentatio concipitur. Quod falsum est, quia argumentator quando<que> mendicat propositiones."

to argue dialectically in their favor, perhaps in part by arguing against alternatives. Thus if the principles of a demonstrative science are called into question, another discipline such metaphysics may be called into action to deal with the problem, but that higher discipline will not demonstrate the challenged principles. Buridan, in the passages in question speaks of natural philosophy "treating" (*tractare, tractanda*), "explaining" (*indigent explanatione per superiores scientias*), and "expositing" (*indigent expositione*) the doubt whether the continuum is composed of indivisibles.[66]

Aristotle himself had suggested in the *Topics* that it was dialectic that had to deal with problems concerning scientific principles of established disciplines:

> We must say for how many and for what purposes the treatise [the *Topics*] is useful. They are three—intellectual training, casual encounters, and the philosophical sciences.... For the study of the philosophical sciences it is useful, because the ability to puzzle on both sides of a subject will make us detect more easily the truth and error about the several points that arise. It has a further use in relation to the principles used in the several sciences. For it is impossible to discuss them at all from the principles proper to the particular science in hand, seeing that the principles are primitive in relation to everything else: it is through reputable opinions about them that these have to be discussed, and this task belongs properly, or most appropriately, to dialectic; for dialectic is a process of criticism wherein lies the path to the principles of all inquiries.[67]

What Buridan and Albert of Saxony drew out of the *Posterior Analytics* was arguably an accurate representation of Aristotle's position. In the *Posterior Analytics* Aristotle describes the ideal structure of a

---

66) Interestingly, in his *De continuo* Thomas Bradwardine attacks the problem of the composition of the continuum from every discipline before raising the issue whether he has committed a *petitio principii.* See John Murdoch, "Thomas Bradwardine: His Mathematics and Continuity in the Fourteenth Century," in *Mathematics and its Applications to Science and Natural Philosophy in the Middle Ages*, eds. Edward Grant and John E. Murdoch (Cambridge, 1987), 103-137; Edith Sylla, "Thomas Bradwardine's *De Continuo* and the Structure of Fourteenth-Century Learning," in *Texts and Contexts in Ancient and Medieval Science*, eds. Edith Sylla and Michael McVaugh (Leiden, 1997), 148-186.

67) Aristotle, *Topics*, trans. W.A. Pickard-Cambridge in *The Complete Works of Aristotle*, ed. Jonathan Barnes, vol. I (Princeton, 1984), Book I, ch. 2, 101a25-b4.

scientific discipline. He also describes how one might go about creating such a discipline. It has often been remarked that Aristotle's own scientific writings do not seem to conform to his ideal model as presented in the *Posterior Analytics*. In a typical natural philosophical work, Aristotle begins by examining previous theories on a given topic and then uses a critique of these theories to argue for what he considers the most reasonable position. In light of the passage from the *Topics* just quoted, it becomes clear that Aristotle's actual scientific practice follows the model he set out in the *Topics* for scrutinizing the principles of an already established science. Commentators on the opening of Book I of the *Physics* (where Aristotle argues that the natural philosopher need not argue against those who deny that all or some things move because it is a first principle of natural philosophy) reacted in various ways to Aristotle's suggestion that, if not the physicist, then the practitioner of some higher discipline might be called upon to debate the principles of physics.[68] In the special case of a subalternate science whose principles had become open to doubt, there was the suggestion that the subalternating science might prove (*probare*) the principles of the subordinate science.[69] Walter Burley, in his *Questions on Aristotle's Posterior Analytics* seems to countenance this possibility.[70] Others, including someone Burley called the "new expositor," and William of Ockham said that the proper principles of the subalternate science were only known by experience. Otherwise, all that a higher discipline can do is to argue, not demonstrate:

68) Sylla, "The A Posteriori Foundations of Natural Science."

69) Based on Markowski, "Dialectische und Rhetorische Argumentation," 582, perhaps more should be made of the use of the word "*probare*": "Die *probatio* als eine der Begründungsformen kommt schon bei Johannes Duns Scotus und Wilhelm Ockham vor; eine intensive Beschäftigung mit ihr erfolgte erst in der Buridanischen Schule; seit dieser Zeit nahm die *probatio* einen zentralen Platz in dem Zweig der Logik ein, der Allgemeine Dialektik gennant wurde. Gleichzeitig mit dem Buridanismus verbreitete sich auch die *probatio* in Mitteleuropa und war auch an der erneuerten Krakauer Universität populär."

70) Sylla, "Status of Astronomy," 274 and n. 31.

> Second it should be noted that, against someone who denies the principles of a subalternate science, the subalternating science can argue; similarly both against someone who denies the principles of the subalternated science and against one who denies the subalternating science, both the metaphysician and the dialectician can argue. And thus he [Aristotle] says that it is not for geometry to argue against someone who denies geometrical principles, but for another science, that is a subalternating science, if it has a subalternating science, or for what is common to all sciences, that is metaphysics and dialectics. However, it should be known that first principles cannot be made evidently known by any common science, because the first principles of any science either are known per se (*per se nota*), and consequently cannot be demonstrated nor made known with an evident knowing (*notitia*) by others. Or they are not known per se, and then they cannot be made evidently known except by experience, which is not gained from any common science whatever. And thus no first principle of any subalternating science can be demonstrated. But against those denying it, it can be argued by the principles of a common science which they concede.... Similarly what they say is false, namely that all science can be resolved into principles known in themselves, because not all first principles in the special sciences are known in themselves, but some are known only by experience, without which they cannot be evidently known.... And thus that there is motion is not known except by experience, which nevertheless is assumed in natural science, as experience should be presupposed in natural science. Whoever wants to learn natural science without experience, strives after the impossible, because it is impossible to acquire natural science without experience, although it is possible to acquire natural faith without experience by believing what other people say.[71]

Perhaps, then, a medieval scholastic, faced with anomalies involving the principles of a demonstrative science, would turn to the method of dialectic to untangle the difficulties. Certainly as far as convincing others that a theoretical novelty was justified, the use of topical arguments might well be the best one could do. Still, it was the practitioner of the subalternate discipline who would take the initiative to suppose or borrow principles as undemonstrated starting points for his own work. The other alternative would be to gather new data from experience to establish new principles *a posteriori* as Ibn al-Haytham recommended doing in optics.

---

[71] Ibid., 274, n. 30; William of Ockham, *Expositio in Libros Physicorum Aristotelis*, in *Opera Philosophica*, vol. IV, eds. Vladimir Richter and Gerhard Liebold (St. Bonaventure, 1985), 44-46.

Buridan and Astronomy

Buridan himself did not claim to be either a geometer or an astronomer.[72] He declared himself (presumably as a natural philosopher) incompetent to resolve astronomical problems such as the problem that eccentrics and epicycles reproduce a given motion equally well.[73]

We know that astronomers not only reasoned mathematically about moving spheres or circles, but that they also took evidence from experience. It was astronomers who had collected the evidence of the motions of the stars and planets that they later used to determine their principles *a posteriori*, including the principle that the heavenly bodies move in uniform circular motion. Buridan knew that astronomers in antiquity had held theories different from those of modern astronomers.[74] There was a difference of theory, for instance, concerning the precession of the equinoxes. Depending on the observation of precession, some had added an extra celestial sphere or extra motions to the same sphere. Aristotle, Buridan said, did not

---

72) For Buridan and mathematics, see J.M.M.H. Thijssen, "Buridan on Mathematics," *Vivarium* 23 (1985), 55-78.

73) John Buridan, *Quaestiones super libris quatuor De caelo et mundo*, ed. Ernest Moody (Cambridge, Mass., 1942), bk. II, q. 18, 211: "Utrum astra moveantur per se vel ad motum suae sphaerae.... Postea etiam mihi videtur quod difficile est demonstrare quod luna non habeat epiciclum; et difficile est etiam demonstrare quod luna habeat epiciclum; quia apparentia quae de luna salvantur per epicyclum et eccentricum, possunt salvari sine epiciclo per plures eccentricos quorum unus deferret alterum. Et hoc est considerandum in astrologia de motibus. Neutrum igitur est mihi demonstrabile, saltem per me; ideo etiam non est mihi demonstrabile utrum luna tali motu circumgirationis moveatur." Buridan makes a similar comment in *In Metaphysicen*, bk. 12, q. 11, f. 74ra: "Et ita etiam qui vellet ponere ecentricos sine epiciclis oportet ponere duos ecentricos, unum in altero, sicut ponimus ecentricum in concentrico. Et tunc salvaretur totum per illos duos eccentricos sicut salvabatur per ecentricum et epiciclum. Et breviter loquendo non apparet michi quod aliqua istarum viarum sit bene demonstrabilis nec etiam bene reprobabilis."

74) Buridan, *In Metaphysicen*, bk. XII, q. 9, f. 72ra: "Ista questio est valde difficilis et aliter esset respondendum secundum Aristotelem et Commentatorem, et aliter secundum fidei veritatem. Et adhuc etiam esset aliter respondendum tenendo principia Aristotelis secundum antiquos astrologos qui erant tempore eius et aliter secundum modernos."

believe that the eighth sphere moved with two motions, as other astronomers did. But to track these motions it had been necessary to observe celestial positions and to preserve the written records of these observations for as much as five hundred or a thousand years. During that time the records might have been falsified.[75]

In his questions on the *Physics* IV, q. 14, in the course of answering whether time is the measure of any motion, Buridan embarked on an extended disquisition about measurement and its problems, in the course of which he said that we cannot measure any motion precisely or to a point, because differences between two magnitudes may be too small to perceive.[76] Not long after Buridan, Nicole Oresme made a strong case for the probable incommensurability between the periods of the various planets, meaning that exact ratios would never be known.

As a natural philosopher, Buridan had asked what the number of intelligences is—whether, for instance, it is more than, equal to, or less than the number of celestial spheres. Is there one intelligence for every sphere, so that there would be intelligences for epicycles as well as the main orbs? Or might there be just one intelligence for the daily components of the motions of all the planets as well as the stars? It has been noticed, in this context, that from the viewpoint of theology Buridan suggested that the heavenly spheres might

---

75) Ibid., bk. XII, q. IX, f. 72vb: "Unde apparet quod Aristoteles non credidit octavam sphera<m> moveri duplici motu, nec forte illud est bene certificatum, quia non potest certificari bene nisi per observationes seu reservationes scriptorum a longissimis temporibus, ut a quingentis aut mille annis. Et possibile est talia scripta fuisse falsificata. Immo aliqui secundum tales reservationes credidissent quod illa octava sphera moveretur contra motum diurnum in centum annis uno gradu. Visum est postea aliis quod post transitum quinque vel sex graduum ipsa revertebat. Et ideo talia non sunt bene certificata. Et ideo adhuc si illi antiqui vidissent illum secundum motum octave sphere ipsi posuissent preter spheram in qua fixe sunt stelle aliam ferentem ipsam vel revolventem sicut fecerunt de planetis."

76) Buridan, *Questiones super libros Physicorum*, f. 81ra: "Et notandum est etiam quod non possumus motus naturales omnino precise et punctualiter mensurare, scilicet secundum modum mathematice considerationis. Non enim possumus per stateram scire si precise libra cere sit libre plumbi equalis. Potest enim esse excessus in ita parva quantitate quod non perciperemus excessum sed sufficit sepe mensuratio ad prope iuxta illud quod de modico non est curandum."

be moved by impetus put into them by God at the time of their creation.[77] But, also from a theological point of view, he said that God might move all the spheres, and indeed all stones, by himself, without the aid of any intelligences. And, of course, he said, there are many legions of angels, not just the movers of the celestial spheres.

If it was appropriate to Buridan as a natural philosopher to ask about the number of celestial intelligences moving the heavenly spheres, it was not necessary for astronomers to deal with this question, because they only needed to know where the heavenly bodies were, the kinematics so-to-speak, in order to make astrological predictions, which was their main goal. In Buridan's words:

> And therefore it suffices for them to accept the easier imagination, according to which (if it were true) the celestial bodies would move with as many motions and with such velocities as they now move. And they do not have to care whether it is so in reality as they imagine. Now if the outermost sphere were posited to drag with it all the others and that sphere were posited to carry the epicycle, the same would result as far as concerns the various aspects of stars with respect to each other and with respect to us as now happens according to the position that we put forth. And therefore it is licit for them to posit another imagination, nor is it against their science, nor against the truth that they intend.[78]

---

77) Buridan, *In Metaphysicen*, f. 734a. He introduces this idea as "una imaginatio nescio an fatua."

78) Ibid., f. 72va, "dico sicut dicit Commentator quod astrologi non habent curare a quibus corporibus corpora celestia moveantur, scilicet an a seipsis an ab intelligentiis: nec utrum ab uno motore an a pluribus: nec etiam utrum una sphera moveat aliam an non, sed sufficit eis scire quod tot motibus moventur et secundum tales velocitates quia solum per hoc volunt scire habitudines situales astrorum adinvicem et ad nos. Et ideo sufficit eis accipere faciliorem imaginationem secundum quam (si esset vera) corpora celestia moverentur tot motibus et talibus velocitatibus sicut nunc moventur et non debent curare utrum sit ita in re sicut imaginantur. Modo si ultima sphera poneretur secum trahere omnes alias et quod sphera poneretur ferre epiciclum, idem omnino proveniret quantum ad diversos aspectus stellarum adinvicem et ad nos sicut nunc provenit ex positione quam posuimus. Et ideo ipsi licite possunt ponere aliam imaginationem, nec illud est contra illorum scientiam nec contra veritatem quam intendunt. Sed de talibus imaginationibus eorum et aliorum philosophus habet inquirere que sit vera et que non."

For Buridan, attempting to speculate about astrophysics, the Ptolemaic system was not at all easy to supplement with physical causes—even if they were intelligences—of the observed motions. It does not appear that creating an astrophysics for a heliocentric system would have been hugely more difficult. In any case, Buridan the natural philosopher, observing mathematical astronomy from outside, thought that astronomers had every right to propose new imaginations—which is the same as new astronomical principles—when they thought it appropriate. But concerning their imaginations and those of others, the philosopher would have to inquire which were true and which not. Perhaps by this Buridan meant that it was the task of the astronomer to know that (*quia*) the planets move with certain velocities are are found in certain places, whereas it was the task of the natural philosopher to explain the cause, *propter quid*, of their motions.

The *Posterior Analytics* and Astronomy

In 1159, in his *Metalogicon,* John of Salisbury wrote:

> The science of the *Posterior Analytics* is extremely subtle, and one with which but few mentalities can make much headway. This fact is evidently due to several reasons. In the first place, the work discusses the art of demonstration, which is the most demanding of all forms of reasoning. Secondly, the aforesaid art has, by now, practically fallen into disuse. At present demonstration is employed by practically no one except mathematicians, and even among the latter has come to be almost exclusively reserved to the geometricians. The study of geometry is, however, not well known among us, although this science is perhaps in greater use in the region of Iberia and in the confines of Africa. For the peoples of Iberia and Africa employ geometry more than do any others; they use it as a tool in astronomy. The like is true of the Egyptians, as well as some of the peoples of Arabia.[79]

In this paper I have tried to examine the theories of science that appear in the questions on the *Posterior Analytics* of John Buridan,

[79] John of Salisbury, *The Metalogicon* , V.vi, trans. Daniel McGarry (Berkeley, 1962), 212. Cf. Jens Høyrup, *In Measure, Number, and Weight. Studies in Mathematics and Culture* (Albany, 1994), 116-117.

with glances at Buridan's other works, and at the *Questiones subtilissime* on the *Posterior Analytics* of Albert of Saxony. I suggest that Buridan fine-tuned his theory of science as he attempted to deal with what he knew of the relatively successful scientific discipline of mathematical astronomy. Perhaps Buridan learned something, as John of Salisbury seems to propose, by looking at astronomy as a model of what a proper scientific discipline should be. In this case, the picture of separate scientific disciplines developed by William of Ockham, John Buridan, Albert of Saxony, and the *moderni* on the basis of Aristotle's *Posterior Analytics* and the actual practice of mathematical astronomers might have reinforced each other. On this model, astronomers might be willing to "recommence the inquiry into the principles and premises, beginning our investigation with an inspection of the things that exist and a survey of the conditions of visible objects," in Ibn al-Haytham's terms. In Buridan's view it was licit for them to do so without going against their science or the truth that they intended.

The Significance of "Chymical Atomism"

William R. Newman*
Indiana University, Bloomington

Abstract
The historical treatment of atomism and the mechanical philosophy largely neglects what I call "chymical atomism," namely a type of pre-Daltonian corpuscular matter theory that postulated particles of matter which were operationally indivisible. From the Middle Ages onwards, alchemists influenced by Aristotle's *Meteorology*, *De caelo*, and *De generatione et corruptione* argued for the existence of robust corpuscles of matter that resisted analysis by laboratory means. As I argue in the present paper, this alchemical tradition entered the works of Daniel Sennert and Robert Boyle, and became the common property of seventeenth-century chymists. Through Boyle, G.E. Stahl, and other chymists, the operational atomism of the alchemists was even transmitted to Antoine Laurent Lavoisier, where it became the basis of his claim that elements are simply "the final limit that analysis reaches."

Keywords
atomist, corpuscularian, negative-empirical principle, operational indivisibility, homoeomerous, matter theory, alchemy, chymistry, Aristotle, Geber, Daniel Sennert, Robert Boyle, Antoine Laurent Lavoisier, Georg Ernst Stahl, Kurd Lasswitz.

Introduction

The history of pre-modern atomism has two convenient bookends. At one end we find the ancient origination of the atomic school by Democritus and Leucippus. Aristotle tells us that the ancient atomists viewed their indivisible particles as being composed of pure being, differentiated by their *schēma, thesis,* and *taxis*, or shape,

* Department of History and Philosophy of Science, Goodbody Hall 130, Indiana University, Bloomington, IN 47405, U.S.A. (wnewman@indiana.edu).

position, and arrangement (Metaphysics I, 4, 985b4 *et seq.*).[1] These particles could be conjoined, separated, or rearranged, thus accounting for generation, corruption, and alteration. The other bookend is provided by the mechanical philosophies of the seventeenth century, only some of which were professedly atomistic, but which nonetheless engaged in a similar move of trying to reduce the phenomena of the world to the size, shape, motion, and texture of microstructural corpuscles.

Historians of science and philosophy are familiar with the important and perennial questions that swell the historiography of atomism and the mechanical philosophy. Atomism and its explicit descendants have provided a particularly rich area of investigation for their philosophical, scientific, and religious implications. Questions focusing on the possibility of a strict reduction of secondary to primary qualities, on the legitimacy of physical versus mathematical indivisibility, on mechanism versus teleology, on the reality of the void, and on the religious or ethical implications of a denial of hylomorphism, flow in an uninterrupted stream from antiquity and the Middle Ages up to the work of Newton. We may or may not accept Richard Westfall's elegantly stated claim that the net result of the mechanical philosophy was that the "world is a machine, composed of inert bodies, moved by physical necessity, indifferent to the existence of thinking beings."[2] But I doubt that many would deny that the picture given by Westfall closely corresponds to what the history of science as a scholarly enterprise has led us see as the salient features of the mechanical philosophy and its forerunners.

The present paper will discuss a very different sort of atomism, which did not dwell on the primary/secondary quality distinction, had no bearing on the issue of mathematical indivisibility, was indifferent to the distinction between teleology and mechanism, could coexist with either plenist or vacuist views, and presented little if any problem to the opponents of Epicurean materialism. As I will

---

1) For other passages describing the atomism of Democritus and Leucippus, see G.S. Kirk, J.E. Raven, and M. Schofield, *The Presocratic Philosophers* (Cambridge, 1983), 402-433.

2) Richard Westfall, *The Construction of Modern Science* (Cambridge, 1977), 33.

argue, the primary historical significance of this hitherto little acknowledged atomism lies elsewhere, and its philosophical implications may at first seem jejune when compared to the wealth of issues raised by its ancient namesake. I have in mind what I will call "chymical atomism," a theory whose origins lie in an unusual fusion of alchemical, medical, and philosophical concepts already present in the thirteenth century if not before. I use the archaically spelled "chymical" as a means of simultaneously linking this form of atomism to the chemical atomism of the nineteenth century, which is its descendant, and freeing it from too strict an association with the chemistry of John Dalton and his heirs. The subject of chymical atomism has preoccupied me for a number of years—certainly since the publication of my edition of Geber's *Summa perfectionis* in 1991. But the research for my recent book, *Atoms and Alchemy*, has made me see the subject in a new light.[3] Whereas I once maintained the distinction already present in Kurd Lasswitz's work between corpuscularism and atomism, where the Geberian theory ranked as a corpuscular rather than an atomistic one, I now propose that the concept of "atomism" itself be re-examined.[4] Such a re-examination

---

3) William R. Newman, *The Summa Perfectionis of Pseudo-Geber: A Critical Edition, Translation and Study* (Leiden, 1991), and Newman, *Atoms and Alchemy: Chymistry and the Experimental Origins of the Scientific Revolution* (Chicago, 2006). My discussion of Boyle in the following pages owes a heavy debt to *Atoms and Alchemy*, chapter six.

4) Kurd Lasswitz, *Geschichte der Atomistik vom Mittelalter bis Newton* (Hamburg, 1890), 1:211-228. Lasswitz treated a wide range of late antique thinkers as corpuscularian rather than atomist. Among the former he ranked technical thinkers who accepted corpuscular modes of explanation but, unlike Democritus and Leucippus, did not derive their belief in a particulate micro-world from metaphysical principles and did not engage in the radically reductionist program of ancient atomism. Hero of Alexandria for example, maintained that bodies were made up of λεπτομερῆ σώματα at the micro-level, and used the interaction of these particles to explain a variety of macro-level phenomena, ranging from the compression of air to the formation of a partial vacuum. But in the analysis of Lasswitz, Hero's corpuscles were in no sense indivisible or intransmutable—they were merely microscopic bits corresponding to Aristotle's four elements. In short, the emphasis of Hero and other corpuscularian thinkers lay in providing a plausible explanation of certain sensible phenomena by means of a physical micro-structure, not in any sort of postulated immutability or resistance to division on the part of the atoms. This, in Lasswitz's view, was the primary distinction between

will lead us to include much of what passed as alchemical corpuscular theory as a type of atomism.

The Case of the Early Robert Boyle

A good entrée into what I have in mind can be gained from examining Robert Boyle's posthumously published *Essay of the Atomicall Philosophy*. The work was written in the mid-1650's, before Boyle had explicitly dissociated himself from the atomism of antiquity. After paying tribute to a list of his modern atomistic forebears, in which Boyle includes Gassendi and Descartes among others, the young virtuoso proceeds to give the details of an atomism that is tacitly borrowed from the Wittenberg medical professor Daniel Sennert. After Boyle's brief introductory remarks, he jumps into a series of experimental examples that are intended to demonstrate the probability of his atomistic hypothesis. These demonstrations rely on the persistence of ingredients during laboratory analysis and synthesis. They were called "reductions to the pristine state" by Boyle's source, namely Sennert. But what I want to focus on here are the theoretical comments, also heavily reliant on Sennert, that precede the experiments themselves. Let us therefore give particular care to the first part of Boyle's quotation, which I have placed in italics:

> *In similar bodyes ^that are really such for wine milke &c. that seeme so are not^ their being constituted by Atomes is very probable, since it is so that their particles are very small & of the same nature with the whole they compose,* thus sylver being dissolv'd in Aqua fortis & that Menstruum so well filter'd that the dissolv'd silver & it will both passe thorough Cap paper all the invisible particles of ye Metall which are so small that they hinder not the Diaphaneity of the Menstruum are yet each of

corpuscularism and atomism. As he put it (at 1:211), "Die unmittelbare Erfahrung der Wechselwirkung wird auf die Atome übertragen, weil dieselben nicht mehr als metaphysische Substanzen, sondern als physikalische Partikeln der Körper selbst vorgestellt werden. Das ist die Geburtstätte der Korpuskularphysik." Unfortunately, Lasswitz's bifurcation leaves no room for chymical atomism, which *did* make resistance to physical division a criterion, but did not engage in the primary-secondary quality distinction that originated out of the reductionist program of ancient atomism.

> them true silver as appeares by precipitating them to the bottome (by a little resolv'd salt of Tartar) in the forme of a subtle powder which is easily reducible into the same numericall silver that was at first corroded & so in the mixture of Metalls there is an union per minima that is Atomes, as if gold & silver be duly melted together each part of the masse has an equall proportion of the respective Metalls, & any part of it being cast into Aqua fortis (which by reason of the virtue wee are now going to ascribe to it is by the French often call'd eau de depart) or water of separation, the Menstruum will corrode and imbibe the Atomes of <the> silver & let those of the gold fall in the forme of powder to the bottome, instances of this Nature might be easily multiplyed if I judg'd them requisite.[5]

A close inspection of this passage reveals interesting surprises. Boyle addresses himself first to the genuine "similar bodyes," among which his text includes the metals. These similar bodies are the homoeomerous substances of Aristotelian natural philosophy, even though Aristotle does not view them as consisting of atoms.[6] Aristotle had argued in *De generatione et corruptione* (I 10 328a 10-12) that true mixture (*mixis*) comes about only when substances are uniformly combined—"the mixture ought to be uniform throughout, and, just as any part of water is water, so any part of what is blended should be the same as the whole" (Forster's translation).[7] As we know, Aristotle contrasts such true mixture with *synthesis* or juxtaposition, which consists merely of placing bodies side-by-side. Stock examples in the Aristotelian corpus of substances that have undergone true mixture include the metals, flesh, blood, wine, and milk.[8]

---

[5] Robert Boyle, *Essay of the Atomicall Philosophy*, in *The Works of Robert Boyle*, ed. Michael Hunter and Edward Davis (London, 1999-2000), 13:228. After inspection of the microfilm of Royal Society, Boyle Papers 26, ff. 162r-175r, I have altered the authorial insertion to read "seeme so" where the editors read "some so." In addition, where the editors use pointed brackets to indicate Boyle's insertion, I use carets.

[6] In the quoted passage, Boyle started by discussing "similar" or homoeomerous materials in general, but subsequently realized that he needed to exclude wine and milk—both of which had been considered homoeomerous by Aristotle—precisely because they lack the resistance to analysis that would make them satisfy his concept of homoeomerity. Boyle is influenced here by Daniel Sennert, *Hypomnemata physica* (Wittenberg, 1636), 113-114.

[7] Aristotle, *On Sophistical Refutations, On Coming-to-Be and Passing-Away*, ed. and trans. E.S. Forster (Cambridge, MA, 1978), 259.

[8] See Aristotle *Meteorologica* IV 383a21-22, 384b30-34, 385a8-10, 388a13- 388b10, 389a7-23, 390b14-22, *et passim*.

In such mixed substances, according to *De generatione et corruptione,* every part will be alike, and the resulting mass will therefore be homogeneous. In short, to Sennert and the young Boyle, each "atom" making up the whole is itself a mixt, bearing the essential characteristics that brand the whole as a substance. The first surprise lies in the fact that Boyle, the professed "atomist," is willing in principle to accept that there are such homoeomerous substances marked by different essences. At this early phase of his career Boyle abandons the ontological simplicity of the Democritean atom, as Sennert had done before him, and treats individual atoms as if they were Aristotelian mixts. This move decisively distinguishes Boyle's early atomism both from the corpuscularism of Descartes with its undifferentiated first matter, and from the adherence to a "uniform catholic matter" of Boyle's maturity.[9]

A second and perhaps greater surprise arises from Boyle's causal assertion linking the homoeomerity of some substances with the fact that they are "atomic." It is "very probable," he says, that some substances are composed of atoms, "since it is so that their particles are very small & of the same nature with the whole they compose." The reader may justly wonder at the sense of this. It is perhaps obvious that atoms would have to be small, but why should the uniform character of a substance that is homoeomerous in the Aristotelian sense support a belief in atomism? After all, Aristotle had, in *De generatione et corruptione* (I 10 327b33-328a18), argued against atomism precisely because it *violated* the principle of uniform, perfect mixture. How then can uniformity of mixture support the claim of atomism? The answer here lies in Boyle's use of the term "Atomes," which he derives directly from Sennert. In this passage of the *Atomicall Philosophy*, the primary sense of "atom" does not refer to a body that is incapable of further physical division into smaller bodies. Boyle is not thinking here in the spatial, even geometrical, terms of classical atomism, even though he has just argued in favor of Democritus that Aristotle was wrong in representing the Abderite as a point-atomist. Instead, "atom" here refers

9) See my *Atoms and Alchemy*, chapter seven, for Boyle's mature corpuscular theory.

primarily to a substance that is incapable of decomposition into more primitive substances.

Like Sennert, the Boyle of the *Essay of the Atomicall Philosophy* is reasoning within the framework of what historians of chemistry have named the "negative-empirical concept." [10] According to this analytical ideal, a substance is viewed as elementary if it cannot be decomposed by the tools of the chymist. Hence the concept is negative in that it defines an element solely in terms of what chymistry cannot do, and empirical in that it relies on the experience of the laboratory. As Boyle would later put it in *The Sceptical Chymist,* elements are "those primitive and simple Bodies of which the mixt ones are said to be composed, and into which they are ultimately resolved."[11] Within the framework supplied by Boyle's *Essay of the Atomicall Philosophy*, "elementary" and "atomic" can be seen as coextensive terms—both imply resistance to decomposition.[12] For this reason, then, the ability of some corpuscles to withstand the corrosive *menstrua* of the seventeenth-century chymist makes it probable that they are operationally indissoluble, and hence atomic. The identity between the smallest available particles and the substance as a whole provides direct evidence for such atomism, since an obvious consequence of this uniformity is that no decomposition products are present. If there were perceptible decomposition products, they would not be materially identical to the original substance

---

10) Bernadette Bensaude-Vincent and Isabelle Stengers, *Histoire de la chimie* (Paris, 1993), 51-52.

11) Robert Boyle, *The Sceptical Chymist,* in *Works of Robert Boyle,* 2, 220.

12) It is not the case, of course, that semi-permanent material corpuscles are really required in order for something to be irreducible and intransmutable (hence "elementary"). One need only think of the spectral colors. Despite Isaac Newton's personal commitment to a corpuscular ontology, his claim for the elementary character of the spectral colors required only that he demonstrate them to be caused by rays of unequal refrangibility, not that they be corpuscular. See Alan Shapiro, "The Evolving Structure of Newton's Theory of White Light and Color," *Isis* 71 (1980), 211-235, and A.I. Sabra, *Theories of Light from Descartes to Newton* (Cambridge, 1981). The phenomena of the chymical laboratory, however, provided many clues to indicate that matter really was corpuscular at the microlevel—clues that ranged from the discrete arrangement of sublimed "flowers" collecting on the inside of a sublimatory to the powders that collect at the bottom of a flask when acid solutions of metals are precipitated out by bases.

being analyzed, and would therefore introduce heterogeneities into the controlled space of the analysis. Therefore homogeneity after attempted analysis is a warrant for the claim that the material at hand is atomic.

Aristotle and Geber

If we accept the legitimacy of Boyle's—and Sennert's—use of the term "atom" to mean a substance that is incapable of division by means of laboratory analysis, then a host of further implications arise. First, it is quite evident that Boyle's source, Sennert, did not invent the negative-empirical approach. Aristotle himself says at *De caelo* 3 302a15-18 (in Guthrie's translation), "Let us then define the element in bodies as that into which other bodies may be analysed ... and which cannot itself be analysed into constituents differing in kind."[13] This would go on to form the basis of the scholastic maxim *in quae dissolvi possunt composita, ex iisdem coaluerunt:* "The things into which composites can be dissolved are the things out of which they are made."[14] Yet the Aristotelian and scholastic claim for analysis admitted, of course, that the elements themselves could be mutually transmuted even if they could not be further analyzed. Hence the putative conversion of water into air by means of heat, for example, did not imply that water was made of air. It is therefore difficult to see how the Aristotle of *De caelo* or *De generatione et corruptione* could have made any hard claims

---

13) Aristotle, *On the Heavens* (Cambridge, MA, 1971), 283. See the treatment of this Aristotelian *locus* in Daniel Sennert, *De chymicorum cum Aristotelicis et Galenicis consensu ac dissensu* (Wittenberg, 1619), 283. Sennert renders the maxim as "Ex ijs corpora naturalia constant, in quae resolvantur," and attributes it to "Arist., 3 De coelo, cap. 3, text 35," and to "Galen, lib. 1. c.3 de elementis." In Bekker notation the Aristotelian passage cited is *De caelo* 3 302a15-18.

14) Thomas Erastus, *Explicatio quaestionis famosae illius, utrum ex metallis ignobilioribus aurum verum & naturale arte conflari possit,* (appendix to *Disputationes de nova Philippi Paracelsi medicina* [Basel, 1572]) 35. For parallel passages in Erastus's *Disputationes de nova Philippi Paracelsi medicina*, see Pagel, *Paracelsus*, 320. See also Allen G. Debus, *The Chemical Philosophy: Paracelsian Science and Medicine in the Sixteenth and Seventeenth Centuries* (New York, 1977), 1:133-134

about the specific elemental constituents of more complex materials based on analysis. The products of analysis were themselves capable of transmutation, making the whole process unreliable at best. In order to pass beyond this impasse, it was necessary to grant some degree of fixity to the end-products of analysis.

This is precisely the move that Aristotle—if the author of *Meteorology* IV really is identical to the Stagirite—makes in chapter seven of that work.[15] When *Meteorology* IV speaks of blood and milk, both professedly homoeomerous materials that can be separated into solid and liquid components, the text (at IV 7 384a16-33) attributes the coagulation of milk to a pre-existent earthy component (γεῶδες) and that of the blood to the presence of fibers (ἶνες) within the blood.[16] In both cases, the coagulation of the liquid is attributed to the separation of these pre-existent solid ingredients from a watery component rather than being explained by transmutation. The congealment of blood is effected when its fibers congregate and its serous liquid is allowed to mass together or pass off. In the case of milk, furthermore, the separation into whey and cheese is something that physicians intentionally do "when they curdle it" (οἷον οἱ ἰατροὶ ὀπίζοντες) by adding rennet and heat.

---

15) My thesis is not at all dependent on the authenticity of *Meteorology* IV as a work by Aristotle. Since medieval and early modern authors routinely viewed the work as authentic, faithfulness to its views was as "Aristotelian" as adherence to any other part of the Stagirite's corpus. For the modern question concerning the genuineness of *Meteorology* IV, see Aristotle, *Meteorology*, ed. and tr. H.D.P. Lee (Cambridge, MA, 1952), xiii-xxi; David Furley, "The Mechanics of Metorologica IV. A Prolegomenon to Biology," in *Zweifelhaftes im Corpus Aristotelicum*, eds. Paul Moraux and Jürgen Wiesner (Berlin, 1983), 73-93; Pierre Louis, *Aristote: Météorologiques* (Paris, 1982), xii-xv. But see H. Strohm, "Beobachtungen zum vierten Buch der aristotelischen Meteorologie," in *Zweifelhaftes im Corpus Aristotelicum*, eds. Moraux and Wiesner, 94-115. Strohm considers *Meteors IV* to be a "Bearbeitung." A fairly recent *status quaestionis* may be found in Carmela Baffioni, *Il IV Libro dei "Meteorologica" di Aristotele* (Naples, 1981), 34-44.

16) For the ἶνες in blood, see also *Meteorology* IV 10 389a19-23. It is very difficult to see how *Meteorology* IV's "fibers" can be present in blood if the latter is construed as a strictly homoeomerous mixture in the sense that Aristotle presents in *De generatione et corruptione* I 10 328a 10-12. The resolution of this problem cannot form part of the present paper, but one obvious likelihood is that *Meteorology* IV is using the term "homoeomerous" in a less strict sense than that of *De generatione et corruptione*.

The author also demonstrates the fact that the whey left behind after the coagulation of the cheese is of a watery rather than an earthy nature by showing that the separated whey cannot be thickened further by heat, but merely boils away. It seems that *Meteorology* IV 7 20-24 has in mind a process of induced analysis, explicitly invoked to show that milk, despite being a homoeomerous substance, contains watery whey and earthy cheese within itself before their separation.

Although it may be difficult to square *Meteorology* IV's rather loose concept of homoeomerity with the stricter notion expressed in *De generatione et corruptione*, the idea of a homoeomerous material that could still contain robust and distinct ingredients found a solid welcome in medieval alchemy. In Geber's *Summa perfectionis*, probably composed at the end of the thirteenth century, the four elements are treated as minute particles that come together in a "very strong union" (*fortissima compositio*) to make up the two principles of metals, mercury and sulfur. These in turn retain their robust being within metallic corpuscles, while the metallic particles themselves are also hard or in some cases impossible to decompose. Thus Geber erected a hierarchical corpuscular system, with elements combining to form particles of mercury and sulfur, and the latter coming together to make up corpuscles of the different metals. The same idea, with modifications, can be found later in the works of Sennert, Boyle, and Stahl, to name only some obvious representatives of early modern chymistry.[17]

The *Summa perfectionis* also considerably develops *Meteorology* IV's attempts to assay the components of materials by means of analysis. Whereas *Meteorology* IV speaks of rather humble tests such as boiling milk with rennet, Geber employs the full panoply of the medieval alchemical laboratory, with its assortment of stills, sublimatories, crucibles, cupels, and specialized furnaces. Geber's approach

---

17) See Newman, *Atoms and Alchemy*, chapter one, for Geber's hierarchical matter theory. Ursula Klein has questioned my reading of Geber as a proponent of compositional hierarchy, in her essay review, "Styles of Experimentation and Alchemical Matter Theory in the Scientific Revolution," *Metascience* 16 (2007), 247–256. See my forthcoming response, "Alchemical Atoms or Artisanal 'Building Blocks'? A Response to Klein," to appear in *Perspectives on Science*.

was marked by the extensive application of laboratory analysis to the determination of underlying material principles. For example, the *Summa perfectionis* explicitly states that the principles of the metals, mercury and sulfur, are known to be homogeneous because they cannot be resolved into their components by means of sublimation in a flask.[18] As Geber says of sulfur, "[it] has a very strong composition, and is likewise uniform and homoeomerous in its particles, because it is homogeneous. Thus its oil is not borne away from it by distillation, as it is from other things having oil."[19] The negative-empirical concept is the same here as in Sennert and Boyle. What is different is merely the tool of analysis—writing before the discovery of the mineral acids, Geber's attempt to analyze sulfur relied on the sublimation vessel as his analytical tool rather than those much more powerful chemical solvents. Hence if we accept the term "atomism" for Sennert and the early Boyle, we must also extend this terminology backwards to include the tradition of Geberian alchemy.

To summarize what we have detailed so far, the tradition of chymical atomism was not wedded to the absolute homoeomerity of *De generatione et corruptione,* but its inspiration was ultimately no less Aristotelian for that. Following the approach of *Meteorology* IV, Geber and his heirs retained a modified concept of homoeomerity. In the *Summa perfectionis* and its heirs, materials were deemed to be of one unified nature in this modified sense if they passed the test of laboratory analysis. Whether the test involved a mere analysis by fire or a more impressive dissolution in one or more of the

---

18) Newman, *Pseudo-Geber*, 663 (Latin, 323): "each of these [principles] *in genere* is of very strong composition and uniform substance. This is so because the particles of earth are united through the smallest particles [*per minima*] to the aerial, watery, and fiery particles in such a way that none of them can separate from the other during their resolution. But each is resolved with the other on account of the strong union that they mutually have received through the smallest [*per minima*]...." See also 646 (Latin, 280-281): "We see a manifest example (*experientia*) of this in the sublimation of spirits. For when a sudden resolution comes about in them by means of sublimation, the humid is not separated from the dry, nor the dry from the humid so that they be totally divided into the parts of their mixture. Instead, their whole substance ascends, or only a little of their components is disunited."

19) Newman, *Pseudo-Geber*, 666 (Latin, 328).

mineral acids, the absence of heterogeneities at the end of the test determined whether the material undergoing examination was fundamental or not. The early modern chymists following in the tradition of Geber accepted that the matter that resisted division by laboratory analysis was corpuscular in structure, whether they employed the traditional term *partes* for the particles or adopted the language of Democritus and Leucippus, as Sennert and the young Boyle did. In short, this was a tradition of operational atomism, much like that espoused by chemists in the nineteenth century, with the obvious caveat that earlier chymists had no concept of atomic weight. [20] We are therefore left with an interesting question. What was the relationship between the chymical atomism that preceded Lavoisian chemistry and the matter theory to which alchemy gave way? Can we detect a measure of continuity between the two, or must we continue in traditional wise to view the Chemical Revolution as a complete rupture with the past? There is perhaps no better way to answer this question than to examine the work of Antoine Laurent Lavoisier himself.

A Lavoisian Coda

In this brief paper, I cannot do more than make some hand-waving gestures to the chemistry of the late eighteenth century. Nonetheless, approaching Lavoisier with a clear image of chymical atomism in mind will cast a very revealing light on his theory of matter. Lavoisier, of course, is renowned as a debunker of the four Aristotelian elements and of that putative principle of flammability, phlogiston. Among historians of chemistry, Lavoisier is also famous for basing his elemental theory firmly on the negative-empirical

[20] Alan Rocke's important book *Chemical Atomism in the Nineteenth Century: From Dalton to Cannizaro* (Columbus, 1984) rightly emphasizes the importance of atomic weight for the development of nineteenth-century chemical atomism. Obviously this factor distinguishes the chemistry of John Dalton and his heirs from the chymistry of the seventeenth century, but the operational consideration of an atom as that which resists laboratory analysis remains the same. See Rocke, *Chemical Atomism*, 1-15. See 10-14 for Rocke's definition of chemical atomism, 21-47 for his treatment of Dalton, and 66-78 for Berzelius.

principle. It was chemical analysis that provided Lavoisier with his new system of operationally indecomposable elements that would displace the old transmutable ones of the scholastics. Intent on distancing himself from the fanciful speculations of the Cartesians with their corpuscles equipped with hooks, eyes and other contrivances, Lavoisier explicitly contrasted his approach to that of microstructural theorists, as in this passage from his 1789 *Traité élémentaire de chimie*:

> All that one can say about the number and nature of the elements is limited, according to me, to purely metaphysical discussions. These are indeterminate problems that one attempts to solve, which are capable of an infinity of solutions, but it is very probable that no particular one is in agreement with nature. Thus I will content myself with saying that if by the name "elements" we mean to designate the *molécules simples et indivisibles* that make up bodies, it is probable that we do not know them. But if, on the contrary, we attach to the name "elements" or "principles" of bodies the idea of the final limit that analysis reaches, all the substances that we have not yet decomposed are for us elements.[21]

Two things immediately stand out here—first, Lavoisier wants to avoid the speculative approach to material composition that he refers to as "metaphysical," which results in unwarranted claims about simplicity and indivisibility. Second, his means of avoiding speculation is to base his compositional assertions on chemical *decomposition*,

[21] Antoine Laurent Lavoisier, *Traité élémentaire de chimie* (Paris, 1864), 7 (accessed electronically on 27 September 2008 via *Les oeuvres de Lavoisier*, at http://www.lavoisier.cnrs.fr/): "Tout ce qu'on peut dire sur le nombre et sur la nature des éléments se borne, suivant moi, à des discussions purement métaphysiques: ce sont des problèmes indéterminés qu'on se propose de résoudre, qui sont susceptibles d'une infinité de solutions, mais dont il est très-probable qu'aucune en particulier n'est d'accord avec la nature. Je me contenterai donc de dire que, si par le nom d'éléments nous entendons désigner les molécules simples et indivisibles qui composent les corps, il est probable que nous ne les connaissons pas: que, si, au contraire, nous attachons au nom d'éléments ou de principes des corps l'idée du dernier terme auquel parvient l'analyse, toutes les substances que nous n'avons encore pu décomposer par aucun moyen sont pour nous des éléments." The translation is my own, since the translation by Robert Kerr is too free for scholarly usage. Kerr, for example, translates "molécule" as "atom." See Antoine Laurent Lavoisier, *Elements of Chemistry*, tr. Robert Kerr, ed. Douglas McKie (New York, 1965), xxiv.

which he says will lead to "the final limit that analysis reaches," namely the elements. The unwary reader may come away from this with the idea that Lavoisier was opposed to *any* discussion of *molécules.* But in fact what Lavoisier really does not want to countenance is the claim of *absolute* simplicity and indivisibility. Later in the *Traité élémentaire* he explicitly affirms that chemical agents, unlike physical ones, divide a body into its *molécules primitives et élémentaires.*[22]

It is worth looking at Lavoisier's use of such terms as *molécules primitives* more carefully. The term *molécule,* of course, is derived from the diminutive of the Latin *moles*, or mass (in Latin, a *molecula*). Thus a *molécule* in an unqualified sense is simply a small mass of matter in Lavoisier's language, not a modern chemical molecule. Frequently Lavoisier uses the term *partie* to describe the same thing, namely a small bit of matter.[23] When speaking of the distinction between physical grinding and chemical dissolution, Lavoisier has the following to say about *molécules*:

> Trituration, porphyrization [i.e. grinding on porphyry], and pulverizing, properly speaking are only preliminary mechanical operations whose object is to divide and separate the *molécules* of bodies, and to reduce them into very fine particles. But however far one may push these operations, they can never resolve a body into its *molécules primitives et élémentaires*: properly speaking they do not even break up their aggregation. Hence each *molécule*, after trituration and porphyrization, still forms a whole similar to the original mass that one intended to divide, unlike the case in truly chemical operations, such as dissolution, which destroys the aggregation of the body and separates the *molécules constitutives et intégrantes* composing it from one another.[24]

---

22) In addition to the passage quoted in the text of my article, Lavoisier expresses much the same idea at *Traité élémentaire* (Paris, 1864), Troisième Partie, 305.

23) See for example, Lavoisier, *Traité élémentaire*, 6, 104, 121, 367, 415 for *parties constituantes* and 27, 143 for *molécules constituantes*—the two terms are used interchangeably.

24) Lavoisier, *Traité élémentaire de chimie*, 294: "La trituration, la porphyrisation et la pulvérisation ne sont, à proprement parler, que des opérations mécaniques préliminaires, dont l'objet est de diviser, de séparer les molécules des corps, et de les réduire en particules très fines. Mais, quelque loin qu'on puisse porter ces opérations, elles ne peuvent jamais résoudre un corps en ses molécules primitives et élémentaires: elles ne rompent pas même, à proprement parler, son agrégation; en sorte que chaque

Here Lavoisier uses *molécule* in two senses—first as a unit of mechanical decomposition in the case of grinding, and second, as a unit of chemical decomposition. In the case of mechanical decomposition, the size of the finest particles depends on the force that one applies to the grinding. As Lavoisier says elsewhere, a heavier pestle falling from a fixed height may grind the matter to a fineness more subtle than that attained by a lighter pestle, but in any event the resulting particles will be huge when compared to those arrived at by chemical decomposition. In the latter case, unlike that of physical grinding, Lavoisier is clearly thinking of the *molécule* as a fixed terminus of decomposition. The *molécules primitives et élémentaires* are simply the smallest bits of matter that Lavoisier's eighteenth-century chemical technology has allowed him to attain—they are "the final limit that analysis reaches."

From our discussion so far, it is clear that Lavoisier is treating "elements" as the smallest bits of matter into which a substance can be divided by currently available means of analysis. To divide the *molécules primitives et élémentaires* still further would mean splitting them into their own elemental constituents.[25] Lavoisier's approach is readily identifiable as an offshoot of the chymical atomism to which this paper has been devoted. The identification of "elements" with chemical bits that are operationally indecomposable—whether the latter be named "atoms," "parts," or "molecules"—is in full accord with an alchemical tradition that was already flourishing in the High Middle Ages, whose own roots lay in the Aristotelian tradition of *Meteorology* IV. Lavoisier's language of *molécules* or *parties* that are *constitutives et intégrantes* and whose existence is revealed by analysis derives directly from the tradition of chymical corpuscular theory espoused by George Ernst Stahl and his followers, a tradition that

molécule, après la trituration et la porphyrisation, forme encore un tout semblable à la masse originaire qu'on avait eu pour objet de diviser, à la différence des opérations vraiment chimiques, telles, par exemple, que la dissolution, qui détruit l'agrégation du corps, et écarte les unes des autres les molécules constitutives et intégrantes qui le composent."

25) From Lavoisier's perspective, the chemist of course does not know whether the *molécules primitives et élémentaires* that he has arrived at by analysis are themselves simple or not. All that he can say is that they represent the limits of the analytical tools currently available.

in turn drew both on Boyle and on older alchemical sources.[26] Lavoisier's reliance on laboratory analysis to arrive at the *molécules primitives* of bodies therefore belongs to a chymical tradition extending back through Stahl, Boyle, and others well into the Middle Ages.

Conclusion

To conclude, let me reaffirm the distinction between what I am calling chymical atomism and other atomic or corpuscular theories. What I view as having been most unusual about chymical atomism was its reliance on the negative-empirical principle as applied to laboratory analysis for determining the material constituents of bodies, its demonstrative use of paired analysis and synthesis, as in the case of the reduction to the pristine state, and its emphasis on particles that retain their substantial identity rather than deriving all of their macro-level properties from the structural characteristics such as size, shape, and motion insisted upon by the stricter mechanical philosophers. This is not to say that chymical atomism made no use of structural explanation, of course. To the contrary, writers from Geber to Sennert extensively employ the relative sizes and distribution of pores and particles to explain a host of phenomenal properties. But what they do not do, by and large, is to speculate about the shapes of their micro-level corpuscles in the Cartesian fashion, or to engage in an explicitly reductionist program of leading secondary qualities back to primary ones. Instead, they employ a sort of theoretical parsimony. In Lavoisier's case, this parsimony is an explicit product of choice, but I tender that it was a deliberate move on the part of many earlier authors as well, even when it is not expressed as an outright credo. When modern historians of science and philosophy treat this sort of atomism as "weak," theoretically underdeveloped, vague, or confused, they are simply missing the point.[27] It was precisely the obstinate refusal to engage in

[26] See the important article by J.B. Gough, "Lavoisier and the Fulfillment of the Stahlian Revolution," *Osiris*, second series 4 (1988), 15-33, especially 23-29.

[27] I have in mind especially Klein, "Styles of Experimentation," 251 *et passim*.

excessive microstructural theorizing that allowed Lavoisier and his heirs to focus on the atom (or *molécule primitive et élémentaire*) as the end-point of analysis. As such, the chymical atom, and its nineteenth century heir, the chemical atom, was an operationally indestructible bit of matter endowed with determinate chemical properties, not a putative entity composed of uniform catholic matter formed into shapes corresponding to those of macro-level machines. When chymical authors *do* make those moves, as Robert Boyle tentatively did in his maturity, it is usually the product of fusing different traditions, as when Boyle combined parts of the Cartesian and Gassendist programs with his Baconian and Sennertian roots. The result, especially in Boyle, can be an uneasy marriage of structure and essence, the latter often disguised under the rubric of "seminal principles" which come in by the back door after having been denied entrance at the front. As Lavoisier showed, there were genuine benefits to limiting one's assertions to the fruits of analysis, an approach in which he was, however, a belated and unwitting heir of the aurific art.

Conjecture, Probabilism, and Provisional Knowledge in Renaissance Meteorology

Craig Martin*
Oakland University

Abstract
For Renaissance Aristotelian natural philosophers, ideally knowledge was certain and based on syllogistic demonstration. Many Italian scholars, such as Agostino Nifo, Pietro Pomponazzi, and Niccolò Cabeo, considered this ideal as inapplicable to the field of meteorology. Rather, because of the accidental nature of meteorological phenomena and the inherent irregularity of the weather, they believed that causal explanations of meteorology were largely conjectural, provisional, and probabilistic. Several of these natural philosophers applied the standard of "saving the appearances" to the field of meteorology because of the difficulties involved in making accurate observations. This lower epistemological standard contributed to the willingness of Aristotelians to revise meteorological theories and deviate from Aristotle's own positions.

Keywords
meteorology, imperfect mixtures, Pietro Pomponazzi (1462-1525), earthquakes, demonstration, conjecture, dialectic, probabilism, "saving the appearances," Aristotelian natural philosophy, Renaissance

Introduction

The two dominating views of Renaissance Aristotelian epistemology or "scientific methodology" entail positions that are incompatible,

---

* History Department, Oakland University, 372 O'Dowd Hall, Rochester, MI 48309, U.S.A. (martin@oakland.edu). I thank Matt Klemm, Bill Newman, Darrel Rutkin, and Edith Sylla for their advice, comments, and corrections. Research and writing of this article was supported by a Gladys Krieble Delmas Foundation Fellowship, an Oakland University URC Faculty Fellowship, and a Dibner History of Science Program Fellowship at the Huntington Library.

if not contradictory. The first view holds that Aristotelians believed that their knowledge was complete and certain. Holders of this position believe that Aristotle, and his later medieval and Renaissance commentators, equated natural philosophical knowledge with the logical methods found in the analytics, that used syllogism for a model of unequivocal demonstration. These accounts begin with analyses of Aristotle,[1] extend from Aristotle's supposed view to his scholastic followers,[2] and find that the rejection of belief in the certainty of knowledge of the natural world to be crucial to the Scientific Revolution.[3] Thus according to this view, seventeenth-century English experimentalists, in a climate of latitudinarianism, overturned their predecessors' epistemology that used the demonstrative sciences as a model "to attain to the kind of certainty that compelled absolute assent."[4] Moreover, the insistence on certainty purportedly endorsed by Aristotelian natural philosophers is considered opposed not just to the new natural philosophies of the seventeenth century but also to the rhetorical traditions of Renaissance humanists that emphasized observation and contingent knowledge.[5]

---

1) James Franklin, *The Science of Conjecture* (Baltimore, 2001), 166: "Aristotle argued for a rational search for certain causes in medicine as everywhere else."

2) Peter Dear, *Revolutionizing the Sciences: European Knowledge and Its Ambitions, 1500-1700* (Princeton, 2001), 5-6: "The lure of demonstrative certainty drew scholastic natural philosophers to believe that they could make knowledge that was analytically solid." Richard G. Olson, *Science and Religion, 1450-1900: From Copernicus to Darwin* (Baltimore, 2004), 90: "According to most Christian Aristotelian philosophers including Thomas Aquinas, one should demand absolute certainty of any religious or natural scientific statement that commands assent."

3) James R. Jacob, *The Scientific Revolution: Aspirations and Achievements, 1500-1700* (Atlantic Highlands, NJ, 1998), xiii: "The shift from an Aristotelian theory of knowledge, which was confident of the truth of what we perceive, to a modern skepticism that doubts our capacity to know truth but nonetheless finds the resources, intellectual and cultural to overcome pessimism and to insist that scientific knowledge is still attainable … ."

4) Steven Shapin and Simon Schaffer, *Leviathan and the Air-Pump: Hobbes, Boyle, and the Experimental Life* (Princeton, 1985), 23-24

5) Barbara Shapiro, *Probability and Certainty in Seventeenth-Century England: A Study of the Relationships between Natural Science, Religion, History, Law, and Literature* (Princeton, 1983), 3-14.

The second view, which stems from the groundbreaking work of first Ernst Cassirer and then John Herman Randall, Jr., stresses the importance of the method called *regressus*, which combined new observationally based discovery of causes (*demonstratio quia* or *methodus resolutiva*) with the certainty of syllogism in producing theoretical explanations (*demonstratio propter quid* or *methodus compositiva*).[6] Using this model of method, natural signs, experience, and observations lead to the hypothetical determination of the causes of effects, by what Averroes termed *demonstratio signi* or "demonstrations from signs." Then by an intellective *negotiatio*, confidence in the hypothetical causes increases. Finally, the causes can then be utilized to explain the effects in a certain demonstration. Thus, *regressus* method allows for the possibility of change and progress in natural philosophical knowledge and a place for hypothetical understandings of nature without rejecting the possibility of syllogistic demonstration. The possibility of change means that the results are not certain but rather provisional, especially for knowledge that has resulted only from the first stage of the procedure, the *demonstratio signi*.

While the combination of all three parts of the *regressus* was intended to bring about certainty or near certainty, the realization of this goal was not always possible because of the contingent, i.e., accidental, nature or difficulty of the subject. Paolo Palmieri notes for example that Giacomo Zabarella, one of the leading proponents of and contributors to the *regressus* theory, did not apply the theory in his textbook, *De rebus naturalibus*.[7] Zabarella, however, and many other Renaissance natural philosophers followed the first stage of the *regressus* in their discussions of meteorology, a field that recommended itself to *demonstratio signi* because of Aristotle's explicit use of signs in his own considerations of meteorological phenomena. The first stage of the *regressus* was not taken to be definitive or certain but rather was provisional. Agostino Nifo, a natural philosopher

6) Ernst Cassirer, *Das Erkenntnisproblem in der neueren Zeit* (Berlin, 1922), 136-144; John Herman Randall, Jr., "The Development of Scientific Method in the School of Padua," *Journal of the History of Ideas*, 1 (1940), 177–206; John Herman Randall, Jr., *The School of Padua and the Emergence of Modern Science* (Padua, 1961).

7) Paolo Palmieri, "Science and Authority in Giacomo Zabarella," *History of Science* 45 (2007), 404-427, esp. 418.

who wrote on logic as well as meteorology, described the discovery of causes through natural effects, a method he took to be typical of natural philosophy, as "conjectural."[8]

An examination of Italian Renaissance Aristotelian meteorology, from Agostino Nifo to Niccolò Cabeo, demonstrates that Aristotelians saw many theories about nature as provisional and at times instrumental. Just as Robert Boyle and other later experimentalists are said to have believed, they thought that their "physical hypotheses were provisional and revisable" in addition to being "removed from the realm of the demonstrative."[9] Thinkers including Nifo, Pietro Pomponazzi, and Lodovico Boccadiferro applied the ideal of "saving the appearances" not just to astronomical theory but also to explanations of changes within the sublunary region. While epistemic goals and standards varied among thinkers, a number of them consciously followed at least the first stage of the *regressus* theory, the demonstration of the fact, in their attempts to establish hypothetical understandings of the causes of weather. Many were fully aware of the uncertainty of their findings and enunciated the provisional nature of their theories. This does not mean, however, that they were anticipating or applying the modern scientific method or Baconian induction. Rather their views were firmly based in Renaissance ideals of demonstration and dialectic based on experience, Aristotelian textual analysis, and the utilization of signs.

Aristotle's Meteorology

Aristotelian meteorology, as opposed to astro-meteorology, was primarily causal in its goals, not predictive. For Aristotle, knowledge of the causes of many meteorological subjects was based not on syllogistic deduction but was the result of interpreting signs that helped to confirm the hypothesis that meteorological phenomena result from two exhalations that circulate in the sublunary region.

---

8) Agostino Nifo, *Expositio super octo Aristotelis Stagiritae libros de physico auditu,* (Venice, 1552), fol. 6v.

9) Shapin and Schaffer, *Leviathan and the Air-Pump*, 24.

Aristotle himself was aware of the impossibility of a unified method for all learning, and proposed a prohibition on μετάβασις, the use of content from one field to explain another.[10] Thus, he contended that while mathematics was appropriate for discussions of astronomy and optics, in other messier fields, the objects of science could not be known through such tools. In the *Nicomachean Ethics*, Aristotle specified that all fields of philosophy cannot be expected to achieve the same level of precision (τὸ ἀκριβής). Rather, the degree of certainty is determined, at least partially, by the nature of the subjects under scrutiny. Thus fields that rely on opinion and uncertain premises can at best provide probable conclusions and approximations to the truth (τύπῳ τἀληθὲς).[11]

While in the *Ethics* Aristotle urged investigators to expect degrees of exactness proportional to the subject at hand, in the *Topics* he detailed two kinds of logical proof that differ in terms of certainty: demonstration (ἡ ἀπόδειξις) and dialectic. Demonstration, the more certain of the two, is a deductive argument based on "true and primitive" premises that are convincing in and of themselves; dialectic starts with reputable opinions, that is, opinions that if not acceptable to everyone at least are endorsed by most wise people. Demonstration provided *scientia*, proven knowledge, while dialectic yielded verisimilitude. In the words of recent commentators, dialectic "could offer only conjectural premises and probable conclusions," no matter how accurate or truthful they seemed.[12] Thus Renaissance thinkers typically understood dialectic as a discussion of probability, that is, a consideration of likelihoods. By the sixteenth century, scholars had added a third category of argument, namely persuasion, which was rhetorical in its nature and even less certain than dialectic. Acknowledgement of these three categories, demonstration, probability, and persuasion, was so widespread that it was common to, in the words of Richard Serjeantson, "every natural

[10] Aristotle, *Posterior Analytics*, 75a38-75b6.

[11] Aristotle, *Nicomachean Ethics*, 1094b13-27.

[12] Jean Dietz Moss and William A. Wallace, *Rhetoric and Dialectic in the Time of Galileo* (Washington, D.C., 2003), 16.

philosopher educated to any level beyond that of rudimentary Latin grammar."[13]

While recognition of the categories of demonstration, probability, and persuasion was nearly universal, it was less clear which category should be applied to which subjects. Meteorology, however, was a particularly good candidate for dialectical or probable arguments for a number of reasons. For one, Aristotle's approach was dialectical throughout the *Meteorology*. He considered and largely rejected the opinions (τὰ ἔνδοξα) of the Presocratics in addition to admitting the relative strengths of the adopted positions.[14] Second, some Renaissance Thomists believed that certain knowledge (*scientia*) was restricted to metaphysical topics while opinion (*opinio*) applied to the physical world, and thus also to meteorology.[15] The lower epistemic standard was used for natural knowledge in part because the regularity of nature is only "for the most part," according to Aristotle. In particular, matter and its accidents, as opposed to substances, are contingent, that is, are not necessary, accidental, and "capable of being otherwise than as it for the most part is."[16] Aristotle went on to explain in the *Posterior Analytics* that non-essential attributes, or accidental qualities, because of the absence of their necessity, are not subject to necessary proofs or demonstrative knowledge (ἐπιστήμη ἀποδεικτική).[17] The contention that the non-essential does not admit to demonstration but rather to dialectic or probable argumentation is crucial to Renaissance meteorology because of the widely-held belief that most meteorological phenomena involve not substances or essential natures *per se* but rather imperfect mixtures of the elements. If meteorology deals with

---

13) R.W. Serjeantson, "Proof and Persuasion," in *The Cambridge History of Science: Volume 3 Early Modern Science*, ed. Lorraine Daston and Katharine Park (Cambridge, 2006), 139-140.

14) Cynthia Freeland, "Scientific Explanation and Empirical Data in Aristotle's *Meteorology*," *Oxford Studies in Ancient Philosophy* 8 (1990), 62-107.

15) Serjeantson, "Proof and Persuasion," 139.

16) Aristotle, *Metaphysics*, VI.2, 1027a12, Trans. W.D. Ross.

17) Aristotle, *Posterior Analytics*, I.6, 75a19-22.

accidental formations of the elements then it stands to reason that any understanding of the field is only probable.[18]

Aristotle realized the difficulty of making comprehensive and accurate determinations of meteorology. According to Aristotle, meteorology strives to explain changes in the sublunary world, including the area above the earth as well as below. The key word is "strives." The phenomena of this region are natural, but necessarily irregular, particularly in comparison to the celestial region.[19] Because of their natural irregularity, a limited confidence in the ability to give true explanations is called for. Aristotle wrote: "Of all these phenomena, some we find inexplicable, others we can to some extent understand."[20] Gaining a good account of meteorological phenomena is hampered not only by their irregularity but also by their inaccessibility to the senses. Before giving his explanation for comets, Aristotle, in a statement that can reasonably be thought to apply to large portions of the *Meteorology*, wrote: "We consider that we have given a sufficiently rational explanation of things inaccessible to observation by our senses if we have produced a theory (λόγος) that is possible."[21] Thus the goal was to provide a "possible" account rather than a "certain" one.

Matter Theory, Uncertainty, and Meteorology

A number of Renaissance treatments of Aristotle's *Meteorology* not only accepted his claims that meteorological knowledge is limited but used matter theory to explain why. These explanations were based on two principles: 1) meteorological phenomena result from imperfect mixtures, which do not possess substantial forms; and 2) the matter of the terrestrial or sublunary realm is unstable and relatively unknowable.

---

18) For similar arguments as applied in Renaissance medicine see Ian Maclean, *Logic, Signs and Nature* (Cambridge, 2002), 132.

19) Aristotle, *Meteorology*, 338b1-3.

20) Ibid., 339a3-4. Tr. H.D.P. Lee (Cambridge, MA, 1952), 5.

21) Ibid., 344a5-7. Tr. Lee, 49.

The idea that the first three books of the *Meteorology* treated imperfect mixture goes back at least to Albertus Magnus who described the subject of these books as phenomena that are in the state of becoming a simple substance.[22] John Buridan was one of the earliest to use the term "imperfect mixtures" to categorize meteorological substrata in contrast to "perfect mixtures" such as flesh, blood, milk, and metals, for which *Meteorology* IV gives an account.[23] Renaissance Aristotelians followed Buridan's phrasing almost uniformly. The acceptance of meteorological substrata as "imperfect" contributed to the idea that meteorological knowledge is probable. If meteorology considers objects that are without their own substantial forms, knowledge of formal causes will be limited. Indeed meteorology most often referred to the forms of the elements that composed the two exhalations, which were the material and efficient causes of meteorological effects. Francesco Piccolomini, a professor at Padua during the last decades of the sixteenth century, explicitly expressed this view: "The form that is a substance is not properly suited to meteors, because they are imperfect mixtures, which are of such a type, that they will not create new forms; therefore their substantial form is not distinct from those of the elements."[24] As a result, scholars were limited in the kinds of formal and final causation that could be used to understand meteorology.

Although Renaissance commentators did not uniformly claim there were no formal or final causes for meteorological phenomena, the rejection of the existence of those causes was frequent. For example, Francesco Vimercati and Jacob Schegk specifically argued that the two causes of meteorological phenomena are efficient and material causes; and the Coimbrans' commentary contended that "meteors" do not have their own formal causes and made little mention

---

22) Albertus Magnus, *Liber quartus meteororum*, in *Opera*, ed. A. Borgnet (Paris, 1890), IV: 705.

23) John Buridan, *Expositio libri meteororum*, Ms. Vaticana lat. 2162, fol. 103r.

24) Francesco Piccolomini, *Librorum ad scientiam de natura attinentium pars quarta. In qua Meteorologica explicantur et connexa cum eis* (Venice, 1596), fol. 4v "Forma, quae Substantia sit, proprie Metheoris non competit, quia sunt mixta imperfecta, quae huius sunt conditionis, ut nova non prodierint, ideo forma Substantialis non est distincta ab ea Elementorum."

of final causes.[25] Effectively, the idea that the end of a substance was the realization of its form was ruled out, leaving only external final causes available for meteorology for those who believed that final causes do indeed exist for meteorology. The character of meteorological phenomena thus eliminated the possibility of a deep knowledge of their formal and final causes, the two most privileged types of causes for Aristotelians.

The Renaissance conception of the imperfection of meteorological phenomena was an amalgam of the Aristotelian ideas of potency and act and Platonic views of matter, that had held currency throughout the Middle Ages. In the *Timaeus*, Plato contrasted the matter of the sublunary world with the forms. Forms are perfect, unchanging, and the source of knowledge. To the contrary matter is the source of imperfection in the world because of its instability. Material necessity is the "errant" cause.[26] As a result, understandings of the material world are necessarily as transitory as its contents, and the *Timaeus* puts forth only a "likely" explanation, which is the best we can hope for.[27] Plato's belief that matter is the source of uncertainty or natural imperfection ran through the Augustinian tradition and emerged in other *loci* of medieval and Renaissance scholasticism. Averroes in his commentary on the *Posterior Analytics* wrote, "matter is the reason that that which is *per accidens* is found in the sciences. And what is *per accidens* is a great distance from truth … Therefore, it is necessary that the matter is less true."[28] Thomas Aquinas expressed a similar view in his commentaries on Aristotelian logic, where he contended that uncertainty is correlative

---

25) Jacob Schegk, *In reliquos naturalium Aristotelis libros commentaria* (Basel, 1550), 335; Francesco Vimercati, *In quatuor libros Aristotelis Meteorologicorum commentarii* (Paris, 1556), 18; Collegium Conimbricense, *In libros Meteororum Aristotelis Stagiritae* (Cologne, 1603), 5.

26) Plato, *Timaeus*, 47e-48e.

27) Plato, *Timaeus*, 29b-c.

28) Averroes, *In librum Aristotelis de demonstratione maxima expositio*, in *Aristotelis Opera cum Averrois commentariis* (Venice, 1572-1576 ; reprt. Frankfurt am Main, 1962), I, 2, fol. 375r: "materia est causa, quod inveniatur in scientiis id, quod est per accidens: et quod est per accidens est magna elongatio a veritate … oportet igitur ut materia sit minus vera … ."

to matter's instability: "Uncertainty is caused by the transmutability of sensible matter; thus however much it approaches it [matter], knowledge (*scientia*) is less certain."[29] These views were influential in logical works even up to the 1590s, when Zabarella cited them to give authority to his similar position.[30] The practical ramifications of this Platonic conception of matter, however, are evident not in the works on logic and method but in meteorological discourse, because of its overwhelming concern with material causes.

Simone Porzio, a professor at Pisa in the 1540s and 50s best known for his defense of the materiality of the human intellect,[31] gave an account of the epistemological basis for the natural sciences in his treatise, *De rerum naturalium principiis*. In the section that considers meteorology, Porzio considered the immediate efficient causes of simple and composite substances. Simple substances can be known according to a formula (*ratio*) because they are constant by nature. Difficulties, however, emerge in attempting to explain particular composites or mixtures because they arise out of matter that is indeterminate. These mixtures are accidental and lack a specific cause; their necessity is internal rather than the necessity of universal efficient causes, such as the sun. Thus, Porzio contended that natural philosophical arguments that move "from efficient causes to their effects are not the most powerful (*potissimae*) because the effects can be impeded by the indetermination of the matter, which can receive multiple and various forms."[32] For Porzio, the plasticity

---

29) Thomas Aquinas, *In Aristotelis Peri Hermeneias et Posteriorum analyticorum expositio*, ed. Raimondo M. Spiazzi (Rome,1964), lib. I, lect. 41, n. 358: "... incertitudo causatur propter transmutabilitatem materiae sensibilis; inde quanto magis acceditur ad eam, tanto scientia est minus certa."

30) Luigi Olivieri, *Certezza e gerarchi del sapere: Crisi dell'idea di scientificità nell'aristotelismo del secolo xvi* (Padua, 1983), 67-69.

31) Simone Porzio, *De humana mente, disputatio* (Florence, 1551).

32) Simone Porzio, *De rerum naturalium principiis* (Naples, 1553), fol. Riiv-Riiir: "Verum respectu aliquorum, ut puta quarundam elementorum affectionum, quas in libro Meteorum alias enarravimus, est causa particularis, hoc est, est proxima & immediata efficiens causa: ... Atque iccirco in rebus naturalibus demonstrationes a causis efficientibus ad suos effectus non sunt potissimae: quia effectus impediri queant, ob materiae indeterminationem, quae varias multiplicesque formas recipere potest."

of matter makes knowledge of composites uncertain and proofs stemming from efficient causes are less than certain. Even though Porzio's conclusion potentially applies to many fields within natural philosophy, it is especially pertinent to meteorology because of the significant role that matter holds in explanations of this field.

Saving the Appearances: Probable, Possible, and Provisional Knowledge

The lack of certainty about meteorological phenomena did not render the field hopeless. Rather honest assessments of the character of meteorological theory were needed, as was recognized well before the Renaissance. Avempace (Ibn Bājja), for example, compared the field of meteorology to mathematical astronomy, which he believed had progressed in recent years, "we have now more principles from mathematical astronomy than in any preceding time." In comparison, meteorological subjects are "difficult" because "the principles we have are not sufficient," but that should not lead us to reject natural science altogether, rather "there is no reason to give no account at all; therefore, let us give an account insofar as we are enabled to do it by the principles we have found."[33] A modest assessment of the possibility and degree of knowledge in meteorology is required until better principles are uncovered. While a critic of a number of aspects of Avempace's natural philosophy, Averroes in his meteorological works agreed concerning the limited epistemic character of meteorology. In his discussion of the Milky Way, Averroes argued that because there are doubts about the *genus* of the Milky Way, our knowledge of its causes should be considered "possible" and our understanding diminished (*cognitio diminuta*).[34] The acceptance of limitations was no hindrance to speculation about

---

33) Paul Lettinck, ed. and trans., *Aristotle's Meteorology and its Reception in the Arab World: With an Edition and translation of Ibn Suwār's Treatise on Meteorological Phenomena and Ibn Bājja's Commentary on the Meteorology* (Leiden, 1999), 399

34) Averroes, *In quatuor Meteorologicorum Aristotelis libros* in *Aristotelis Opera,* V, fol. 414r: "Accidit autem huic cognitioni in substantia Galasiae esse cognitionem diminutam: propterea quod genus illius est ignotum esse per se."

causes as long as we recognize that they are only possible; thus, Averroes offered two separate possibilities for the causes of the flickering Milky Way: the weakness of our eyes causes it because of the distance of small stars; or, a place in the skies receives and multiplies the light of the stars.

The views of Averroes resonated during the Renaissance. Pietro Pomponazzi expressed skepticism about the possibility of complete knowledge of the natural world, using our inability to understand meteorological phenomena as evidence for the limitation of human knowledge. In his *De incantationibus*, a work dedicated to giving potential explanations for strange and seemingly miraculous events, for prodigies, including fountains and statues that drip blood, or for bizarre meteorological events chronicled in histories, such as the time it rained wool, he conceded that an epistemological standard below certainty was appropriate. In the introductory pages of *De incantationibus*, he claimed that some Peripatetics used demons to explain the intractable, not only because demons are posited by "ecclesiastical decrees," but also because their presumed existence allows us to "save many phenomena."[35] The employment of demons in natural philosophy is thus instrumental, parallel to epicycles and eccentrics that save the phenomena in astronomy. While Pomponazzi rejected demons, he did not reject the idea that naturalistic explanations are meant to save the appearances. Citing Averroes' commentary on the *De caelo* and Aristotle's *Topics,* Pomponazzi argued "that in difficult and hidden matters, the answers more removed from inconveniences, and more consonant with sensations and reason, are to be better received than contrary arguments."[36] As a result, Pomponazzi's controversial claim that the miraculous events

---

35) Pietro Pomponazzi, *De naturalium effectuum causis sive de incantationibus* (Basel, 1567; reprint Hildesheim, 1970), 6. For saving the appearances in Pomponazzi see Ian Maclean, "Heterodoxy in Natural Philosophy and Medicine: Pietro Pomponazzi, Guglielmo Gratarolo, Girolamo Cardano," in *Heterodoxy in Early Modern Science and Religion,* ed. John Brooke and Ian Maclean (Oxford, 2005), 1-29, esp. 15-16; Franco Graiff, "I prodigi e l'astrologia nei commenti di Pietro Pomponazzi al *De caelo,* alla *Meteora* e al *De generatione,*" *Medioevo* 2 (1976), 331-361, esp. 331-332.

36) Pomponazzi, *De incantationibus*, 130-131. For raining wool see Pliny, *Historia naturalis*, II, 57.

recounted in scripture "can on the surface be reduced to natural causes"[37] need not be taken as an endorsement of the view that the events actually were the result of natural causes alone but rather as support of the more modest claim that natural causes can give an explanation that potentially conforms to our experiences and reasoning.

Pomponazzi further considered epistemology and the natural world in his *In libros Meteororum* (ca.1522), a work that discusses these difficult and hidden matters. In this work he contended that Aristotle at times adopted the epistemological standard of verisimilitude and employed rhetorical arguments in natural philosophy. As a result, Pomponazzi conceded that saving the appearances was an appropriate ideal for natural philosophy.

In a *dubium* dedicated to *Meteorology* 2.1, he addressed the degree of knowledge and certainty that can be ascribed to meteorological subjects as he tried to make sense of a passage of the *Meteorology* that confounded and contradicted his own experience of the natural world. In this passage, Aristotle attempted to explain the sources that create various kinds of bodies of water. He divided the kinds of bodies of fresh water into two types: standing and running. Running water, such as rivers and streams, comes from sources that are higher than the stream or river. Standing water, however, is of two types. Typically, standing water naturally comes from the accumulation of rain water and is static; the relevant examples are lakes and swamps. According to Aristotle, standing water that comes from underground sources only does so when artificially created, such as in the case of wells. This last statement, which Pomponazzi found problematic, does not appear to have been a slip of the Stagirite, as it appears twice. Aristotle wrote: "Some [standing water] springs from sources, but is always made to do so artificially (χειρόκμητα), as for instance the water in wells." And two lines later wrote: "Hence water in streams and rivers runs of its own accord (αὐτόματα), but well-water needs an artificial construction (τέχνης ἐργασομένης)."[38]

---

37) Pomponazzi, *De incantationibus,* 81.

38) Aristotle, *Meteorology,* 353b25-29. Trans. Lee, 127.

Pomponazzi found the contention that all standing springs or wells must be artificial untenable: "this is *contra experimentum*, as many from the schools have told me they have seen, and I myself have seen, many springs and natural wells."[39] After rejecting the interpretation attributed to Thomas Aquinas, which was shared by Gaetano of Thiene and "all of the Latins [he] had seen," because this view, although in agreement with Aristotle, was not in accordance with reason and experience, and after dismissing Alexander's view, which he said was also false, Pomponazzi tried to determine why Aristotle believed as he did. He speculated that perhaps "in Aristotle's country all stationary bodies of water are man-made," although in his own "regions this is not the case."[40] But his conclusion was that Aristotle's view is "probable and does not demonstrate."[41] More strongly he contended that "in my judgment Aristotle's theory is without value." The only way Pomponazzi found to make sense of Aristotle was to lower the epistemological standard that Aristotle used to reach this theory. It is not one of demonstration as found in the *Analytics* but rather closer to the standards of a rhetorical argument:

> I do not wish to seem rash for I do not say this because I wish to reprehend the words of Aristotle, yet without blaming it is conceded to me and licit to doubt about this matter. For in my view Aristotle's position is without value. I know how to respond to this position [i.e., Aristotle's] even if I do not know it [i.e., the correct position] and I would like to respond to it. Perhaps it should be said that Aristotle does not proffer this position as a demonstration and necessary proof but in order that it be some likely persuasion. For it seems that the sea could not arise from springs by the method that Aristotle posits but does not demonstrate. I should say that Aristotle did not posit it as a demonstration. [His position] is valid according to verisimilitude, not by [his] advancing [it] absolutely. For in advancing

39) Pomponazzi, *In libros Meteororum*, Ms. Biblioteca Ambrosiana R. 96 sup., fol. 49r: "Et hoc est contra experientiam; nam, ut multi ex scolaribus mihi dixerunt vidisse, et ego ipse vidi multos fontes et puteos stantes naturales naturaliter neque aliquo artis ministerio."

40) Ibid., fol. 50r: "sed forsan in patria Arist[otelis] ita erat quod omnes aquae stationariae erant manufactae. In nostris tamen regionibus non ita est."

41) Ibid., fol. 49v: "Hoc autem est probabile, et non demonstrat."

> it absolutely, it would be false. But Aristotle has assumed it not absolutely, but for the most part.[42]

Pomponazzi thus contended that Aristotle for many topics argued not absolutely but with likely persuasion or according to verisimilitude. These many topics included others discussed in the *Meteorology*, where Aristotle seems to have made the claim that he could not use demonstrative proof because of the inaccessibility and irregularity of many meteorological effects.

Pomponazzi resumed this discussion of the epistemology of meteorology in another *dubium* in which he addressed Aristotle's theory of the formation of typhoons. According to this theory typhoons result when two winds collide in a cloud. The weaker wind is thrust aside and begins to move in circular motion, just as an eddy does in water. Pomponazzi asked simply: How did Aristotle know typhoons are formed in such away "when he was neither above the clouds nor was he able to see them generated?" The answer depends on the relatively lower standard of demonstration for natural phenomena inaccessible to the senses. Citing the first book of the *Nicomachean Ethics*, where Aristotle held that one cannot expect the same level of demonstration in ethical matters as one finds in mathematics, Pomponazzi suggested "that as much should be expected by the auditor as the treated material allows." The natural world defies complete explanation, while metaphysics can admit certain demonstration:

> In natural things we cannot have demonstration *semper sic* … since no one can know these *diurna superiora*. For we are like manual workers, and God truly is as

42) Ibid., fol. 50r: "Nollem ego videri temerarius. nam haec non dico ut velim dicta Arist. reprehendere. sed tamen circa illa absque reprehensione mihi concessum et licitum est dubitare; nam meo iudicio ratio Aristo. nullius esţ valoris. Ad hanc rationem scio ego respondere nec eam novi, et [ut written above] respondere velim. forsan dicendum est, quod Ar. non protulit hanc rationem pro demonstratione et necessaria probatione, sed ut esset quaedam persuasio. Verisimile enim videtur quod non possit mare a fontibus ortum trahere. Ratione illa quam ponit Ar[istoteles] non autem demonstrat. id dicerem Aristotelem non potuisse pro demonstratione. Valet ergo a verisimili, non absolute proferendo. absolute enim proferendo falsa esset. Sed Ar[istoteles] assumpsit eam ut in pluribus non absolute."

> Architect, we can make artificial things and not natural, and in this science [i.e., meteorology] we can have demonstration and certitude in so far as we can, not however as [we can] in metaphysics.

Rather the standard for much of the impermanent natural world is "saving the appearances." Pomponazzi thus concluded that: "but because Aristotle is further from contradiction, therefore his words are tested in natural things."[43] Many discussions of the natural world and therefore discussions of meteorology are evaluated not by strict standards of syllogistic proof but rather by their consistency with experience and free from internal contradiction.

Agostino Nifo agreed with Pomponazzi, his rival and contemporary, that theories in meteorology should attempt only to save the appearances because of the indetermination of matter combined with concerns over the possibility of accurate sensation and the limits of human comprehension of natural subjects. In a short treatise on the causes of catastrophes printed in 1505, he considered and rejected the possibility of accurately predicting meteorological catastrophes through diagnoses of air because of the near infinite number of mixtures, each with a different temperament. "To explain how so many signs from mixtures occur is laborious and perhaps beyond human capabilities." Thus only an empiricism based on observation, not on matter theory, can aid in predicting future natural

---

[43] Ibid., fol. 84r: "Alia dubitatio satis trivialis: quomodo scivit Aristoteles tiphonem illo modo contrarium, cum non fuerit supra nubem, nec viderit eum generari. Notetis quod tantum petendum est ab auditore, quantum concedit materia tractata: et clare habetur primo Aethic. [i.e., *Nicomachean Ethics*] et in meth. [i.e., the *Metaphysics*]. In rebus naturalibus non possumus habere demonstrationem semper sic, ut diximus in lib. de Anima, meth., autem prius, Quoniam nemo possit scire illa diurna superiora. Nos enim sumus ut manuales, Deus vero ut Architectus, possumus enim nos facere artificialia non autem naturalia, unde in hac scientia possumus habere demonstrationem, et certitudinem eo modo quo possumus, non autem ut in meth. Sed quia Aristoteles remotior est a contradictione, ideo sua dicta in naturalibus probantur." For Pomponazzi's claims of using an uncertain method in other works see: Stefano Perfetti, "Docebo vos dubitare. Il commento inedito di Pietro Pomponazzi al *De partibus animalium* (Bologna 1521-24)," *Documenti e studi sulla tradizione filosofica medievale* 10 (1999), 439-66; Francesco Paolo Raimondi, "Pomponazzi's Criticism of Swineshead and the Decline of the Calculatory Tradition in Italy," *Physis* n.s. 37 (2000), 311-358, esp. 326.

disasters.[44] Similar doubts were common to critiques of medical diagnosis and astrological prediction, fields that potentially considered an infinite number of combinations of complexions and influences.[45] In any case, in his commentary on the *Meteorology* that was first printed in 1523, Nifo went even further, conceding that the field of meteorology is not a science in itself (*in se*) because of the irregularity of meteorological phenomena and "also because the elementary causes, from which these [meteorological] phenomena arise, act contingently and are acted upon contingently."[46] The indeterminate nature of matter coupled with the difficulty of observing many meteorological phenomena led Nifo to conclude that the standard of meteorological knowledge required that the accounts were coherent rather than certain in a strict sense.

> Because we think we have sufficiently given demonstration about effects unclear to the senses according to theory (*secundum rationem*) if we reduce those appearances to the possible, that is, to such a certainty to which what is impossible does not follow.[47]

In his treatment of Aristotle's *Physics*, Nifo, basing himself on his reading of the *Meteorology*, expanded this judgment to all of natural science, which he says is not a *scientia simpliciter* such as mathematics because "it does not treat the true causes of natural effects, but only in so far as they are possible through conjecture."[48]

---

44) Agostino Nifo, *De nostrarum calamitatum causis liber ad Oliverium Carafam* (Venice, 1505), fol. 18r: "Explicare quot e mixtionibus contingant significationes laboriosum est: et fortasse supra captum humanum: quapropter observatoribus pensitanda relinquimus."

45) For medicine, see Maclean, *Logic, Signs and Nature*, 134.

46) Agostino Nifo, *In libris Aristotelis Meteorologicis commentaria* (Venice, 1540), fol. 2r: "Tum etiam quia cause elementarie e quibus, hec proficiscuntur, contingenter agunt, et contingenter patiuntur."

47) Nifo, *In Meteorologicorum,* fol. 26v: "Quoniam autem de effectibus sensui immanifestis putamus sufficienter demonstrasse secundum rationem si ostensa de his reduxerimus ad possibilem, hoc est ad talem certitudinem, ad quam non sequantur impossibilia."

48) Agostino Nifo, *Expositio super octo Aristotelis Stagiritae libros de Physico auditu* (Venice, 1552), fol. 6v: "Dicendum, scientiam de natura non esse scientiam simpliciter,

A similar skepticism toward the possibility of certain knowledge of meteorology is found in Tiberio Russiliano's *Apologeticus adversus culcullatos* (1519), a work that applied physical causes to both the foundations of Christian dogma and to what was traditionally understood as miraculous. Russiliano, a former student of Nifo, in a *quaestio* in which he put forth the argument that according to philosophical arguments there must have been an infinite number of universal floods, considered his argument "demonstrative and unassailable," but only if his suppositions are accepted. He admitted that not all of his suppositions were necessarily true even if they were clear in themselves (*ex se patent*); rather they were conjectural, based on common agreement and sensible signs.

> The suppositions are clear of themselves; first they derive out of common agreement and experience, since signs and traces of an inundation appear in mountainous regions, such as seashells and oysters, so that we should reasonably arrive at the conjecture that when there was a universal flood it covered and surpassed all of the mountains.[49]

The key is that signs and traces, based on experience, lead to conjectures that then become the basis for premises in deductive arguments. The deductive arguments, however, are only as sound as the conjectures upon which they are based.

---

qualis est scientia mathematica, est tamen scientia propter quid: quia inventio causae, quae habetur per syllogismum coniecturalem, est propter quid effectus. Per haec delentur obiectiones, quae contra haec fieri solent: Prima quidem delentur ex eo, quia non est circulus in demonstratione, cum primus processus sit tantum syllogismus, secundus vero demonstratio propter quid. Deletur etiam secunda obiectio, quia effectus semper est notior ipsa causa in genere notitiae quia est. Nunquam enim causa potest esse ita certa quia est, sicut effectus, cuius esse est ad sensum notum. Ipsum vero quia est causae, est coniecturale, utrum tale esse coniecturale est notius ipso effectu, in genere notitiae propter quid. Nam posita inventione causae, semper scitur propter quid effectus. unde & Aristo[teles], in libro *Meteororum* concedit se non tradidisse veras causas effectuum naturalium, sed quo erat sibi possibile coniecturabiliter."

49) Tiberio Russiliano, *Apologeticus adversus cucullatos*, ed. Luigi De Franco (Cosenza, 1991), 154: "Suppositiones autem ex se patent; prima quidem ex communi consensu et experientia, cum in locis montuosis signa adhuc inundationis appareant atque vestigia, ut conchae et ostrea, ut deveniamus iure in coniecturam quandoque diluvium fuisse universale, montuosa quaecumque cooperiens ac tegens."

Since the time of Pierre Duhem, the epistemological standard of "saving the appearances" has been used to distinguish the astronomical science from cosmology and terrestrial physics.[50] As Peter Barker and Bernard Goldstein have argued, the epistemological goal of "saving the appearances" was meant to give *demonstratio quia* while admitting the impossibility of *demonstratio propter quid.* [51] While for astronomy the limitations on certainty result from the mathematically indistinguishable nature of starting principles and the inability to observe accurately celestial bodies, for meteorology the limitation comes from the accidental nature of meteorological phenomena. Lodovico Boccadiferro, a professor of natural philosophy at Bologna in the period 1527-1545, who had studied with Pomponazzi, perhaps gave the clearest summation: "This law must be observed: that when the causes of some effects are unknown to us, we must accept suppositions, or principles, from which nothing impossible, nothing contrary to the senses, and nothing repugnant to the appearances follows." Boccadiferro admitted that this "contingent possible proposition is that which is not true, but could be true."[52] Thus the epistemological standard for meteorology fell far below that of certain truth; and the goal of "saving the appearances" should not be taken to be characteristic only of astronomy but for some parts of natural philosophy as well.

50) Pierre Duhem, *SOZEIN TA PHAINOMENA, essai sur la notion de théorie physique de Platon à Galilée* (Paris, 1908).

51) Peter Barker and Bernard R. Goldstein, "Realism and Instrumentalism in Sixteenth-Century Astronomy," *Perspectives on Science* 6 (1998), 232-258.

52) Lodovico Boccadiferro, *Lectiones super primum librum Meteorologicorum Aristotelis* (Venice, 1565), fol. 45v-46r: "Et lex observanda est ista, quod cum causae alicuius effectus sunt nobis ignotae, debemus accipere suppositiones, aut principia, ex quibus nihil impossibile sequatur, neque contra sensum, neque apparentibus repugnans … . Propositio possibilis contingens est illa quae non est vera, sed potest esse vera, ut habetur primo priorum. & Averr. hoc contingens appellat inventum, & contingens & possibile … ."

Observations, Theory, and Revisions

The contention that meteorological theory was conjectural and at times only capable of saving the appearances was widely known among those concerned with this field in the sixteenth century. Moreover, this contention justified the application of new observations to the field, which could be used as signs to correct Aristotle's own theory. For example, a number of Aristotelians as early as the 1520s used the observations of sailors to amend Aristotle's position that there was an uninhabitable torrid zone in the area around the equator.[53] Moreover, Vimercati contended that, contrary to Aristotle, Portuguese sailors and Columbus had observed flows in the Atlantic Ocean, which had caused their return trips to be of different lengths than their departing voyages.[54] Zabarella utilized his experience of being upon Monte Venda, outside of Padua, on a day when it rained in the lowlands but did not on the mountaintop to conclude that he had observed the "middle region" of air, i.e., the region above the clouds. From his observations, he concluded, against Aristotle, that this region is composed of normal air, not exhalations.[55]

One of the most common topics where authors took into account the application of newly observed signs was in explanations of earthquakes. Not only had Aristotle's treatment included an extremely lengthy discussion of signs, but a number of his observations and generalization did not conform to the experiences of those who lived in the sixteenth century. Lucio Maggio, a self-described Bolognese *gentil'huomo* who authored a dialogue dedicated to explaining the

---

[53] Craig Martin, "Experience of the New World and Aristotelian Revisions of the Earth's Climates during the Renaissance," *History of Meteorology* 3 (2006), 1-15.

[54] Aristotle, *Meteorology*, 354a20; Francesco Vimercati, *In quatuor libros Aristotelis Meteorologicorum commentarii* (Paris, 1556), 176: "Observatus vero est & alius in mari fluxus, tum in Mediterraneo, tum in Oceano, quo videlicet fluit ab ortu ad occasum, & in Mediterraneo rursus ad ortum refluit, quomodo etiam in sinu illius Adriatico. Quem fluxum, etsi non evidentem, observaverunt tamen nautae ex itineribus, quae breviori tempore conficiunt, cum ab ortu ad occasum navigant, quam cum ab occasu ad ortum, aquae fluxu navium motum aut adiuvante, aut impediente."

[55] Giacomo Zabarella, *De rebus naturalibus libri xxx* (Venice, 1590), 386-387. See Charles B. Schmitt, "Experience and Experiment: A Comparison of Zabarella's View with Galileo's in *De Motu*," *Studies in the Renaissance* 16 (1969), 80-138, at 98-100.

earthquakes that struck Ferrara in 1570, followed Aristotle's general explanation based on underground exhalations and winds but disagreed with particular details of Aristotle's account. He credited Aristotle himself with having developed a limited epistemological goal for meteorology: "Aristotle did not tell a lie saying that of those things that are hardly clear to the senses, we demonstrate appropriately if by reasoning about these matters we deduce some causes that are not contrary to the effects."[56] Maggio's primary intention was to demonstrate the natural origins of the Ferrarese earthquakes in order to dispute Pope Pius V's claim that the earthquakes resulted from divine anger over Ferrara's hospitable treatment of its large Jewish population,[57] so he attempted to demonstrate that all of the particulars of this earthquake were not beyond the understanding of natural philosophy even if they did not correspond to exactly what was found in Aristotle. The fact that the Ferrarese earthquakes began in November is taken as evidence that Aristotle's contention that earthquakes most often occur in winter during tranquil weather at night should not be understood as a uniform rule.[58] Other observations were put to the service of upholding Aristotle's contention that earthquakes occur in locales where the earth is porous and hollow.[59] Maggio asserted that the Ferrarese terrain contains "holes, pores, caverns, and subterranean passages" that are typically filled with water. The untypically warm summer and autumn, however, dried out these pores, which Maggio wrote he had observed. The heat rendered the passages "deprived of their usual humor," thereby allowing the earthquake-inducing exhalation to enter and subsequently to cause tremors.[60]

---

56) Lucio Maggio, *Del terremoto dialogo del Signor Lucio Maggio gentil'huomo bolognese* (Bologna, 1571), fol. 56v: "ho havuto sommo piacere di vegghiare, & considerare, che Aristotele non disse la bugia dicendo che di queste cose poco manifeste al senso convenientemente demostriamo se ragionandone adduciamo alcune cause che non siano agli effetti contrarie."

57) Emanuela Guidoboni, "Riti di calamità: Terremoti a Ferrara nel 1570-1574," *Quaderni Storici,* n.s. 55 (1984), 107-135.

58) Maggio, *Del terremoto,* fol. 44r.

59) Aristotle, *Meteorology,* 366a24-26.

60) Maggio, *Del terremoto,* fol. 34v-35r: "Ma tornando là donde ci partimmo dico, che il terreno Ferrarese come molti altri, contiene in se meati, pori, caverne, & vie sotter-

Discussions such as Maggio's remained present within the commentary tradition on Aristotle's *Meteorology* during the seventeenth century. Niccolò Cabeo, a Jesuit originally from Ferrara, emphasized experience and chymical experimentation in his 1646 commentary on Aristotle's *Meteorology*.[61] Similarly to Maggio, he noted approvingly that Aristotle used the first stage of the *regressus* method in the analysis of earthquakes, "He [Aristotle] began well, as I have said, in trying to show *a posteriori*, or rather by the *methodus resolutoria,* what is the cause assigned to earthquakes."[62] For Cabeo, however, Aristotle's attempts were insufficient and his theory was unable to explain the real cause of earthquakes. Experience and observations taken from recent earthquakes support a chymical explanation. Instead of maintaining that winds or the eruption of subterranean exhalations provoked tremors, Cabeo believed that earthquakes result when veins of flammable substances such as sulfur, bitumen, and niter ignite.[63] This theory is supported by a number of observations, or arguments *a posteriori.* The earth in the area around Ferrara, which had been struck by earthquakes again in 1625 and 1636, for example, is full of niter, confirmed by the red

ranee, dove si generano l'acque, & per le quali l'acque delle pioggie, & del Pò, vanno penetrando, & lo rendono humido… . Hora essendo stato grandissimo caldo, & siccità del mese di Luglio fino à i dodici di Novembre, che non cadde mai tanta acqua dal cielo, che se bagnasse la polvere, & essendo passato tutto queltempo, senza che mai piovesse, si come in quei tempi essendo il Signor Mattio in Ferrrar per servigio dell'Illustrissimo Signor Cardinale Paleotti, & io del Signor Prencipe osservammo insieme, & più volte discoremmo sopra quel la straordinaria sicittà, per la quale erano secchi i fiumi, i pozzi, & ogni humore, si sono sempre per lo secco fatte maggiori le caverne, i pori, & meati della terra, & restando asciute, & prive del solito humore, doveano empirsi di alcuna cosa, percha la natura, come sapete, non comporta il vacuo … ."

61) Craig Martin, "With Aristotelians like These, Who Needs Anti-Aristotelians? Chymical Corpuscular Matter Theory in Niccolò Cabeo's Meteorology," *Early Science and Medicine* 11 (2006), 135-161.

62) Niccolò Cabeo, *Commentaria in libros Meteorologicorum* (Rome, 1646), II, 246: "Incipit, ut dixi, probare a posteriori, seu potius Methodo resolutoria, ostendere, bene assigatam esse causam terraemotus."

63) Ibid., II, 243. The idea that the causes of earthquakes were chymical did not originate with Cabeo, see: Georgius Agricola, *De ortu et causis subterraneorum libri V* (Basel, 1558), 31-32.

color of the local well-water.[64] The 1636 earthquake provided Cabeo the opportunity to witness the flow of these waters that were accompanied by burning fumes.[65] No winds, however, were seen escaping from the newly created gashes in the ground. Moreover the numerous earthquakes that occur in the region around Vesuvius, a region famed for volcanic eruptions, spontaneous fires, and bituminous soil, are signs of the correctness of Cabeo's general theory.[66] Thus according to Cabeo, Aristotle's general method of arguing from effects to causes leads to chymical theories that Aristotle himself did not endorse.

Conclusion

The distinction between theories of scientific knowledge and practice is necessary for historical understandings of Renaissance natural philosophy. While epistemological ideals played a role in the presentation of concepts and guided the direction of natural philosophy, they should not be mistaken for actual methods of research. Palmieri rightfully distinguishes Zabarella's logic from the paths followed in his treatises on cosmology and nature. Nevertheless, the *methodus resolutiva*, with its emphasis on using experience to modify and create provisional theories, guided Renaissance thinkers. While certainty was the ideal for syllogistic knowledge, Renaissance commentators on the *Meteorology* recognized that the intractable nature of the field, the difficulty of accurate observations, the inaccessibility of many of the subjects, and the accidental nature of material causes rendered much of meteorology conjectural. Observations could lead to new theories, departures from Aristotle's own positions, which would not be certain but strive to correspond to observable natural effects.

---

64) Cabeo, *Commentaria,* II:243.

65) Ibid., II:248.

66) Ibid., II:249. The sulfuric, bituminous, and nitric nature of the area around Vesuvius was widely noted by observers of its 1631 eruption and before. E.g., Giovanni Francesco Porrata Spinola, *Discorso sopra l'origine de' fuochi gettati dal Monte Veseuo* (Lecce, 1632), 4; Simone Porzio, *De conflagratione agri Puteolani… epistola* (Florence, 1551), 6.

The intractability of meteorology led authors to proofs within the field, not as demonstrative in a strict sense, but as dialectical, or, for Pomponazzi, even as rhetorical. Recognition that Aristotle's arguments were probable or approached verisimilitude extended not just to the portions of the field that were especially difficult but its foundations as well. For example, two Franciscans, Bartolomeo Mastri and and Bonaventura Belluti, while agreeing with Aristotle that there were four elements, found Aristotle's arguments not demonstrative but rather probable in their 1640 disputations on meteorology.[67] Natural philosophers in the Aristotelian tradition realized that while *scientia* required syllogism, the probable arguments of dialectic were at times the only available solution for discussions of the natural world.

While natural philosophers of the sixteenth and seventeenth centuries might have found new impetus in humanism and latitudinarianism for endorsing the belief that natural philosophy was composed of probable knowledge, that belief was not foreign to Aristotelian natural philosophy. When Claude Bérigard, a professor at Pisa and then Padua, wrote in his 1661 *Circulus pisanus* that Aristotle spoke correctly that every opinion of natural philosophy was "nothing other than a hypothesis," that could help solve difficulties just "as was the custom in the science of the stars, which puts forth epicycles, concentrics, and eccentrics," he was following a long Aristotelian tradition rather than undermining it by casting doubts.[68] Thus Athanasius Kircher's probabilism found in the *Mundus*

---

67) Bartolomeo Mastri and Bonaventura Belluti, *Disputationes in libros de celo et Metheoris* (Venice, 1640), 203: "Numerus elementorum, quamvis varius fuerit apud Antiquos, ut dicitur in Phys. communiter tamen censetur esse quaternarius; verum est tamen, quod non datur ratio aliqua evidenter demonstrans elementa non esse plura, nec pauciora, & praesertim rationes Arist. sunt probabiles... ."

68) Claude Bérigard, *Circulus pisanus... veteri et peripatetica philosphia in Aristotelis libros de octo Physicorum. Quatuor de Coelo. Duos de Ortu & interitu. Quatuor de Meteoris, & tres de Anima* (Padua, 1661), 19: "Haec ratio iam nihil concludit, sed nos remittit ad discutiendas totius Physicae difficultates, ut tandem constet an eas dirimat accuratius Aristoteles, an antiqui: & vere opinio cuiuscunque philosophi nihil est aliud quam hypothesis, qua posita videndum est an facilius enodentur omnes difficultates scientiae naturalis, ut fieri solet in Astrologia quae varios statuit epicyclos, concentricos & eccentricos, ut iis quae apparent in coelo respondeatur: sic arbitrandum est de

subterraneus (1665), a work that is largely meteorological, at least partially exhibits continuity with the Aristotelian tradition in addition to being an example of Jesuit endorsement of this epistemic standard.[69] Aristotelians recognized that natural philosophy dealt with contingencies and that its theories were dialectical and therefore uncertain. The imperfections of the matter of the sublunary regions and our inability to observe accurately the effects of this matter, made Aristotelians recognize that knowledge of nature was incomplete and probable. If indeed there was a major shift during the seventeenth century, it was the broader endorsement of the uncertainty or inconclusiveness of metaphysics rather than the adoption of the idea that natural philosophy provided probable knowledge.

opinione Aristotelis & antiquorum, nec ante sententiam ferre oportet, quam praecipuis dubitationibus responsum fuerit." Cf. Paolo Marangon, "Aristotelismo e cartesianesimo: Filosofia accademica e libertini," in *Storia della cultura veneta* (Vicenza, 1984), IV.2, 95-114.

69) Mark A. Waddell, "The World as it Might Be: Iconography and Probabilism in the *Mundus subterraneus* of Athanasius Kircher," *Centaurus* 48 (2006), 3-22.

Setting up Copernicus? Astronomy and Natural Philosophy in Giambattista Capuano da Manfredonia's *Expositio* on the *Sphere*

Michael H. Shank*
University of Wisconsin-Madison

Abstract
In 1499, while Copernicus studies in Bologna, the commentary on Sacrobosco's *Sphere* by the Padua master Francesco Capuano da Manfredonia first appears in print. It will be revised and reprinted several times thereafter. Like Copernicus, Capuano has a high view of astronomy and mingles astronomical and physical considerations (flies moving on wheels, men on ships, impetus, comets, *raptus*). Also, Capuano offers a flawed argument against a two-fold (diurnal and zodiacal) motion of the Earth. Multiple thematic resonances between Capuano's commentary and *De revolutionibus*, I, 5-11, suggest the hypothesis that Copernicus is answering Capuano, whose work was owned by Joachim Rheticus, if not Copernicus himself.

Keywords
Capuano da Manfredonia, Medieval natural philosophy, Padua, Bologna, Sacrobosco, Nicholas Copernicus, *raptus*, Joachim Rheticus, *De caelo* II, 14

No one doubts that Nicholas Copernicus and his contemporaries saw his proposal for a rotating Earth that also revolved around the Sun as antithetical to the established world order. Friends and foes alike worried about the novelty of *De revolutionibus orbium coelestium.* Copernicus himself was reluctant to publish the work and

* Department of the History of Science, University of Wisconsin-Madison, Social Science Building 7143, 1180 Observatory Dr., Madison, WI 53706-1320, U.S.A. (mhshank@wisc.edu). I thank Florence Hsia, Richard Kremer, William Newman, and Edith Sylla for comments on earlier drafts.

did not do so before Joachim Rheticus's *Narratio prima* (1540) had successfully "tested the waters." While seeing *De revolutionibus* through the press (1543), Andreas Osiander was so worried about its physical claims that he undercut them by inserting a "note to the reader" intended to shield its author from the likes of Martin Luther, who by 1539 had already called Copernicus a fool.[1]

Our own contemporaries also believe that *De revolutionibus* had revolutionary features, but for different reasons. Many historians of science see the heart of Copernicus's innovation in his disrespect for allegedly sharp boundaries between mathematical and physical (natural philosophical) approaches to the cosmos. In Westman's widely cited classic formulation, Copernicus "violated Aristotle's prohibition of *metabasis*, a prohibition that disallowed the transfer of the principles of one discipline into those of another." He dared to use arguments from mathematical astronomy to draw physical conclusions about the location and motion of the Earth.[2]

Severe space limits allow only one point about the extensive historiography on the subject.[3] The received wisdom focuses mostly on the pronouncements of natural philosophers, largely neglecting what writers more concerned with astronomy said and did around the time of Copernicus. Accordingly, the university context, in which natural philosophy occupied a preeminent place, is judged to be relatively static and fruitless for the minority activity of astronomy. Some natural philosophers did dispute the legitimacy of bringing astronomical claims to bear on their discipline, which they considered superior. But it is crucial to understand that the matter was contested. Whatever the natural philosophers thought, several astronomers, both before and in Copernicus's era, shared his notion that

---

1) Cf. Alexandre Koyré, *La révolution astronomique: Copernic, Kepler, Borelli* (Paris, 1961), 76-78.

2) Robert S. Westman, "The Astronomer's Role in the Sixteenth Century: A Preliminary Study," *History of Science*, 17 (1980) 105-147; and "Proof, Patronage and Poetics: Copernicus's Preface to *De Revolutionibus*," in David C. Lindberg and Robert S. Westman, eds., *Reappraisals of the Scientific Revolution* (Berkeley, 1990), 167-205, esp. 182-83.

3) Most work to 2003 is summarized in Michael Kokowski, *Copernicus's Originality: Towards Integration of Contemporary Copernican Studies* (Warsaw/Cracow, 2004).

astronomy stood at the apex of the sciences, engaged in causal inquiries, and could pass judgment on physical matters relevant to its subject. While a graduate student of John Murdoch's more than twenty-five years ago, I began to document this attitude in Johannes Regiomontanus (d.1476), Copernicus's leading Latin predecessor. More evidence has accumulated in recent years.[4] Regiomontanus and Copernicus both belonged to a subculture that took for granted that astronomers (*mathematici, astronomi, astrologi*) could raise and answer questions in natural philosophy. The extent of that subculture remains to be charted, but we can be sure that it included at least three people.

The third man was Francesco (later Giambattista) Capuano da Manfredonia (fl. 1494-1505), a master at Padua ca. 1494-1495 who left commentaries on Georg Peuerbach's *Theoricae novae planetarum* (Venice, 1495, 1499, etc.; Paris, 1515) and on Johannes de Sacrobosco's *Sphaera* (Venice, 1499 *et alibi*). Evidently begun as university lectures, these works first appeared contemporaneously with Copernicus's studies in Bologna (1496-1500), were available when both men probably overlapped in Padua (1501-1503), and were reprinted in various editions during his lifetime, the last ones in 1531. Earlier treatments of Capuano have already noted that his work was mixing astronomical and physical material, indeed "redefining astronomy as primarily a part of natural philosophy."[5]

As I illustrate below in preliminary fashion, Capuano's *Expositio* of the *Sphere* is considerably more significant, both on its own terms and in relation to Copernicus, than the modest earlier scholarship

[4] The earlier literature is cited in Michael H. Shank, "Mechanical Thinking in European Astronomy (13th-15th Centuries)," in Massimo Bucciantini, Michele Camerota, and Sophie Roux, eds., *Mechanics and Cosmology in the Medieval and Early Modern Period* (Florence, 2007), 3-27; idem, "Regiomontanus as a Physical Astronomer: Samplings from the Defence of Theon against George of Trebizond," *Journal for the History of Astronomy*, 38 (2007) 325-49.

[5] Pierre Duhem, *Études sur Léonard de Vinci: Ceux qu'il a lus et ceux qui l'ont lu*, 3 vols. (Paris, 1906-13), II:351-54; Pierre Duhem, *To Save the Phenomena: An Essay on the Idea of Physical Theory from Plato to Galileo*, trans. E. Dolan and D. Maschler (Chicago, 1969), 52-53, 63; Christe Ann McMenomy, "The Discipline of Astronomy in the Middle Ages" (Ph.D. dissertation, History Department, UCLA, 1984), 227-247 (transcription problems), quotation on 294.

on him has realized.[6] It reveals a late-fifteenth-century university master grappling with the motions of the Earth and their physical consequences, freely mingling natural philosophical arguments with the astronomical material he was teaching. While commenting on Sacrobosco's elementary text of the mathematics curriculum, Capuano also brings together salient arguments from the *Physics* and *De caelo* commentaries of the previous century, including the possibility of mixed circular motions and of combined rectilinear and circular motions. When Capuano examines hypothetically the arguments for the motions of the Earth, he of course discusses its rotation, a familiar hypothesis in fourteenth-century natural philosophy. Surprisingly, he also attempts to refute a hypothetical two-fold motion of the Earth. Although Capuano defends the Earth's immobility, his commentary significantly enriches our understanding of academic discussions of astronomy at the time of Copernicus.

Two decades ago, I noted several formal resemblances between Capuano's themes and arguments and those of Copernicus in book I of *De revolutionibus*, and I found more when I returned to Capuano for another project several years ago. The number, quality, and specificity of these resonances suggest the hypothesis that Capuano's commentary on the *Sphere* may have served as a sparring partner for Copernicus, who is notoriously silent about his debts to contemporaries. These intriguing echoes are probably not coincidental, for this text is certainly connected to Copernicus's circle, and probably to Copernicus himself.

Introducing Capuano

While not exactly familiar, Giambattista Capuano's name has surfaced in scholarship at least since the late nineteenth century.[7] Pierre Duhem, who famously thought that Copernicus's claims about

---

6) McMenomy, "The Discipline of Astronomy," 247: "With the exception of his prologue, however, Capuanus contributes little new material to the exposition of the sphere."

7) See note 5 and also Antonio Favaro, *Galileo e lo studio di Padova*, 2 vols. (Padua, 1883, 1966), I:96-7; Antonio Favaro, "I lettori di matematiche nell'Università di Pa-

physical reality reflected a serious misunderstanding of physical theory, already saw a similar trend in Capuano's commentary on Peuerbach's *Theoricae novae planetarum* (1495). Named Francesco when he taught astronomy in Padua in 1494-1495 as "doctor of arts and medicine," he later took the name "Giambattista" when he became an Augustinian canon regular of the Lateran Congregation.[8] Although he died in Naples, Capuano probably did not leave the Veneto immediately. The Lateran Congregation had a house in the Venetian lagoon, on the small island of San Giorgio in Alga (1404-1668) between the Giudecca and the mainland town of Fusina.

Before he entered the order, Capuano published the second commentary on Peuerbach's *Theoricae novae* (Venice, 1495). The first, by Adalbert of Brudzewo (1482), the Cracow master who was probably Copernicus's teacher, was printed in Milan in April 1495.[9] Notable in its own right, Capuano's commentary defended the physicality of Peuerbach's partial-orb models in the *Theoricae novae* and sparred anonymously with the Averroist physician and natural philosopher Alessandro Achillini, who was then teaching in Bologna.[10] The controversy continued while Copernicus studied there: in 1498, Achillini published his work on concentric celestial spheres, *De orbibus*, which objected forcefully to Ptolemy's non-concentric astronomy and to his "error in natural philosophy of trying to save the causes of phenomena with fictitious bodies."[11] Although Capuano and Achillini both held degrees in arts and medicine, their intellectual

dova dal principio del secolo XIV alla fine del XVI," *Memorie e Documenti per la Storia della Università di Padova* (Padua, 1922), I, 46.

8) Pierre Duhem, *Les Origines de la statique*, 2 vols. (Paris, 1905-06), II:60-61, 343-45; G. Gliozzi, "Capuano, Francesco," in *Dizionario Biografico degli Italiani,* vols. 19 (Rome, 1976), 255-57; Duhem, *To Save the Phenomena*, 52-53, 63.

9) Albertus de Brudzewo, *Commentariolum super theoricas novas planetarum Georgii Purbachii*, ed. L.A. Birkenmajer (Cracow, 1900), xlii-xliv, lvi. Birkenmajer claims that Capuano used Brudzewo, but offers little evidence.

10) On Peurbach's model, see Michel-Pierre Lerner, *Le Monde des sphères: Genèse et triomphe d'une représentation cosmique*, 2nd ed., 2 vols. (Paris, 2008), I:121-30, esp. 128-29 (and notes).

11) Alessandro Achillini, *De orbibus* in *Opera omnia* (Venice, 1551), 33ra-rb: "error est Ptolemaei in naturalibus in reddendis apparentium causis ex fictis corporibus." For a recent sketch of the Bolognese and Paduan contexts, see Peter Barker, "Copernicus

emphases differed markedly. Capuano identified with mathematicians, including Ptolemy, whom he defended against Averroist attacks ("This opinion of Averroes is against all mathematicians and therefore the truth").[12]

An imperfect version of Capuano's commentary on Sacrobosco first appeared in a 1499 Venetian astronomical compendium, which also reprinted his Peuerbach commentary. His title in that edition suggests that he was still teaching at Padua as a layman. By the time Capuano had revised his commentary on the *Sphere* in 1505, he had become an Augustinian canon. In later editions, he mentions a lunar eclipse of 15 August "in the current year of 1505," which "we saw with all the associates and fathers" (36va).[13] This is evidently a reference to members of his order, to which he had introduced the mathematical sciences, teaching for at least twelve years, presumably in the Veneto.[14] The revised version of the commentary on the *Sphere* (with a new preface) did not appear until the astronomical compendium of 1508. At least two editions of both commentaries appeared in 1518 and two more in 1531. Capuano's commentary on Sacrobosco was perhaps the longest of the genre until Christopher Clavius's a century later.[15]

---

and the Critics of Ptolemy," *Journal for the History of Astronomy,* 30 (1999), 343-58, esp, 348-52.

12) Throughout I cite the commentary as "Capuano, *Expositio,*" with foliation from the Giunta astronomical compendium *Sphera mundi noviter recognita cum commentariis* ... (Venice, 1518), in which the commentary is entitled: *Joannis de Sacrobusto Astronomi celeberrimi Sphericum opus cum expositione Domini Joannis Baptiste Capuani de Manfredonia, Canonici regularis ordinis sancti Augustini episcopi congregationis lateranensis feliciter incipit* (24ra). Capuano, *Expositio,* 46rb: "Hec opinio <de immobilitate polorum> Averrois est contra omnes mathematicos et consequenter veritatem."

13) Capuano, *Expositio,* 36va: "hoc presenti anno 1505 quando fuit eclipsis 15ª die Augusti quam vidimus cum omnibus sociis et patribus...."

14) Favaro, *Galileo,* 97; Gliozzi, "Capuano," 255.

15) Duhem, *Léonard de Vinci,* II, 351-54; Gliozzi, "Capuano," 257; James Lattis, *Between Copernicus and Galileo: Christopher Clavius* (Chicago, 1994).

A High View of Astronomy

Capuano's preface discusses the status of astronomy as a science and its relation to mathematics, concluding that astronomy (he uses both *astronomia* and *astrologia*) belongs to physics (natural philosophy) rather than mathematics.[16] In Capuano's hierarchy of the sciences, astronomy stands near the apex. It is better than all other sciences except the divine, because it deals with ingenerable and incorruptible substance—the noblest of all natural entities and bodies. In addition, the celestial body is the cause of all inferior things. This is a crucial point. Already in Aristotle, but especially in the post-Aristotelian astrological worldview, the sublunar and the supralunar are tightly linked in a causal relationship. Astronomy/astrology thus deals with causes. In addition, *astrologia* is more certain in its conclusions than all other sciences except mathematics, precisely because it uses geometrical demonstrations that rank first in their degree of certainty. Capuano argues that *astrologia*'s use of demonstration places it above both *scientia naturalis* and *scientia divina* (24ra-rb), an outlook analogous to Ptolemy's own in *Almagest*, book 1. Capuano reiterates the same point later in the preface, concluding "since the subject of astrology is natural/physical and its mode of demonstrating mathematical, it participates in the nobility of natural science and in the certainty of mathematics...."[17] His answers to the question, "whether astrologia is a mathematical science?" encapsulate his position:

> Therefore I say that *astrologia* is not a science that is simply mathematical, since a science that does not abstract from sensible matter is not mathematical.... Secondly, I say that by reason of its principles, [*astrologia*] is in some way natural/physical and in another way mathematical.... Third, I say that by reason of its end, astrology is more natural [=physical] than mathematical.... Fourth, I say that no science is simply mathematical unless it is arithmetic or geometry.[18]

---

16) McMenomy, "Discipline of Astronomy," 238-39.

17) Capuano, *Expositio*, 25vb: "cum subiectum astrologie sit naturale et modus demonstrandi mathematicus, quod participat nobilitatem scientie naturalis et certitudinem mathematice."

18) Ibid., 26vb: "Ideo dico quod astrologia non est scientia simpliciter mathematica, quia scientia que non abstrahit a materia sensibili non est mathematica.... Secundo

Capuano does cite Aristotle, *Physics* 2, to the effect that astronomy is a mean between physics/natural philosophy and mathematics. In the end, however, he qualifies this judgment by making astronomy more natural/physical than mathematical (26ra, 26vb).

In both the preface and the body of his work, Capuano sees astronomy as a form of causal inquiry, for good physical and astrological reasons. He argues not only that the celestial body (*corpus celeste*) causes all changes in the sublunar world (24ra), but also that the celestial motions cannot be uniform with respect to the Earth for astrological reasons. Complete uniformity in the celestial motion would always produce the same uniform effect—not the manifold terrestrial physical effects we see (32rb). Within this framework, Capuano broaches explicitly, on separate occasions and at some length, arguments for the Earth's motion. Although he goes on to refute them, what is striking is his non-perfunctory approach to them. He does not dismiss them summarily as absurd, as some predecessors had done; rather, he deals with them singly and repeatedly. We may perhaps see here the effects of the fourteenth-century Parisian discussions of the Earth's possible rotation.[19]

Capuano's commentary on the *Sphere* illustrates the confluence and interaction of the astronomical and natural philosophical traditions, as he often argues at the interface between astronomy and local motion (what we call mechanics). Capuano brings into his discussion a host of physical notions, including impetus, mixed motions (rectilinear and circular, and bi-circular), men walking on moving ships, as well as the important but neglected late-medieval mechanical notion of *raptus* and its cognates. This last term includes

dico quod ratione principiorum est aliquo modo naturalis et aliquo modo mathematica, quia aliqua principia accipit a naturali, ut quod celum sit rotundum probat per principium naturale, scilicet quod non sit vacuum nec penetratio corporum nec corpus sine loco.... Tertio dico quod ratione finis astrologia est magis naturalis quam mathematica. Nam finis scientie naturalis est cognitio rei naturalis; finis autem mathematice est cognitio proprietatum quantitatis absolute vel abstracte; modo astrologus intendit cognoscere corpus celeste inquantum est quoddam naturale.... Quarto dico quod nulla scientia est simpliciter mathematica nisi Arithmetria (!) et Geometria...."

19) Capuano's views on the Earth's shifting center of gravity echo Albert of Saxony's (Duhem, *Origines de la statique*, 343 ff.), sometimes while citing Albertus Magnus (Duhem, *Léonard de Vinci*, II:352-53; see Capuano, *Expositio*, 41va).

and extends Aristotle's concept, in the *Meteorology*, of the lunar sphere's mechanical action on the spheres of the "fire-like" element and air, resulting in their rotation. As used and diffused by Sacrobosco, *raptus* also designates the action of the daily westward motion of the primum mobile as it effectively overpowers or drags the other planetary spheres below it, all of which have a proper natural tendency in the opposite direction (eastward).[20]

> Fifth, it should be noted that all elements except earth move sensibly with local motion. For fire moves circularly with the diurnal motion, as Aristotle demonstrates in *Meteorology*, book 1, since <fire> is carried along on account of (*rapitur ex*) the celestial orbs surrounding it and carrying it with them, as demonstrated by comets and other fiery impressions generated from the hot and dry vapor contained in fire; which, since they are moved circularly, demonstrate the circular motion of the fire itself. The upper part of the air is also moved with the same motion, as is known by experience; the lower part of the air is moved chiefly laterally with the blowing of the wind to various different positions. Water is also moved with flux and reflux [= the tides], for it flows for six hours and in the same time flows back, the same time in which the Moon with diurnal motion covers individual quarters of the sky, as Ptolemy has it in *Centiloquium* 56.[21]

In this picture, not only the celestial spheres below the primum mobile, but also some of the elements participate in a motion that is not proper *to them*. This view, derived from the *Meteorology*, departs from our stereotype of Aristotle's canonical distinction between

---

20) Michael H. Shank "Rough Edges: *Raptus* and the Medieval Disturbances of the Celestial-Terrestrial Boundary," unpublished paper, European Science Foundation Workshop: Mechanics and Cosmology, Max-Planck-Institut für Wissenschaftsgeschichte, Berlin, November 2006.

21) Capuano, *Expositio*, 30va: "5° est advertendum quod omnia elementa preter terram moventur motu locali sensibiliter. Nam ignis movetur circulariter motu diurno, ut demonstrat Aristoteles primo *Metheorum*, quia rapitur ex orbibus celestibus ipsum amplectentibus et secum ducentibus: sicut demonstrant Comete et alie impressiones ignee generate ex vapore calido et sicco in igne contente, que cum circulariter moveatur, demonstrant motum ipsius ignis circularem. Eodem motu movetur etiam pars aeris superior, ut notum est per experientiam; pars autem inferior movetur maxime lateraliter flatu ventorum ad diversas differentias positionis. Aqua etiam movetur motu fluxus et refluxus per motum lune, nam sex horis fluit et totidem refluit, quo tempore eadem luna motu diurno transit singulas celi quartas, ut vult Ptolemeus *Centiloquium*, 56."

rectilinear sublunar motions and circular supralunar ones. On this account, at least the spheres of fire and of upper air move circularly while the lower part of the sphere of air moves laterally (i.e., not simply up or down). It is the *raptus* of the lunar sphere that causes the rotation of the spheres of fire and of the upper air, along with the comets that the latter occasionally carries.[22] Like his fourteenth- and fifteenth-century predecessors from Buridan onward, Capuano calls this motion "preternatural" (neither natural nor violent).[23]

Medieval natural philosophy had extended this concept to the interaction of the celestial spheres. Capuano takes the enormous speed of the primum mobile to make perfect sense since its task (via *raptus*) is to cause all the lower celestial spheres to move (31rb). In short, much more is at stake in these discussions than simply the relative motions of points of light. *Raptus* brings movers, and therefore a physical theory, to bear on the celestial observations.

Multiple Motions

Capuano's discussion also illustrates the extent to which the late medieval science of motion had moved beyond Aristotle's association of one simple body with one simple motion, and accordingly offered resources to cosmology: "I say first that it is possible for two motions to exist in the same sphere or for a sphere to move with two motions" (31vb). He goes on to talk about three ways in which a body can move with several motions. The last two deserve

---

22) The association of comets with the rotations of the spheres of fire and air also appears in the commentary on the *Sphere* by Robertus Anglicus (citing al-Bitruji); Lynn Thorndike, *The Sphere of Sacrobosco and its Commentators* (Chicago, 1949), p. 150; Jamil Ragep, "Tusi and Copernicus: The Earth's Motion in Context," *Science in Context*, 14 (2001) 145-63; Shank, "Regiomontanus as a Physical Astronomer," 342

23) On preternatural motion in Buridan and Copernicus, see Dilwyn Knox, "Copernicus's Doctrine of Gravity and the Natural Circular Motion of the Elements," *Journal of the Warburg and Courtauld Institutes*, 68 (2005) 157-211, esp. 163-64; in fifteenth-century Vienna, see Shank, "Regiomontanus as a Physical Astronomer: Samplings from the Defence of Theon against George of Trebizond," *Journal for the History of Astronomy*, 38 (2007) 325-49, esp. 331-33.

our attention, for they clearly illustrate Capuano's belief in mixed motions:

> In the second way, something can be understood to move with several motions, since it moves with a motion mixed of several motions, which motion is called mixed since it has the properties of several motions. For example, if someone were to descend through an arc whose chord is the radius of the world, that motion would be oblique and a descent, and this is possible, as is clear from the posited example. Likewise, some spiral circles or spirals can be imagined from the surface of the Earth to its center, by which means someone could move <starting> from the surface of the Earth; such a motion would be circular and also a descent, as is clear.[24]

These abstract examples are hypothetical, in the tradition of *per imaginationem* reasoning that such fourteenth-century natural philosophers as Buridan and Oresme used. But they are also associated with several striking examples with a more mechanical cast.

> Third, it can happen that something is moved with several motions, not equally primarily (*eque primo*), nor by the same mover, but by different ones, namely by one proper and intrinsic, and by another separate and extrinsic, or better by another body. And this is possible, as examples show clearly: suppose first that a ship is moved westward, and that a man on the ship also is moved westward with the motion of the ship (which motion is not proper to the man but extrinsic, as is clear). Given this case, the same man can be moved with a proper motion from within, in the same ship, in the contrary direction, namely from west to east. Doubtless this [latter] motion is proper, from within, and belongs primarily to him. The first motion, however, was not proper, but extrinsic, which is why the same man is moved with two motions, not equally primarily.[25]

---

24) Capuano, *Expositio*, 31vb: "Secundo potest intelligi aliquid moveri pluribus motibus, quia movetur motu mixto ex pluribus motibus, qui motus dicitur mixtus, quia habet proprietates diversorum motuum; verbi gratia, si aliquis descenderet per arcum cuius chorda esset semidiameter mundi, motus iste esset obliquus et descensus; et hoc est possibile, sicut patet in exemplo posito. Similiter imaginentur a superficie terre quidam circuli gyrativi vel gyre usque ad centrum terre per quos aliquis moveatur a superficie terre; talis motus esset circularis et etiam descensus, ut patet."

25) Ibid., 31vb: "Tertio potest contingere aliquid moveri pluribus motibus non eque primo nec ab eodem motore sed a diversis: ab uno scilicet proprio et intrinseco, et ab alio separato et extrinseco, vel potius a corpore alio. Et hoc est possibile, quod patet per exempla; primo, dato quod navis moveretur versus occidentem et homo existens

In his thirteenth-century commentary on the *Sphere*, Campanus of Novara had mentioned a simple moving ship in an argument for the Earth's rotation, which he quickly attributed to a perverse turn of mind (*propter pravam dispositionem intellectus*).[26] In the quotation above, however, Capuano uses a complex ship argument similar to that in Oresme's *Livre du ciel et du monde*.[27] It shows a good grasp of the compounding of motions: witness his appeal to mixed motion and his conclusions that the man's proper motion is not identical to the ship's and that these two motions can be contrary. Capuano does not say explicitly whether, like Galileo in Day Two of the *Dialogue*, he considers the ship's motion circular (i.e., moving on the spherical surface of the water).[28] But it seems likely that he does, for his second example is analogous to that of a ship's circular motion on a spherical Earth. Not only does this discussion occur in the context of two motions in the same sphere, but also his appeal to a rotating wheel adds precisely this element of complexity to the compounding of motions:

> Another example: let a wheel be imagined rotating westward in which a fly is moved according to the said motion, and yet from the outside. And with this, let the same fly be moved in the wheel with a proper eastward motion [i.e., on its own]. It will not be moved by each motion equally primarily, as is clear. Note, however, that one of these motions is faster than the other. Therefore, although the man is moved eastward with respect to the ship, since the ship nevertheless moves westward with a faster motion, the man is said to be moved westward

---

in navi etiam moveretur hoc motu ab oriente in occidentem ad motum navis, qui motus non est proprius homini sed est ab extrinseco, ut notum est; cum hoc stat quod homo idem moveatur motu proprio et ab intra in eadem navi econtrario, scilicet ab occidente in orientem; verum hic motus est proprius ab intra competens ei primo; primus vero non erat proprius, sed extrinsecus, quare idem homo movetur duobus motibus non eque primo."

26) Campanus of Novara in *Sphera mundi noviter recognita cum commentariis*..., f. 154v.

27) Nicole Oresme, *Le livre du ciel et du monde*, ed. Albert Menut and Alexander J. Denomy (Madison, 1968), 522-25.

28) Oresme imagines "Sortes" moving southward on a ship that is moving eastward, so that he "describes a diagonal straight line" (*lineam rectam transversalem*). See Garrett Droppers, "The *Questiones de spera* of Nicole Oresme: Latin Text with English Translation, Commentary and Variants" (Unpublished Ph.D. dissertation, History of Science, University of Wisconsin, 1966), 270-71.

> simply (*simpliciter*), since he approaches the west simply [i.e. his net motion is westward]; but toward the east <he is moved> not simply, but only with respect to the ship.[29]

Insects on wheels appear in earlier commentaries on the *Sphere*, notably those of Robertus Anglicus (a fly) and, *mutatis insectis*, Cecco d'Ascoli (an ant). In each case, the example illustrates how the westward motion of the primum mobile (the wheel) carries along the planets (the insects) while the latter retain their own proper eastward motion.[30] Cecco thus states that the ant moves with a "two-fold motion" (*duplici motu*). Oresme uses a fly crawling up and down the spoke of a revolving wheel as a counterexample to Ptolemy's claim that a stone thrown upward on a rotating earth would not fall to the spot from which it had been thrown.[31]

Capuano's language and its immediate context suggest that he had in mind a picture similar to Oresme's. Capuano was thinking not about a freely revolving wheel or a horizontal mill stone like that of Vitruvius in *De architectura*,[32] but about a vertical wheel moving west to east—whether a mill wheel or a cart wheel is unclear—with the fly walking against the direction of rotation, presumably on the inside rim ("*in* the wheel"). The coordinates of the wheel, like those of the man and the ship, are earth-bound, not astronomical. For Capuano, the dominant motion is determined not ontologically, but by comparing relative speeds:

---

29) Capuano, *Expositio*, 31vb: "Aliud exemplum: imaginetur rota circumvoluta versus occidentem in qua sit musca que ad motum rote moveatur dicto motu, tamen ab extra. Et cum hoc moveatur eadem musca in rota motu proprio versus orientem; non tamen movebitur utroque motu eque primo, ut patet. Nota tamen quod unus istorum motuum est velocior altero; ideo licet homo respectu navis moveatur versus orientem, quia tamen navis movetur ad occidentem velociori motu, dicitur homo simpliciter ad occidentem moveri, quia ad occidentem simpliciter appropinquat: ad orientem vero non simpliciter, sed solum respectu navis." Oresme discusses the compounding of rectilinear and circular motions with the example of a mill wheel that is descending as it rotates; Droppers, "The *Questiones de spera* of Nicole Oresme," 270-71.

30) Thorndike, *Sphere*,153, 360.

31) Claudia Kren, "The *Questiones super De celo* of Nicole Oresme" (Unpublished Ph. D. dissertation, History of Science, University of Wisconsin, 1965), 683-84.

32) Bernard R. Goldstein, "The Origin of Copernicus's Heliocentric System," *Journal for the History of Astronomy*, 33 (2002), 219-35, at 230-31.

> these two motions are possible in the supercelestial body so that it can be moved with two motions—not, however, equally primarily. For the primum mobile is moved westward, and since it embraces all the lower spheres, it moves and leads them with the same motion and an equal velocity, precisely as the moved ship moves the man in it. This motion is not proper to the lower orbs, but <comes> from the uppermost body, namely the primum mobile; it is therefore necessary that <the lower orbs> have another natural motion proper to them and <originating> from the form intrinsic to them, and this is the second motion, namely the one by which they are moved eastward.[33]

These two possible motions in the same sphere not only exist, but also are necessary:

> the first necessity is because the primum mobile is the greatest of all natural bodies, moved most rapidly, with a velocity than which no greater is found naturally, for in twenty-four hours, it completes an entire revolution. Therefore the primum mobile is moved by such a velocity and (as I might say) impetus that it drags with it all lower bodies down to the air; therefore, unless there were a second motion contrary to the first and resistant to it, not only the spheres of fire and air, but also the water and earth would be carried away (*raperentur*) and moved; everything would be moved, everything would be jumbled, and consequently the world would be destroyed. The Earth also would be moved, and consequently the heavens would not, since every motion requires something at rest, as it says in *On Animals.* Lest therefore the heaven by its impetus, which it exerts by its diurnal motion, move and jumble everything, ingenious nature has made a second motion toward the opposite part by which the first motion is modified and regulated.[34]

---

33) Capuano, *Expositio*, 31vb-32ra: "Ad propositum: hi duo motus sunt possibiles in corpore supercelesti, ita quod moveatur duobus motibus, non tamen eque primo. Nam primum mobile movetur ab oriente in occidentem, et quia complectitur omnes spheras inferiores, movet et ducit eas eodem motu et equali velocitate precise sicut navis mota movet hominem in ea existentem, et hic motus non est proprius orbibus inferioribus, sed est a corpore superiori, scilicet, a primo mobili; ideo oportet quod habeant alium motum sibi proprium naturalem et a forma sibi intrinseca; et est motus secundus quo scilicet moventur ab occidente in orientem."

34) Ibid., 32ra-rb: "Tertio dico quod non solum est possibile esse duos motus, similiter non tantum est ita quod sunt duo motus in supercelestibus, immo est necessarium; et prima necessitas est quia primum mobile cum sit maximum omnium corporum naturalium velocissime motum qua naturaliter non est maiorem velocitatem reperire; nam in 24or horis complet totum circuitum; mobile igitur primum movetur tanta velocitate et (ut dicam) impetu quod secum trahit omnia corpora inferiora usque ad aerem; ergo nisi esset motus secundus contrarius primo ei resistens, tunc moverentur et raperentur non tantum sphere ignis et aer, sed etiam aqua et terra; omnia moverentur, omnia

Capuano raises pointedly the question of the natural motion of the lower spheres in the context of the westward *raptus* of the primum mobile. For each lower sphere, both the daily westward motion and the specific eastward motion proper to it belong to it: the eastward motion is natural to it, while the rapid daily motion is preternatural to it. Here Capuano clearly bristles at the suggestion that this terminology connotes violence: "I therefore call this motion preternatural, and if it were said to be natural, this would perhaps not be inappropriate, since the motions that are in the celestial bodies can be nothing but natural." Thus Capuano goes on to soften the connotation of violence that he associates both with *raptus* and with Sacrobosco's mention of the inferior spheres' striving against it (*contra nitentibus*).[35] Suffice it to say for now that, in Capuano, physical concepts are pervasive and in tension with one another—the language of impetus, resistance, and dynamic balance appears alongside that of the natural and the preternatural.

Strikingly, immediately following this passage Capuano turns to the cosmic order, which his argument *contra* sets up as an "incomprehensible" problem, to be solved after dealing "first with the order of all the spheres; and second, more specifically, [with the spheres] of the Sun, Venus, and Mercury."[36]

---

confunderentur, et consequenter mundus corrumperetur. Terra etiam moveretur et per consequens celum non moveretur, quia omne motum indiget quiescente, ut habetur in libro *De animalibus*: ne igitur celum suo impetu quem agit per motum diurnum omnia moveat et confundat, natura ingeniosa fecit motum secundum; quo existente ad partem oppositam, modificatur primus motus et regulatur." .

35) Ibid., 32va: "ideo voco eum motum preternaturalem, et si diceretur naturalis, fortasse non esset inconveniens, quia motus qui sunt in corporibus celestibus non possunt esse nisi naturales. Et si diceretur quod uni corpori simplici est unus motus simplex naturalis, verum est primo, vel a principio activo intrinseco. Et propter hoc motus iste primus respectu inferiorum spherarum dicitur raptus, non quia violentus sed quia ab extrinseco et corpore superiori, unde raptus accipitur improprie. Eodemmodo improprie sphere alie dicuntur niti contra primam, non quia resistent, tanquam si violentarentur ab ea, sed quia moventur contra motum illius, unde ista nixio contra non est nisi motus contra motum primi." Cf. Thorndike, *Sphere,* 79, 119-120.

36) Capuano, *Expositio,* 32va: "Queritur secundo de ordine spherarum quem ponit Auctor, an sit verus. Videtur quod non; primo illud quod est incomprehensibile non debet affirmari; sed nullus sensus potest comprehendere ordinem spherarum, quia sphere non apprehenduntur per sensus, sed solam imaginationem, ut dicit Averroes

The Motion (or Rest) of the Heavens and the Earth

After agreeing with Sacrobosco and attacking Geber on the order of the spheres, Capuano examines whether the heavens move, and begins by offering arguments against their motion. The first argument appeals to the greater nobility of the heavens, and consequently to the greater propriety of their being at rest. His second argument brings the motions of the terrestrial elements to bear on the case, starting with the "circular" motion of fire and the "lateral" motion of the air, which in turn moves the water (i.e., the tides); therefore the Earth should move as well. But every motion requires something at rest, therefore the heavens must be at rest (33va). His third argument asks which of two possibilities is most appropriate: to save the phenomena with moving heavens and an Earth at rest or vice versa. "But it is more suitable to assume that the Earth rather than the heaven moves, since the Earth is better suited to motion, since it is smaller and more easily movable; therefore the heavens are at rest."[37]

In opposition, Capuano argues that to assert that the heavens are at rest would undermine the science of astronomy, which is premised on celestial motion. This argument shows the firmness of the late-medieval mechanical worldview: Capuano effectively finds it impossible to countenance planetary motion if the primum mobile, which moves everything else, rests.

A few lines later, Capuano alludes specifically to the Pythagoreans' claim, which Aristotle had mentioned in *De caelo*, that the rotation of the Earth might save the phenomena of the stars' rising and setting (33va). A ship moving in relation to the shore returns, this time in a case arguing for the plausibility of the heavens at rest: the heavens seem to move because "we move circularly with the Earth." While Capuano concedes that the Earth's rotation saves the diurnal motion, he claims that it fails to save either the

---

secundo *Celi*.... In hac questione, primo videbitur ordo omnium spherarum. Secundo particulariter de Sole Venere et Mercurio."

37) Ibid., 33va: "Tertio de possibilibus, quid convenientius est, poni debet; apparentie autem possunt salvari per motum celi et quietem terre, et econtra per motum terre et quietem celi; sed magis est conveniens ponere terram moveri quam celum, quia est ad motum aptior, quia minor et facilius mobilis, ergo celum quiescit."

proper motion of the fixed stars (precession) or the motions of the planets. Strikingly, for Capuano, it is a lesser problem to concede that the heavens move than that the Earth moves. He therefore rejects the stationary heavens, because in the end the Earth cannot move circularly. "Not only is it more appropriate for the heavens to move than the Earth; rather, it is impossible for the Earth to be moved, especially circularly, since there is no specifying that by which the Earth can move." Admittedly, the Earth is small, but it is also heavy, fixed in the center, and most incapable of motion, "which is why the phenomena are saved more appropriately by the motion of the heavens than that of the Earth."[38]

Refuting a Two-Fold Motion of the Earth

In arguing further for a stationary Earth, Capuano then proposes a reductio ad absurdum premised on a two-fold (diurnal and zodiacal) motion of the Earth:

> To the extent that something moved circularly is lower, to that extent it is moved by more motions, as is clear of all spheres. This is reasonable, since the first sphere, from the fact of its proximity to the being that is simplest and immobile, moves only with a single, most simple, and regular motion. For, however much a sphere is lower <and> more distant from the first, to that extent it is changed by many transmutations, at least by the diurnal motion and by the motion that occurs in the zodiac. But this motion that occurs in the zodiac in no way appears in the Earth, since the fixed stars are always seen to rise and to set out of the same part of the sky, which would not happen if the Earth moved with these two motions.[39]

[38] Ibid., 33vb: "non solum est convenientius celum moveri quam terra; immo est impossibile terram moveri precipue circulariter, quia non est assignare a quo terra moveatur. Etiam terra, licet parva, est tamen gravis, fixa in centro et ineptissima ad motum; quare salvatur apparentia convenientius per motum celi quam terre."

[39] Ibid., 40vb: "Secunda ratio: quanto aliquid motu<m> circulariter est inferius, tanto pluribus motibus movetur, ut patet de omnibus spheris; quod est rationabile, quia prima sphera ex quo est propinquissima enti simplicissimo et immobili, movetur tantum unico motu simplicissimo et regulari; quanto vero sphera est inferior magis distans ab illo primo, tanto pluribus transmutationibus mutatur, cum ergo terra sit ab eo remotissima; si movetur circulariter, deberet moveri pluribus motibus, saltem motu

This argument appears in the context of the question "Whether the Earth is at rest?" (40va), after Capuano has already dealt once with the rotation of the Earth. The question is a typical one not only in the *Sphere* and its commentators, but also in late medieval treatments of Aristotle's *De caelo*.[40] Capuano's strategy here heightens the implausibility of the Earth's motion by requiring, if it is to move circularly, that it move with at least two motions instead of one.

Both this specific argument and the general refutation of it hark back to *De caelo* II, 14 (296a34-b6), in which Aristotle also was arguing against multiple motions of the Earth.[41] The complexities of Aristotle's text deserve close study in their own right, as do the fundamental disagreements of the modern commentators about this passage. Of the medieval Latin commentators on this passage, space allows consideration only of Aquinas, whose exposition of *De caelo* Capuano cites. Aquinas paraphrases Aristotle sufficiently to clarify his own understanding of the latter's sparse discussion.[42] Aquinas's explanation reiterates the uniqueness of the sphere of fixed stars, which has (before the discovery of precession) only one motion. All of the other circularly moved bodies lag behind the sphere of fixed stars and thus move with several motions. If the earth moved, it too "would necessarily be moved with several motions." Aquinas understands the latter to be motions like those of the planets: the motion of the fixed stars around the poles of the celestial equator (the daily motion, not named) and "some other proper motion around the poles of the zodiac." On this account, the latter motion would cause "changes and turnings" of the stars (*mutationes et versiones*,

diurno et motu qui fit in çodiaco; sed hic motus qui fit in çodiaco nullo modo apparet in terra quia stelle fixe semper videntur oriri et occidi ex eadem parte celi, quod non contingeret si terra moveretur his duobus motibus."

40) Ernest A. Moody, ed., *Iohannis Buridani quaestiones super libris quattuor de caelo et mundo* (Cambridge, Mass., 1942), book II, qu. 22 (226 ff.); Albert of Saxony, *Quaestiones subtilissimae in Aristotelis de caelo et mundo* (Venice, 1492; Hildesheim, 1986), Bk II, qu. 10, E6ra-va.

41) *Aristotelis Opera cum Averrois Commentariis* (Venice, 1562-74), V:162rF.

42) See Edward Grant, ed., *Sourcebook in Medieval Science* (Cambridge, Mass., 1974), 496-97.

the very words of the Latin Aristotle). Aquinas here explains that the Earth's motion in the zodiac would "fall short" (i.e., lag behind the motion of the primum mobile) "just as happens with the planets" (i.e., like them, it would have its own proper motion eastward). But we do not see such changes, therefore the Earth does not move in this manner.[43]

Like Aristotle's argument, Aquinas's exposition is premised on a polarity of motions (single-multiple) that is treated independently of the Earth's hypothetical motions. That is, the argument takes the motion of the primum mobile as a given: it does not consider whether the hypothetical motions of the Earth undermine the premise in any way or make the motion of the primum mobile obsolete. These key features reappear in Capuano's argument, not coincidentally since he earlier had cited Aquinas's *De caelo* commentary, albeit in other contexts (25rb, 31va).

Capuano's own argument starts with the premise that the motions of the celestial spheres are ranked by simplicity. The first sphere has the simplest motion (the daily motion of the fixed stars) explained by its proximity to the first mover ("the being that is simplest and immobile"). As one moves away from the latter, the number of motions increases. Capuano's language ("*quanto...tanto*") suggests a proportionality that he does not elaborate and in fact does not need for his argument. All he needs is the concession that all spheres but the first have more than one motion: the minimal plurality that all the lower spheres share consists of at least two motions. If the Earth moved circularly, it too would have at least two motions (diurnal and zodiacal).

This language strongly echoes Aquinas's exposition. Although Capuano does not directly compare the Earth to a planet, his naming of its two hypothetical motions as diurnal and zodiacal (which Aquinas had implicitly done) effectively gives the Earth planetary characteristics. If the Earth moves circularly, it cannot merely rotate, as the first sphere does; it must have at least one additional motion,

---

43) Thomas Aquinas, *In Aristotelis libros De caelo et mundo, De generatione et corruptione, Meteorologicorum expositio*, ed. P.F. Raymundi and M. Spazzi (Rome, 1952), 261-63.

like all other non-first-spheres. Capuano's terminology has given the Earth the number of motions of the simplest planet, the Sun.

Capuano concludes his argument by denying the Earth's hypothetical motion in the zodiac on observational grounds recycled from the *De caelo* via Aquinas. Unlike his predecessors, Capuano does not specify how the heavens would look if the Earth moved with multiple motions. He simply states that what we see—the unchanging rising and setting of the stars from the "same part of the sky"—is incompatible with the Earth's motion in the zodiac.

Capuano does not explain what he meant thereby, but his account is consistent with Aquinas's exposition of Aristotle. Capuano's argument shares the flaws of its predecessors, even as his naming of the Earth's hypothetical motions as diurnal and zodiacal puts those problems in sharper relief. His premise—that the first sphere has one simple motion, other spheres more than one—makes the hypothetical diurnal rotation of the Earth superfluous. Indeed, Capuano conceded as much in his apocalyptic argument that if the Earth moves, the heavens must rest. Yet in his own argument, the fixed stars and the Earth both have a diurnal motion. Likewise, when Capuano imagined a zodiacal motion of the Earth, he evidently considered its effect only on the stars, not the planets, of whose variation in distance he was aware. His emphasis on our sighting of the stars always from the "same part of the sky" suggests that he imagined how the night sky would appear from an Earth moving on a sphere with a zodiacal pole. On this interpretation, his objection to the zodiacal motion of the Earth around the center of the zodiac would be that this motion is inclined to the equator and yields counterfactual positions of the fixed stars. Like his predecessors, Capuano in effect added the Earth's motions to the traditional celestial ones, but did not compensate for the addition.

What makes Capuano's use of a hypothetical two-fold motion of the Earth, diurnal and zodiacal, particularly intriguing is that, in the astronomy of the day, the only planet with these two motions was the Sun. Yet giving the Earth new Sun-like motions would necessarily require adjusting the Sun's motions and place in order to save the phenomena. But Capuano no more considered the consequences, for the other planets, of the Earth's motion in the zodiac than he did the consequences of the Earth's rotation for his premise of the primum mobile's diurnal rotation.

In any case, the language of Capuano's argument sharpened, no doubt unwittingly, the contradictions in Aristotle's classic argument in a very suggestive way. By requiring that an Earth with *any* motion had to have at least diurnal and zodiacal motions, Capuano introduced not only planetary, but also solar language into the hypothesis of the Earth's motion.

A Few Thematic and Dialectical Resonances between Capuano and Copernicus

Most of these issues show intriguing parallelisms to the front matter and the cosmological sections of Copernicus's *De revolutionibus*, Book I. Most generally, the common outlook between Capuano and Copernicus is obvious. Like Regiomontanus, these two contemporaries see astronomy as a discipline at the apex of the sciences and one with physical consequences. They do not hesitate to use and to resolve, for their own purposes, arguments in natural philosophy. In addition, they both understand astronomy as causal.[44]

I now turn to some dialectical parallels between Capuano and Copernicus. At one level, my suggestion that Copernicus appears to be answering Capuano is trivial. Since Copernicus argues against the set of all defenders of a central, stationary Earth, he is implicitly arguing against every member of the set, hence against Capuano. As I hope to show, however, the issues that Capuano brings up in this one treatise are sufficiently unusual, specific, and numerous to cast doubt on the hypothesis that the similarity is coincidental.

In *De revolutionibus*, I, after summarizing the main arguments for the centrality of the Earth and against its daily rotation (chap-

[44] Copernicus opens the manuscript of Book I with a reference to the "studies that deal with the godlike circular motions of the world" and lists among the topics in it "the causes of the other phenomena of the heaven" (This passage was omitted in the 1543 edition); *The Manuscript of Copernicus' On the Revolutions: Facsimile*, Complete Works, vol. 1 (London, Warsaw, Cracow, 1972), Nicholas Copernicus, *On the Revolutions of the Heavenly Spheres*, trans. A.M. Duncan (Newton Abbot, England and New York, 1976), 35 (modified).

ter 7), Copernicus in chapter 8 specifically takes on the arguments against the motion of the Earth. Instead of dealing piecemeal with the natural or unnatural character of such a rotation, he steps back to make a meta-point: if the Earth rotates, its motion is natural. Naturalness of motion will characterize whichever body is moving, whether the primum mobile or the Earth. Recall that one of Capuano's arguments for keeping the Earth stationary was that, if it moved, "there is no specifying that by which the Earth can move." (33vb). Here, Copernicus's appeal to its natural motion sounds like a direct answer to Capuano. In short, there is no independent criterion of natural motion, only a choice of options: either the heavens stand still and the Earth rotates, or the heavens rotate and the Earth stands still. In either case, the motion of the moving body and the immobility of the stationary one are natural; the question at hand is the determination of which body is in motion. In the process, Copernicus alludes to his own apocalyptic scenario if the heavens move by *raptus*.[45] Capuano had already rehearsed the arguments for the greater nobility of rest for the heavens and the greater ease of motion for the smaller body (the Earth), only to refute them. Copernicus adopted them.

Still in Chapter 8, Copernicus discusses the motion of comets (6r), and air. This passage caught the attention of Jamil Ragep, who was puzzled by its apparent uniqueness in the Latin tradition.[46] Significantly, Capuano had also discussed comets in the context of the *raptus* of the celestial spheres, not surprisingly perhaps, since the motion (or lack thereof) of the elements belongs to the problem of the motion (or lack thereof) of the Earth (30va). Copernicus uses *raptus* or its cognates three times in *De revolutionibus* I, 8 (6r-6v).

As we have seen, Capuano had argued that a sphere could move with mixed circular motion and elemental bodies could move with combined circular and straight motions. The availability of such

45) Copernicus, *Revolutions* (Duncan), 43-44; Nicholas Copernicus, *De revolutionibus orbium coelestium libri VI* (Nuremberg, 1543), facsimile (Brussels, 1966), 5v: "rapietur."

46) Jamil Ragep, "Tusi and Copernicus: The Earth's Motion in Context," *Science in Context*, 14 (2001) 145-63, esp. 146-50.

arguments helps to explain Copernicus's own appeal to mixed straight and circular motion (6r-v) and his rapid march through the traditional natural philosophical problem of a rotating Earth in *De revolutionibus* I, 8. Although such arguments are scattered throughout fourteenth-century natural philosophy, we do not know precisely how they reached Copernicus. In Capuano's revised commentary on the *Sphere*, they were not only readily available, but also put to work in a combined astronomical and physical context. Copernicus put similar arguments to work for the elements, arguing that the apparent rectilinear rise and fall of the latter exhibit a compound of circular and rectilinear motion.[47]

But it is the matter of requiring more than one motion of the Earth that is most striking. In *De revolutionibus* I, 9, Copernicus's argument resonates with the framework of Capuano's discussion, which Copernicus has turned on its head. Having argued in *De revolutionibus* I, 8 that the mobility of the Earth is more probable than its immobility, in the first sentence of Chapter 9, Copernicus asks, "whether several motions are appropriate for it" (7r).[48] The immediate leap to the plural strikingly echoes Capuano's argument rooted in *De caelo*, II, 14. After sketching his theory of gravity,[49] Copernicus explores the consequences of making the Sun and Earth trade places and motions: "if the [yearly circuit] is transformed from solar into terrestrial, immobility being conceded to the Sun, the risings and settings of the zodiacal signs and the fixed stars, by which they become morning and evening stars, will appear in the same way." On the force of this fundamental observational equivalence, Copernicus goes on to assert a stronger one: "Also the stations, retrogressions, and progressions of the wandering stars will be seen to be not their own motion, but the Earth's, which they transform into their appearances. Finally, the Sun itself will be thought to own the center of the universe...."[50] These assertions are a direct denial of Capuano's observational claim, which was probably rooted

47) Copernicus, *Revolutions* (Duncan), 44-45.

48) Ibid. (Duncan), 46.

49) See Knox, "Copernicus's Doctrine of Gravity."

50) Copernicus, *De revolutionibus*, 7r-v; Copernicus, *Revolutions* (Duncan), 46 modified.

in the objections of *De caelo* II, 14 and of Aquinas's commentary on it, that the stars would exhibit "changes and turnings."

If Copernicus had read the work, the incompatible premises in Capuano's argument, generated by the juxtaposition of the simple motion of the fixed stars with the added assumption of "*diurnal* and zodiacal" motions, must have stood out when Copernicus tested the consistency of his insight.

Strikingly, the passage in Capuano that immediately follows this discussion turns to the order of the planets, and discusses extensively the place of Mercury and Venus in relation to the Sun: "In this question, it will be a matter first of the order of all the spheres; second, and more specifically, of the Sun, Venus, and Mercury."[51] Precisely this problem, on Bernard Goldstein's convincing interpretation, motivated Copernicus's move to heliocentrism by causing him to rethink the order of the planets—the subject of *De revolutionibus* I, ch. 10.[52]

Not least, Copernicus evidently agreed with Capuano that giving the Earth only diurnal and zodiacal motions could not save the phenomena. Accordingly, he spent the better part of *De revolutionibus* I, 11 offering two derivations of the third motion he ascribed to the Earth. Copernicus's motion of "declination" keeps the Earth's axis pointed in the same direction on an annual basis and also accounts for the precession of the equinoxes. Equally significant, immediately thereafter, the manuscript includes material omitted in the printed text, to the effect that Philolaus and Aristarchus probably saw the motions of the five planets as more consistent with the mobility than the immobility of the earth "although they were not moved by the reason that Aristotle alleges and criticizes."[53] This is a direct allusion to *De caelo* II, 14, the very passage that informs the arguments of Aquinas and Capuano.

The significance of Capuano's treatise lies less in the specific views he presents, many of which were scattered throughout late-medieval natural philosophy, than in his collection and linkage of them

---

51) Capuano, *Expositio*, 32va: "In hac questione, primo videbitur ordo omnium spherarum. Secundo particulariter de Sole, Venere et Mercurio."

52) Goldstein, "The Origin of Copernicus's Heliocentric System," 219-35.

53) Copernicus, *Revolutions* (Duncan), 53.

in a detailed commentary on the *Sphere*, the basic elementary text of the mathematical curriculum. In recasting Aristotle's argument against multiple motions of the Earth, Capuano called the two hypothetical motions "diurnal" and "zodiacal." In so doing, he effectively raised the possibility of the Earth's being a planet—not simply a body rotating at the center of the universe, but a body with the traditional motions of the Sun. In Capuano's commentary on the *Sphere*, written by a Paduan master whose path he may have crossed in Padua, Copernicus could conveniently find a range of recent and detailed arguments against the motion of the Earth. By 1543, he had formulated arguments that look like answers to them.

But was Copernicus answering Capuano? First, Copernicus very likely knew about Capuano through the controversy with Alessandro Achillini. His studies in Bologna (1496-1500) overlapped with Achillini's tenure and publication of *De orbibus* (1498).[54] A new piece of evidence increases the likelihood of his familiarity with Achillini's work. *De orbibus* contains several mysterious citations from Averroes's *Epitome of the Almagest*, previously thought to be available only in Arabic and Hebrew.[55] Equally intriguing, Copernicus cites this work, calling it *Averroes in Ptolemaica paraphrasi*, when discussing a possible transit of Mercury (*De revolutionibus*, 8v), the topic of one of Achillini's quotations.[56] Second, if Copernicus did not know about Capuano through Achillini in Bologna, he is likely to have heard about his commentaries (the commentary on Peuerbach competed with that of Brudzewo, who was probably Copernicus's teacher in Cracow). Third, Copernicus and Capuano easily could

---

54) Mario di Bono, *Le Sfere omocentriche di Giovan Battista Amico nell'astronomia del Cinquecento* (Genoa, 1990), 62-65; Peter Barker, "Copernicus and the Critics of Ptolemy," *Journal for the History of Astronomy*, 30 (1999), 343-58, esp. 349-50.

55) Achillini mentions Averroes's *Epitome of the Almagest* in *De orbibus*, 26rb, 27vb, 28vb, 33ra, suggesting access to a Latin version in late-fifteenth-century Bologna, perhaps through Mordekhai Finzi; see Tzvi Langermann, "The Scientific Writings of Mordekhai Finzi," *Italia*, 7 (1988), 7-44, 30-31. Averroes's *Epitome* also appears in the Latin excerpts of Levi ben Gerson's *Wars of the Lord*; see José Luis Mancha and Gad Freudenthal, "Levi ben Gershom's Criticism of Ptolemy's Astronomy," *Aleph*, 5 (2005), 35-167, 98.

56) Achillini, *De orbibus*, 30va.

have met: Copernicus's years in Padua (1501-03) are bracketed by the years in which Capuano was publishing in the Veneto and presumably teaching his fellow canons there. Not least, Capuano's *Expositio* of the *Sphere* was known in Copernicus's circle. The Stockholm, Royal Library, copy of the 1518 Venice Giunta astronomical compendium containing Capuano's two works is inscribed to "Ioachimo Rhoetico."[57]

With studies in Cracow, Bologna, and Padua, Copernicus paid his dues as a university man. It should not be surprising if the resources on which he drew to refute the arguments against the motions of the Earth included the academic commentaries, disputations, and controversies of his day. Indeed Copernicus tells us that he searched for anyone who held beliefs about the motions of the spheres other than the ones "that they who taught astronomy (*mathemata*) in the schools posited" (f. iiijr). When drawing up his list of beliefs, he could not have done much better than to start with the collection of astronomical and physical arguments that Capuano tried to refute.

57) L.A. Birkenmajer believed that Copernicus gave the volume to Rheticus (see his *Stromata Copernicana* (Cracow, 1924), 320-21); André Goddu is skeptical ("Copernicus's Annotations: Revisions of Czartoryski's 'Copernicana,'" *Scriptorium*, 58 (2004), 202-226, 207-08). The volume (106A, Fol. RAR) contains annotations in several hands, at least one of which (Goddu's Plate 39), in my view, strongly resembles Copernicus's (166r).

Copernicus's Mereological Vision of the Universe

André Goddu*
Stonehill College

Abstract
Mereology is the philosophical study of part/whole relations. Copernicus, Mästlin, and Kepler addressed explicitly some of the logical issues in their support of heliocentrism. Their emphasis on harmony and commensurability to evaluate theories as more or less likely fits with their use of part/whole relations to argue for the greater reasonability or probability of heliocentrism. The essay summarizes the logical and metaphysical issues that earlier traditions discussed, and it uses those discussions to illuminate features of heliocentric theories that remain otherwise obscure even in Newton.

Keywords
mereology, part/whole relations, topics, Copernicus, Mästlin, Kepler, Newton

Introduction

I am delighted to dedicate this paper to John Murdoch, one of the foremost experts on medieval logic and especially of the metalinguistic techniques of the fourteenth century.

"Mereology" is a major subject of discussion in late antique and medieval philosophy.[1] Derived from the Greek *méros* (part), mereology

---

* Stonehill College, 320 Washington St, North Easton, MA 02357-5620, U.S.A. (agoddu@stonehill.edu). Edith Sylla's penetrating questions led to correction of mistakes and expansion of the context. Christoph Lüthy's objections at the Murdoch conference led me to recognize an oversight in Mästlin's defense of Copernicus. To both I am grateful for their careful attention to details.

[1] Liddell and Scott's *Greek-English Lexicon* (Oxford, 1883) lists "part" as only one of the many meanings of *méros*. Achille Varzi, "Mereology," *Stanford Encyclopedia of Philosophy*, http://plato.stanford.edu/entries/mereology/, 1-5, provides a variety of meanings for "part" and "whole" in both technical senses and in ordinary language.

is the "science" of part/whole relations. Medieval thinkers analyzed relationships between parts and wholes with great acuity.[2] Humanists' diatribes against scholastic logic did not have a noticeable impact on university curricula until the seventeenth century. Even then, the curricula continued to emphasize the more useful parts of logic, especially dialectical inquiry and the use of topics in the construction of arguments.[3] There were and continue to be unanswered questions about the arguments proposed by Nicholas Copernicus, Michael Mästlin, and Johannes Kepler in support of heliocentrism. In this paper I examine in some depth their arguments relying upon dialectical topics involving mereology or part/whole relations.

At Cracow and Bologna, Copernicus's teachers instructed students on the use of dialectical topics for the construction of arguments. Although I think it unlikely that Copernicus knew the most technical literature, he may have had some acquaintance with the discussions at Cracow. Of course, Copernicus's insights may have been intuitive in part, and shaped by his critique of Ptolemaic astronomy, but the structure of his book and the arguments employed there, especially in Book I (probably written in the late 1520s) contain evidence of his knowledge of standard dialectical topics. In my previous treatment of Copernicus's arguments, I concluded that the most important, relevant to his criticisms of Aristotle and Ptolemy, is the topic *from an integral whole*. In two papers I laid out the textual background for the claim that Copernicus held a notion of implication that relied on dialectical topics to argue for the probable truth of his hypotheses.[4] Lacking a decisive observation in astronomy

---

[2] Peter King, "Medieval Mereology," *Stanford Encyclopedia of Philosophy*, http://plato.stanford.edu/entries/medievalmereology/. The bibliography on mereology is extensive and can be found in the following: Desmond Paul Henry, *Medieval Mereology* (Amsterdam and Philadelphia, 1991); Hans Burkhardt and Carlos Dufour, "Part/Whole I: History," in *Handbook of Metaphysics and Ontology*, 2 vols. (Munich and Philadelphia, 1991), 2: 663-673; Peter Simon, *Parts, A Study in Ontology* (Oxford, 1987).

[3] Lisa Jardine, "Humanist Logic," in *The Cambridge History of Renaissance Philosophy*, ed. Charles Schmitt and Quentin Skinner (Cambridge, 1988), 173-198; André Goddu, "The Use of Dialectical Topics in the Sixteenth and Seventeenth Centuries," *Medioevo*, 24 (1998), 301-355.

[4] André Goddu, "Consequences and Conditional Propositions in John of Glogovia's and Michael of Biestrzykowa's Commentaries on Peter of Spain and their Possible

and strict scientific demonstration, Copernicus admitted into astronomy only those conditional propositions that receive their warrant from relations of relevance, containing and contained, inclusion, explaining to explained, and the following of consequents from antecedents without the addition of anything extraneous.

A Brief and Selective Survey of the Tradition

To situate this discussion in the context of Aristotelian and scholastic logic, I begin with an overview that explains my focus on dialectical topics, specifically on the topics *from an integral whole* and *from an integral part.*

Aristotle distinguished demonstrative, dialectical, rhetorical, and sophistical reasoning and arguments. He dedicated the *Prior Analytics* to the study of syllogistic, and challenged readers to transform valid arguments into syllogisms. He treated demonstration and induction in the *Posterior Analytics,* dialectical arguments in *Topics,* rhetorical arguments in *Rhetoric,* and fallacies in *Sophistical Refutations.* The purpose of the last treatise is to train students to recognize fallacies in arguments. Rhetorical arguments rely on techniques of persuasion that appeal to emotions, the moral character of the speaker, and logic for their effectiveness.[5]

The term "dialectic" has many connotations. In the Middle Ages it was often used interchangeably with "logic." When considered as a distinct part of logic, "dialectic" may refer to all logic that is not strict demonstration or, more specifically, to all reasoning that begins from premises that are not necessarily true but probably true, true for the most part, or accepted by experts in their field. In this essay

---

Influence on Nicholas Copernicus," *Archives d'histoire doctrinale et littéraire du moyen âge,* 62 (1995), 137-188; André Goddu, "The Logic of Copernicus's Arguments," *Early Science and Medicine,* 1 (1996), 28-68. Aside from the intrinsic topics *from an integral whole* and *from an integral part,* Copernicus also used the intrinsic topics *from the proper attribute, from efficient cause, from final cause,* and the extrinsic topic *from authority.*

5) Robin Smith, "Logic," in *The Cambridge Companion to Aristotle,* ed. Jonathan Barnes (Cambridge, 1995), 27-65, esp. 47-64. Aristotle, *Rhetoric* I, 1-2.

I refer to "dialectic" in this last sense, and this is the sense that fits with the discussion of dialectical topics and that Aristotle treated in *Topics*.

There is, in fact, a close relationship between dialectic and rhetoric. Many dialectical topics have counterparts among the rhetorical topics. Aristotle himself saw rhetoric as the counterpart of dialectic—rhetoric aims specifically at persuasion while dialectic provides the strictly logical arguments.[6] Throughout the *Rhetoric* Aristotle refers readers to the *Topics*. Later in Book I, Aristotle says of rhetoric that it is an offshoot of both dialectic and ethical studies.[7] The aim of a dialectical argument is primarily to provide reasons in support of a conclusion; a rhetorical argument aims to achieve agreement on the acceptability of a conclusion. Both aim at persuasion, but dialectic relies primarily on logic, whereas rhetoric relies also on emotive language and the moral character or reputation of the speaker to achieve consensus.

Aristotle gave priority to demonstration among the types of reasoning and argument, and scholastic commentators followed him to some extent. Yet the emphasis on demonstration does not seem to correspond to practice where dialectical and rhetorical arguments prevail. The requirements for demonstration, especially scientific demonstration in the strictest sense, are so rigorous that meeting them seems to be a seldom-achieved ideal. In most disciplines the premises from which arguments begin are widely shared assumptions or opinions that are probable and hence the arguments are dialectical, not demonstrative. In a number of contexts, however, medieval philosophers held some conclusions so universally that they were accorded a certainty appropriate to demonstrated conclusions when in fact they did not actually satisfy the strictest requirements. The techniques taught in *Topics* were intended to achieve propositions that are true, necessary, certain, appropriate to the

[6] Aristotle, *Rhetoric* I, 1.1354a1-15. Frans van Eemeren and Peter Houtlosser, eds. *Dialectic and Rhetoric* (Boston, 2002), 3-11.

[7] *Rhetoric* I, 2.1356a20-30. There are at least ten explicit references to the *Topics* in the three books of the *Rhetoric*.

phenomena, and universal, but the results of actual practice suggest again that the achievement fell short of the ideal.[8]

On some accounts there is a hypothetical character to Aristotelian science that tends to get obscured by the emphasis placed on necessity and certainty.[9] Some modern interpreters see in Aristotle's views on deduction a distinction that corresponds to the deductive systems of the mathematical sciences, on the one hand, and to the hypothetico-deductive systems of the empirical sciences, on the other.[10] The demonstrations that derive from premises that hold necessarily and without exception belong to strictly deductive systems. Demonstrations derived from premises that hold usually or for the most part belong to hypothetico-deductive systems. Dialectical method leads the scientist to the principles that are appropriate for the object of the science in question. If the principles hold universally and necessarily, then the system is strictly deductive, corresponding to what is meant by scientific demonstration—demonstration in the strictest sense where the conclusions possess certainty. If the principles hold usually or for the most part, then the system is hypothetico-deductive, and the conclusions possess only greater probability than alternative conclusions.[11] Aristotle adopted scientific demonstration as the ideal, yet he cautioned readers to adopt the principles and methods appropriate to the scientific object in question.[12] Because principles in natural philosophy and cosmology hold usually or for the most part, the following account restricts

---

8) Eileen Serene, "Demonstrative Science," *The Cambridge History of Later Medieval Philosophy*, ed. Norman Kretzmann et al. (Cambridge, 1982), 496-517.

9) Serene, p. 510, expresses the point thus: "Aristotelian science was designed to explain a particular phenomenon by showing it to be a token of a type whose essential nature always necessitates that exact kind of phenomenon under the specified conditions."

10) Demetrios Koutras, "Aristotle and the Limitations of Deduction in Scientific Knowledge," in *Aristotle and Contemporary Science*, ed. Demetra Sfendoni-Mentzou et al., 2 vols. (New York, 2001), 2: 153-160.

11) Koutras, "Aristotle and the Limitations," 154-157.

12) R.J. Hankinson, "Philosophy of Science," in *Cambridge Companion to Aristotle*, 109-139, at 114-115 discusses the logic of "for-the-most-part" arguments, and maintains that Aristotle never worked out the details of the logic of these kinds of arguments.

the discussion to dialectical reasoning and argument. Two cautionary points bear repeating, however. First, the syllogistic form of the arguments creates the impression that the arguments are demonstrative in character. Second, dialectical inquiries and topics are intended to generate the principles from which deductions follow, and may thereby suggest that the principles are more demonstrative in strength than they actually are.

Until scholars began routinely to learn Greek and return to the originals, medieval masters and students were dependent on Latin translations of Aristotle's works. In late antiquity Boethius translated numerous Greek works into Latin, and he wrote his own treatises on logic. Aristotle's *Topics* provides rules that are supposed to make it easier for debaters in dialectical disputation to discover arguments.[13] This notion of discovery influenced the ancients, but Aristotle's text is very long and unwieldy; followers began to systematize the topics in different ways to make them easier to use. In his definitive work on dialectic, *De topicis differentiis*, Boethius regards the finding of arguments as its main task. "Topic" means "place." The notion here is that by means of a topic we can find a term that will mediate between two given terms in an argument.[14] Boethius divides topics into maximal propositions or principles that are self-evident, true generalizations and differentiae under which maximal propositions can be grouped. The differentiae constitute Boethius's main instrument for finding arguments providing a term or intermediate term between two terms of a question. The third term joins the other two together in the form of a syllogism. In specific cases where a differentia is used to find a third term, it may be necessary to validate the argument, and that is the function of the maximal proposition, namely as a sort of generalized rule. In categorical syllogisms, maximal propositions are general premises

[13] Smith, "Logic," 57-62; Niels Green-Pedersen, *The Tradition of the Topics in the Middle Ages* (Munich, 1984), 37-126; Eleonore Stump, "Dialectic," in *The Seven Liberal Arts in the Middle Ages*, ed. David Wagner (Bloomington, 1983), 131; and Eleonore Stump, *Dialectic and its Place in the Development of Medieval Logic* (Ithaca, 1989), 3.

[14] Smith, "Logic," 61-62; Stump, *Dialectic*, 57-66; Jan Pinborg, "Topik und Syllogistik im Mittelalter," in *Sapienter ordinare, Festgabe für Erich Kleineidam*, Erfurter Theologische Studien, 24 (1969), 157-178, at 158-160.

that guarantee the validity of the argument; in hypothetical syllogisms they validate the connection between the antecedent and consequent in the conditional premise of the argument. Hence, differentiae help us find third terms to construct arguments, and maximal propositions validate the argument discovered.[15]

The most important introductory textbook on logic after the thirteenth century was Peter of Spain's *Tractatus*.[16] It is divided into two main parts, the first summarizing Aristotelian logic as interpreted by Boethius, and the second the scholastic terminism of the Middle Ages. The first part, then, contains an introduction to categorical syllogisms, and treatises on predicables, predicaments (categories), syllogisms, topics (*De locis*), and fallacies. The second part includes treatises on supposition, relatives, ampliation or extensions, appellation, restriction, and distribution, all of which derive from the thirteenth-century common theories about properties of terms.[17]

Peter conceived of dialectic as an art rather than as a science. "An argument," he says, "is a reason producing belief regarding a matter that is in doubt, that is, a middle proving a conclusion that needs to be confirmed by means of an argument."[18] Peter followed Boethius in dividing a topic into a maxim (maximal proposition) and a differentia, and also dividing topics into three subcategories: intrinsic, extrinsic, and intermediate. An intrinsic topic appears in an argument that is taken from the nature of one of the terms of the argument; extrinsic in an argument that is taken from what is external to their nature; and intermediate is one that comes partly from the nature of the term and partly from without.[19] For Peter a topic confirms an argument by providing a middle term through

---

15) Stump, "Dialectic," 133-135.

16) Peter of Spain, *Tractatus*, ed. L.M. de Rijk (Assen, 1972).

17) Stump, "Dialectic," 128-131.

18) Peter of Spain, *Topics*, tr. Eleonore Stump, in *The Cambridge Translations of Medieval Philosophical Texts*, ed. Norman Kretzmann and Eleonore Stump (Cambridge, 1988), 226-245, at 226.

19) Stump, *Dialectic*, 139. In dividing topics into intrinsic, extrinsic, and intermediate, Boethius had followed Themistius.

which an enthymeme, an incomplete syllogism, can be transformed into a complete syllogism.[20]

The intrinsicness of topics in the first subcategory does not mean that these topics produce demonstrations.[21] Where they are dialectical, the syllogisms that are produced by means of intrinsic topics are probable, hence even intrinsic topics in such syllogisms are considered probable and only contingently related to a term in the conclusion. In demonstration, by contrast, intrinsic terms are always essentially and necessarily related to a term in the conclusion. As this clarification suggests, there is a metaphysical doctrine at work here. In strict demonstrations (*demonstratio propter quid*) the middle term refers to a real cause of the connection between extreme terms. Even in demonstrations of the fact (*demonstratio quia*) the middle term is an assertion about the essence of the subject by means of which we demonstrate an essential attribute of the subject. In dialectical arguments, however, the middle term does not refer to a cause and it is not an assertion of the essence of a subject but rather a sign of the connections between things and the validity of an inference. As a consequence, dialectical arguments produce at best only highly reliable opinions, not knowledge in the strict sense.[22]

Peter distinguishes twenty-five differentiae, forty-three topics, and fifty-seven maxims. Under intrinsic differentiae he distinguishes between topics from substance and topics from concomitants of substance. Under the latter Peter enumerates *from the whole* (universal whole, a species or subjective part, *integral whole*, whole in quantity, whole in mode, whole in place, and temporal whole), *from a cause*, *from generation*, *from destruction*, *from uses*, and *from associated accidents*. As an example of an integral whole, Peter provides the following example: "A house exists, therefore, a wall exists." The topic is *from an integral whole*, and the maxim is "If an integral whole is posited, then any one of its parts is also posited."

---

20) Ibid., 140-146, where William of Sherwood makes explicit what is admittedly only implicit in Peter.

21) Pinborg, "Topik," 178.

22) Stump, *Dialectic*, 155.

That in brief and with focus on the topic *from an integral whole* is Peter's account of topics. Topics serve a function in arguments. A conclusion is a proposition that is proved by an argument. Before it is proved, however, it is in doubt, that is, we have a question about it. From the conclusion we identify extreme terms, and the exercise is to find a term that mediates between the extreme terms in the conclusion. Topics help us to find middle terms. An argument, then, is a reason that produces belief about a matter that is in doubt.

The relevance of such details to Copernicus's comments about method and to his arguments brings us to discussions at Cracow University in the 1490s. In evaluating conditional arguments and consequences, John of Glogovia in particular came to grips with the conditions for their validity. In commentaries on Peter of Spain's treatises, John claimed that the truth or falsity of a consequence is independent of the truth or falsity of its component statements, but is rather dependent on a *connection* between antecedent and consequent.[23] The connection between antecedent and consequent is "based on the intentions (meanings, concepts) whose relationships are recognized by *topical maxims*" [emphasis added]—and the topics mentioned by John are essential superior to inferior, whole to essential part, *integral whole to part*, cause and effect, cause of following, correlative, inclusion, and containing and contained. Among the criteria for a sound consequence enumerated by professors at Cracow was the following minimal condition: what is signified by the consequent must be included in, contained by, or relevant to what is signified by the antecedent. Conversely, a good consequence admits nothing extraneous or irrelevant.[24]

This background helps to illuminate Copernicus's controversial statement in the dedicatory letter to the Pope:[25]

---

[23] Ivan Boh, "John of Glogovia's Rejection of Paradoxical Entailment Rules," in *Die Philosophie im 14. und 15. Jahrhundert*, ed. Olaf Pluta (*Bochumer Studien zur Philosophie* 10) (Amsterdam, 1988), 373-383, at 381.

[24] Boh, "John of Glogovia's Rejection," 381.

[25] Nicolaus Copernicus, *On the Revolutions*, tr. Edward Rosen, *Complete Works*, Vol. 2 (Warsaw and Cracow, 1978), Preface, 4.

> Hence in the process of demonstration or "method", as it is called, those who employed eccentrics are found either to have omitted something essential or to have admitted something extraneous and wholly irrelevant. This would not have happened to them, had they followed sound principles.

The point is that scholastic philosophers used topics such as *from an integral whole* and conditions such as relevance between antecedents and consequents to evaluate the validity of conditional arguments and consequences.

We may also consider briefly another possibility. If Copernicus attended classes in law at Bologna, then he almost certainly received instruction in the application of dialectical topics to legal cases. In treatises on legal dialectic we encounter references to the standard medieval logical treatises, especially Peter of Spain, and to the same kinds of definitions and distinctions that we mentioned above. In addition, legal scholars developed topics specific to legal jurisprudence, some of which derive from humanist contributions on the editing and interpretation of texts. The topics *from similitude* and *from authority*, as one would expect from jurists, were the most important, but they often employ the topic *from an integral whole*. It is also a striking fact that jurists routinely assumed connection or relevance as a condition of validity for consequences.[26] In short, Copernicus might have first encountered these techniques and forms of argumentation at Bologna, although it is likelier that his education at Cracow prepared him to follow lectures on the law and to apply dialectical topics to legal cases.

I began this part of my account by explaining why I was situating the discussion of mereology within the context of dialectical topics, a point to which I return later in the paper. As the introduction also indicates, however, mereology was a major philosophical subject in its own right. To complete this brief and selective survey, I now turn to philosophical analyses of part/whole relations in the western tradition.

The first explicit analysis of part/whole relations in western thought appears in Plato. In *Parmenides* (137a-145e) Plato pursues the

---

26) Vincenzo Mortari, "Dialettica e giurisprudenza. Studio sui trattati di dialettica legale del sec. XVI," *Annali di storia del diritto*, 1 (1957), 239-401.

consequences of contrary assumptions. If being has parts, then what is the relation of the whole to the parts? Is the whole just the collection of the parts or more than the sum of the parts? In *Theaetetus* (202e-205e) Plato notes the difference between a syllable and the letters in it. A syllable is formed from letters yet is distinct from them, for it has its own peculiar form or arrangement. Plato also concludes that there is a difference between a sum and a whole.[27] In other words, an arithmetical sum is not the same as a whole that possesses unity (*hólon* or *súnolon*).

Aristotle took up these issues in several texts, too many to summarize briefly.[28] I select texts essential for the results that follow, namely, those regarding integral whole and parts and the texts where Aristotle draws conclusions about the order of the universe and the natural motions of bodies. Although Aristotle did not discuss part/whole relations in a systematic way, his discussion of concrete wholes in *Topics* (VI, 13, 150a22-b36) forms the basis for the later dialectical topics *from an integral whole* and *from an integral part*: "The whole perishes when the parts do, but the parts do not necessarily perish when the whole has perished." This assertion is the basis for the consequences, "If the whole is, then the part is; if the part is not, then the whole is not." But the following are not valid: "If the whole is not, then the part is not; if the part is, then the whole is."[29] Aristotle adds that a whole compounded of parts in a particular way is not just a totality, sum, or collection of parts. In other

27) Additional texts from Plato: *Phaedrus* 265d-266b, *Philebus* 16c-17a, *Sophist* 260e-261d. Burkhardt and Dufour, "Part/Whole," 663-664; King, "Medieval Mereology," 2.

28) *Topics* V, 5; VI, 11, 13-14; *Physics* I, 1, 4; II, 4; III, 5-6; IV, 5; *De caelo* I, 3; II, 14; *Metaphysics* I, 5; IV, 2, 5; V, 2-3, 23, 25-27; VII, 2; VIII, 1, 6; X, 1; XI, 10; XII, 8, 10; and XIII, 8; *Parts of Animals* 640b1-5; 646a25-30; *Rhetoric* II, 24.2.1401a23-b5; and *Poetics*, 6 and 12. In the analytical indexes of technical terms in his translations of *Metaphysics* (Ann Arbor, Michigan, 1960) and *Physics* (Lincoln, Nebraska, 1961), Richard Hope provides many more references to *méros*, *mórion* (*pars*, *particula*), *hólon* (*totum*), and *súnolon* (a composite or integral whole, sometimes also rendered *totum* in Latin and *totum integrale*).

29) *Rhetoric* II, 24, 1401a23-b5 supports these qualifications, but also points out examples of syllogisms that are not genuine, for example, from a whole action that is wrong concluding that each part is wrong.

words, a quantity (*posón*) is not what he understands by a composite whole (*súnolon*).

Metaphysics (V, 15-16) clarifies this qualification by using the word "total" (*hólon*) to describe a quantity the parts of which can change position without affecting the aggregate. A total, in turn, is distinguished from organic substances composed of homogeneous parts such as tissue, flesh, and bone, thus generating the distinctions (*Parts of Animals* II, 1, 646a1-647b9) among three sorts of substantial composition: elemental, uniform (bone, flesh, and tissue), and non-uniform (organs). Things in which the whole is not like a heap or lump have some uniting factor (*Metaphysics* VIII, 6, 1045a8-12). Such texts are connected to the teleological dimension in Aristotle's biological thinking but they have echoes in his cosmological views of Earth and the celestial spheres as primary beings—the world is a cosmos, a definite something, all things ordered together around a common center (*Metaphysics* XI-XII). We commonly divide the Aristotelian cosmos into two spheres (celestial and terrestrial), but Aristotle allows for many ways in which the celestial influence the terrestrial. For example, the motions of higher bodies are transmitted to lower bodies even down to the rotation of the upper terrestrial atmosphere, the Sun warms Earth, and many continued to be believe in the astrological influences of the qualities and motions of celestial bodies.

In *Metaphysics* IV, 5, 1010a28-32 Aristotle criticizes his predecessors for having attributed to the whole that which is true only of the part:

> We should add another criticism against those who hold these views: they have been reporting what they observe in only a few sensible things as if it were true of the whole cosmos. For it is only the region of what is sensible round about ourselves that is continually in process of destruction and generation; but this is, so to speak, not even a small part of the whole, so that it would have been more just to acquit this small bit because of the whole than to condemn the whole because of this small bit.[30]

30) Hope translation, 79-80.

Aristotle himself concluded that the heavens were made of an element different from the elements in the sublunar realm. For him, each simple body or element has only one natural motion, and the whole of an element, if united, and the part move naturally in the same direction, as, for example, the whole Earth together and a small clod (*De caelo* I, 3, 269b30-270a13). *De caelo* II, 14, 296b27-297a7 summarizes the argument succinctly:

> If it is inherent in the nature of earth to move from all sides to the centre (as observation shows), and of fire to move away from the centre towards the extremity, it is impossible for any portion of earth to move from the centre except under constraint; for one body has one motion and a simple body a simple motion, not two opposite motions, and motion from the centre is the opposite of motion towards it. If then any particular portion is incapable of moving from the centre, it is clear that the earth itself as a whole is still more incapable, since it is natural for the whole to be in the place towards which the part has a natural motion.[31]

It should be noted, however, that constructive arguments from part to whole depend on the conception of a natural whole thing. By concluding that the heavens are made of a different fifth element, the aether, with a distinctive natural motion in circles around the center of the cosmos, Aristotle cut off arguments that the celestial motion follows the same patterns as terrestrial motion.

Although they departed from Aristotle on their notions of the relation between God and the created world, Neoplatonic and medieval authors devoted much attention to part/whole relations.[32] Their motives varied, but a number of puzzles caught their attention. For Neoplatonists like Proclus, the whole is hierarchically related to the parts, an idea that could be easily adapted to Aristotle's arrangement of the spheres down to the center of Earth. Boethius, sympathetic to such schemes, summarized and interpreted Aristotle's doctrines on topics and division. As we indicated above, he devised in particular the formulas that would be repeated down to the

---

31) W.K.C. Guthrie, tr. *Aristotle on the Heavens* (Cambridge, Massachusetts, 1939), 245-247.

32) King, "Medieval Mereology," 2-3; Annik Charles-Saget, *L'Architecture du divin* (Paris, 1982), 65-89. I owe the latter reference to Dilwyn Knox.

Copernican era. With regard to whole/part topics, Boethius organized the topics into rules for the discovery of arguments. He adopted Cicero's division of topics, two of which are *from the whole* and *from the enumeration of the parts*. In discussing parts, Boethius distinguished between two ways: parts taken as species or as members. By "members" Boethius meant "real parts." He discussed inferences from whole to parts and from parts to whole, often referring to examples of integral whole and integral parts.[33] The scholastics adopted this category of topic, introducing a number of distinctions by way of ever more sophisticated examples. From the existence and non-existence of whole and parts, they developed rules for valid and invalid constructive and destructive inferences respectively.

Already implicit in Aristotle, who, however, had applications to organic substances pre-eminently in mind, the consequences that we owe to Cicero and Boethius are the ones found later in Peter of Spain with the typical example of the house:[34] (1) If the house exists, then the parts (roof, walls, foundation) exist. (2) If the house is not, then the parts are not. (3) If the parts are, then the house is. (4) If the part is not, then the house is not. Everyone agreed that (1) and (4) are valid, although some authors added qualifications. For example, Abelard regarded an integral whole as identical with a unique set of parts including the arrangement that individual parts have. (2) is then qualified to read: (2') If the house is not, then the parts-of-this-house are not (which in this form is valid). Although (3) remains invalid, in the following sense it is valid: (3') If the parts-of-this-house are, then the house is *potentially*. Later scholastics emphasized integral wholes that are actually composed out of their parts, implicitly recognizing that not all parts of a whole (*totum*) are plausibly parts of an integral whole (*totum integrale*), otherwise a monster results. For example, cobbling together hands, feet, and a head does not make up a human being.[35]

---

33) Eleonore Stump, tr. *Boethius's De topicis differentiis* (Ithaca, N.Y., 1978), 37, 51-52, and 65; Anicius Boethius, *De divisione liber*, ed. and tr. John Magee (Leiden, 1998), 38-41.

34) Peter of Spain, *Tractatus*, V, 64.

35) King, "Medieval Mereology," 16-18.

The claim about unique arrangement of parts can generate other puzzles noted by medieval philosophers. If I repair a house by replacing some parts, is it still the same house? Some changes (renovation) may indeed produce a new, different house. Such considerations led medieval philosophers to a variety of solutions. Some distinguished between principal (or vital) parts and less principal (non-vital) parts, one consequence of which is that the destructive application *from the integral part* (If the part is not, then the house is not) holds only for principal parts. For things that change, move, grow, or diminish, the distinction provides a way of preserving identity over time and for distinguishing between principal and less principal parts.[36]

Answers to questions about the composition of homogeneous and heterogeneous substances, and about continuous, contiguous, and discrete things often depended on how authors interpreted Aristotle. Aristotle thought the cosmos *qua* unified thing "has" a nature in the sense that the parts are ordered according to the substantial or qualitative forms that they possess as inferred from their observed motions and behaviors. Where Aristotle observed contrary simple motions (up and down), he concluded that the contrary motions must be due (aside from their generating cause) to contrary qualitative forms that account for a natural directionality in the things possessing those qualities. The observed circular motions of celestial bodies must be due to a form that has no contrary thus revealing a different substance the nature of which is to move in circles. Aristotle's metaphysical analysis also led him to the existence of immaterial intelligences that are the cause of eternal circular motion. The contiguity of the celestial spheres is revealed in the way that the circular motions influence the motions of lower spheres down to comets in the upper atmosphere and even to the circular motion of the Earth's atmosphere.[37]

Even within Aristotle's lifetime, however, astronomers made observations that introduced complications into his picture. A few

36) King, "Medieval Mereology," 24; Henry, *Medieval Mereology*, 54-57, 82-84, 334-336; Simon, *Parts*, 198-204.

37) *De caelo* I, 2; *Metaphysics* XII, 7-9.

astronomers departed radically from the geocentric view, but Plato's insistence on uniform circular celestial motions,[38] Aristotle's physical interpretation of the Eudoxan-Calippan model, and his strong support of commonsense observation effectively constrained alternative theories and interpretations into different versions of geocentrism. On the other hand, Ptolemy's more accurate mathematical geocentrism also successfully resisted criticism. Although some philosophers and astronomers continued to seek and construct alternatives, all of them remained geocentric.

Copernicus himself was committed to the uniform, circular motions of all of the celestial spheres. Ptolemy's failure to preserve the uniform circular motions of the planetary and lunar spheres (the so-called Platonic axiom) without resorting to an equant model and other failures led Copernicus to a major cosmological revision in the service of an otherwise conservative reformation of astronomy.[39]

I turn now to the role of mereology in Copernicus's revision of cosmology.

Copernicus's Use of Mereological Topics

To apply the above summary of the tradition in a way that makes the issues as clear as possible, I resort here to simpler kinds of dilemmas. For example, the topics *from an integral whole* and *from an integral part* seem to imply that if, for example, Socrates loses a hand, then Socrates does not exist. One strategy for dealing with

---

38) Plato, *Republic* VII, 529a-530d; Olaf Pedersen and Mogens Pihl, *Early Physics and Astronomy* (New York, 1974), 26-30 and 65-67.

39) Copernicus did not explain exactly how he arrived at his version of the heliocentric system. I have reviewed the reconstructions in "Reflections on the Origin of Copernicus's Cosmology," *Journal for the History of Astronomy*, 37 (2006), 37-53. Only Copernicus's acceptance of the Platonic axiom and spheres explains his insistence on preserving uniform, circular motions and use of eccentrics and epicyclets in his models. On pp. 45-46, I order the steps in Copernicus's path to his cosmological theory, showing in particular how Copernicus assembled and reflected on irregularities, inconsistencies, and several striking facts that neither Ptolemy nor any geocentrist could explain.

this apparently absurd consequent is to define an integral part as the kind of part that cannot be removed without destroying the whole. As we saw above, authors who applied this strategy made distinctions between vital parts (applicable only to organisms) and less vital parts, or between principal as opposed to less principal parts. Or, to put it negatively, a non-integral part is one that we can remove and still have the same individual. This criterion does not remove all ambiguity. We could attempt to find some non-arbitrary criterion, but we would obscure an important lesson from these examples. The answer that we would give to such a question reveals our view of what it is to be an individual and of what constitutes the identity of an individual.

Copernicus's mereological vision of the universe refers to an intuition about the universe as a composite or integral whole. By analyzing the intuition held by him and his followers, we can reveal features of their heliocentrism that otherwise remain obscure. We will also discover why they believed that their conclusions were more probable and likely than those of geocentrists. Finally, we will also appreciate why Michael Mästlin and Johannes Kepler found Copernicus's answers persuasive, even compelling.

Copernicus finally explained the reasons, already implicit in *Commentariolus*, why he adopted heliocentrism in its clearest form some thirty-two years after he first proposed the heliocentric theory.[40] In the Preface to the Pope, he repeats all of the complaints about the failures of his predecessors, but he expresses there with the greatest clarity to date, perhaps under Georg Rheticus's influence, why his predecessors by their assumptions and methods had failed. They had failed to construct a unique, non-arbitrary, and commensurable structure befitting the exquisite craftsmanship of its architect. They had, as late scholastic authors had suggested, cobbled together

---

40) Note that the reasons for adopting heliocentrism constitute a reconstruction, not a description of the path to his heliocentric cosmology. In the reconstruction, Copernicus provides hints about how he arrived at the heliocentric theory, the astronomical details of which I have discussed in "Reflections" and in André Goddu, "Hypotheses, Spheres, and Equants in Copernicus's *De revolutionibus*," in *Les elements paradigmatiques thématiques et stylistiques dans la pensée scientifique*, ed. Bennacer el Bouazzati (Rabat, 2004), 71-95.

from incompatible and irrelevant parts a thing that resembled a monster. They would not have failed, he alleges, if their method had been sound.[41]

Copernicus uses repeatedly the topics *from an integral whole* and *from an integral part* like a chisel to fashion a heliocentrism that is likelier than geocentrism, and to refashion Aristotelian and non-Aristotelian accounts of natural elemental motion so that they fit into a heliocentric cosmos.

In the Preface Copernicus argues that the hypotheses should be relevant to the structure of the universe, and the resulting composite whole should possess commensurable parts. The topic *from an integral whole* supports an inference from the composition of the whole having parts to the existence and place of a part. He compares the universe to the portrait of a single individual having relevant parts ordered harmoniously or commensurably with a picture with parts taken from different individuals resulting in a monster. In other words, the Copernican system is to the Ptolemaic system as a portrait of a harmonious individual is to the portrait of a monster.[42] He supports this analogy and the inference on which it is based with additional arguments that as "the best and most systematic Artisan of all," God arranged the parts of the world machine in a definite and unique order that we are capable of discovering because God created the system for our sakes. He clinches this series of arguments with the following inference: Because "the order and size of all the planets and spheres, and heaven itself is so linked together," by virtue of the topic *from an integral whole*, it follows "that in no portion of it can anything be shifted without disrupting the remaining parts and the universe as a whole."[43]

41) Rosen translation, *On the Revolutions*, 4-5. This is the argument that relies on dialectical topics to argue for relevance between antecedent and consequent as a condition of validity.

42) That Copernicus means "harmonious" to be taken as "commensurable" is clear from the ratios that he assigns to the planetary spheres.

43) Rosen translation, *On the Revolutions*, 5. Copernicus adopts the same arrangement for the entire book beginning with the general structure of the universe and in the remaining books treating the motions of the parts.

In I, 4 of *De revolutionibus*, Copernicus makes his first application of the topic *from an integral part*. After acknowledging the apparent disagreements between uniformly moving spheres and the observed non-uniform motions, Copernicus cautions us to examine the relation of Earth to the heavens carefully, otherwise we may make the error of attributing to the heavens what belongs to Earth thus violating the rule that permits only a negative inference from integral part to whole. That is to say, the error would be to attribute to the whole that which belongs to one of its parts only. Copernicus, however, soon qualifies this rule.

As he takes up the problem of relative motion (I, 5), the question, as Copernicus sees it, is whether we should attribute motion to a part or to the universe as a whole. In I, 6, Copernicus takes advantage of Ptolemy's eccentric models and his admission that the Earth's diameter is as a point in relation to the size of the universe. It follows, argues Copernicus, that it is impossible to say whether Earth is the exact geometrical center of the universe. If we cannot prove that it is the exact center, we cannot prove that it is always at rest in the middle. This is the reason why the ancients resorted to physical arguments to prove that Earth is at the center. Their arguments relied on claims about "natural" motions and that the elements have simple natural motions. Copernicus's objections rest on what we call "natural." To do so we must identify the nature of a thing correctly. How Earth in fact behaves must be ascertained before we can conclude what its nature is and what is natural to it. This suggests, however, not that all inferences from part to whole are invalid, but that before making such inferences about physical explanations, it is necessary to be sure that one's facts are accurate. Only a careful critique that relies on already well-established features of the whole can guide us. In other words, only inferences from part to whole that result in conclusions inconsistent with established axioms or conclusions about the whole are invalid.

Copernicus almost certainly relied on ancient interpretations of Aristotle and accounts that are alternatives to his.[44] He intends all

---

[44] Dilwyn Knox, "Copernicus's Doctrine of Gravity and the Natural Circular Motion of the Elements," *Journal of the Warburg and Courtauld Institutes*, 68 (2005), 157-211.

of his objections to raise doubts in the mind of readers about the immobility of Earth based on the claim that a rectilinear motion is a simple natural motion. Rectilinear motions of objects seem rather to belong to objects that are not in their "proper condition and not in complete accord with their nature."[45] Circular motion belongs to wholes, rectilinear motion to parts, and hence they are relative to one another. After he draws all of the consequences from consideration of multiple centers of revolutions and the relativity of motions, he appeals to the "principle governing the order in which the planets follow one another" and "the harmony of the entire universe." In short, Copernicus assumed from the outset that we can answer affirmatively a question about a part only by establishing the order of the whole. From the existence of the order of the whole we can infer the order of one of its parts. Copernicus's reasoning depends on architectonic principles that subordinate some natural-philosophical hypotheses to astronomical considerations. His argument from integral whole to part retains a dialectical character for it depends on the greater probability of his architectonic principles.

Copernicus's mereological vision of the universe refers, then, to his belief in a universe that is a unified and integral whole composed of discrete principal parts arranged in a definite commensurable order. A hypothesis that seems counterintuitive or contrary to commonsense perception can be shown to be probable by reference to the whole structure, for we can draw conclusions about the position and motion of a part only from consideration of the whole. Our conclusions about the whole, in turn, derive from a critique of accounts inferred from appearances of a part.

This analysis does not exhaust the logic of Copernicus's arguments, but it does illuminate one of his principal logical resources in support of a heliocentric cosmology. To my knowledge, Michael Mästlin is the only author who noticed this strategy. In a preface

---

See also André Goddu, "Copernicus's Annotations: Revisions of Czartoryski's 'Copernicana'," *Scriptorium*, 58 (2004), 202-226 with eleven plates, where I analyze Copernicus's annotations and reading of sources.

[45] Whether Copernicus means that rectilinear motions are unnatural or merely less natural is unclear.

to the 1596 edition of Rheticus's *Narratio prima* and Kepler's *Mysterium cosmographicum*, Mästlin expresses this principal logical or dialectical warrant in Copernicus's arguments:[46]

> Is not the earth and the air around the earth the seat and the whole dwelling-place of all things that to us are heavy or light? But how are earth and the air around it related to the immense vastness of the whole world? They are as a point or small points, and calling them even smaller, if that were possible, would be more fitting. For those reasons do you not consider the Philosopher's inference from such a little part or point to the whole world to be a weak argument? We cannot, therefore, conclude anything certain about the center of this most spacious world from those things that approach this little point or move away from it. This [tendency] of heavy and light things to seek their proper place, which by testimony of the Philosopher is the perfection of the thing, is really an inclination implanted in [all] things by nature, as Copernicus says with erudition in Book I, chapter 9, [and] belongs credibly also to the sun, the moon, and the other wandering stars, that they retain permanently the shape of a sphere as representing their most perfect form. If any such place is the [unique] center of the world, then it is by accident. Truly, Copernicus's reasoning in astronomy proceeds, not from the minute parts to the whole, but on the contrary from the whole to the parts. But from this is easily recognized which is the more convincing, Copernicus's procedure or that

46) Michael Mästlin, preface, in: Heribert Nobis and Anna Pastori, ed. *Receptio copernicana*, in *Nicolaus Copernicus Gesamtausgabe*, 9 vols. (Berlin, 1974-2004), 8/1: 448-452, at 449-450: "An non omnis sedes et totum domicilium omnium eorum, quae nobis gravia sunt aut levia, Terra, et circa terram Aer est? Sed quid Terra, quid eam ambiens Aer, respectu immensae totius Mundi vastitatis? Punctum sunt, sive punctuli, et si quid minus dici posset, rationem habent. Quod cum sit, an non Philosophum dicturum putas, quod infirma argumentatio a particula sive hoc punctulo ad totum Mundum extruatur? Non ergo ex iis, quae ad hoc punctulum appetunt vel ab eo refugiunt, de spaciosissimi huius Mundi centro certi esse possumus. Locum quidem suum proprium, qui Philosopho teste est perfectio rei, haec nostra gravia et levia a Natura sibi tributum appetunt, quam affectionem, ut Copernicus lib<ro>. I cap<itulo> 9 erudite disserit, credibile est etiam Soli, Lunae caeterisque errantium fulgoribus inesse, ut eius efficacia in ea, qua se repraesentant, rotunditate permaneant: Quod si is locus alicubi simul sit Mundi centrum, id non nisi per accidens contingit. Verum Copernici rationes Astronomicae non a particula, ea quae minutissima, ad totum: sed contra, a toto ad partes procedunt. Sed et ex ipso hypothesium usitatarum et Copernici processu facile agnoscitur, utrae plus fidei mereantur. Etenim Copernici hypotheses omnium Orbium et Sphaerarum ordinem et magnitudinem sic numerant, disponunt, connectunt et metiuntur, ut nihil quicquam in eis mutari aut transponi sine totius Universi confusione possit; quin etiam omnis dubitatio de situ et serie procul exclusa manet."

of the usual hypotheses. Because Copernicus's hypotheses enumerate, dispose, connect, and measure the order and size of all orbs and spheres in such a way that none of them can be changed or transposed without throwing the entire universe into confusion, [for all of these reasons] all doubt about position and order remains far removed.

Mästlin criticizes claims made by Aristotelians that the observations of bodies falling to Earth prove that Earth is the center of the universe. Such observations confirm at best that the observed bodies have a tendency to fall towards the center of Earth. What is the justification for inferring the whole from the part? Copernicus was right to argue from the whole to the parts. Mästlin follows that claim immediately with Copernicus's hypotheses by means of which he enumerates, arranges, connects, and measures the order and magnitude of all orbs and spheres, such that no change can be admitted without throwing the entire universe into confusion. Mästlin fails to notice, however, that the inference about gravity depends on a constructive application of the topic *from an integral part.*

Copernicus's vision of a structured and commensurable heliocentric universe persuaded Kepler. Kepler's astronomy bears an almost ironic relation to Copernicus's, yet he defended Copernicus on logical grounds:[47]

> I particularly abhor that axiom of the logicians [dialecticians] that the true may follow from the false, because people have used it to go for Copernicus's jugular, while I am his disciple in the more general hypothesis concerning the system of the world. I therefore considered it particularly worthwhile now to show the reader how it does happen here that the true follows from the false.
> First, you have already seen that what has followed is not exactly the truth.[48]

---

47) By "ironic" I am referring to his critique of the geocentric remnants in Copernicus's theory.

48) See Johannes Kepler, *Astronomia nova*, ed. Max Caspar, *Gesammelte Werke*, 21 vols. (Munich, 1937-1990), 3: 183: "Cavsa, cvr falsa hypothesis vervm prodat et quatenus? Porro quia ego axioma hoc dialecticorum, ex falso verum sequi, vehementer odi, propterea quod eo Copernici (quem sequor magistrum in hypothesibus universalioribus systematis mundani) jugulum petatur: operae precium putavi lectori ostendere, quomodo his ex falso verum sequatur. Primum jam vidisti, non plane verum sequi." *The New Astronomy*, tr. William Donahue (Cambridge, 1992), 294.

Kepler goes on to explain that his own first supposition does not give the planet Mars the right latitude. "So it is not exactly the truth that follows from this false hypothesis." Although he does not cite Aristotle here and, indeed, expresses his contempt for academic logicians, Kepler was in fact following Aristotle's assertion that a false assumption will sooner or later be exposed by the consequences that follow from it.[49] That was why he found it necessary to test extensively every relevant alternative hypothesis. By contrast, from what is true only the true will follow.

There is an even more profound sense in which Kepler followed Copernicus in logic, but to a conclusion that might have shocked Copernicus and did worry Mästlin.[50] Kepler does not explain the logic of his move, yet it is clear from his procedure. If we can argue validly from an integral whole to one of its parts, Kepler concluded that there must be a unified set of physical principles that account for the motions of all of the planets. Once we have concluded from Copernican cosmology that Earth is a planet, then physical principles that account for motions of Earth must be true of the motions of other celestial bodies as well. Because all of the observable heavenly bodies appear to be spheres, the rectilinear component of falling bodies led Copernicus to speculate that gravity is a tendency that God implanted in all of the observable celestial bodies in the form of a sphere (Sun, Moon, and planets). His cosmos, however, is finite and the spheres of the fixed stars and the Sun are stationary, hence Copernicus did not conclude that all spheres rotate on their axes. Valid inferences from part to whole, then, rely on observation and depend on a critique of existing accounts, eliminating those that do not fit in the already established structure. Kepler concluded that the physical principles of motion discovered on Earth must be the principles of all motion.

Mereology is a vast and highly contentious area of philosophy. This essay offers a mere sketch, and provides only a glimpse of Copernicus's mereological vision of the universe. For all that I hope

---

49) Aristotle, *Nicomachaean Ethics* I, 8, 1098b11-12.

50) Michael Mästlin, "Annotations," in Owen Gingerich, *An Annotated Census of Copernicus' De Revolutionibus* (Leiden, 2002), 223.

that the paper suffices to reveal the power of mereology for Mästlin and Kepler, and, in conclusion, to anticipate its effect on Newton's rules of reasoning in natural philosophy. Rule 3 reads in part:[51]

> The extension, hardness, impenetrability, mobility, and force of inertia of the whole arise from the extension, hardness, impenetrability, mobility, and force of inertia of each of the parts; and thus we conclude that every one of the least parts of all bodies is extended, hard, impenetrable, movable, and endowed with a force of inertia. And this is the foundation of all natural philosophy.

Modern interpreters refer to this concept as "transdiction" and "projection."[52] It depends on the principle that "nature is always simple and ever consonant with itself,"[53] and is, in fact, a version of the constructive application of the topic *from an integral part.*

---

51) Isaac Newton, *The Principia*, tr. I.B. Cohen and Anne Whitman (Berkeley, 1999), 795-796.

52) Cohen, "A Guide to Newton's *Principia*," in *Principia*, 199, note 5.

53) Newton, *Principia*, 795.

From "Dragonology" to Meteorology: Aristotelian Natural Philosophy and the Beginning of the Decline of the Dragon in China

Qiong Zhang*
Wake Forest University

Abstract
The cult of the dragon in China, which expressed itself not only in the ritual sacrifices to the dragon kings during drought and floods but also in the rationalization of the dragon's power to make rain by many serious thinkers from diverse intellectual persuasions, was first subjected to sustained criticism during the early modern era as part of an "enlightenment" drive against popular cults and "superstitions" led by some of the Jesuit-inspired Chinese scholars. This paper examines how these critics drew on Aristotelian conceptions of nature and meteorological theories introduced by the Jesuit missionaries to attack the core ideas of the traditional dragon lore and their underlying cosmology. It argues that the de-animated and rigidly stratified view of nature articulated by this small but clearly discernable group of Chinese critics can be seen as marking the beginning of the decline of the dragon, the allegedly semi-divine aquatic animal which swims, walks, flies, and makes rain.

Keywords
dragon lore; *Qi* and Chinese cosmology; Aristotle; *meteorologica*; Jesuit mission in China

"On the fifteenth day of the twelfth month, in the eighth year of Zhizheng reign (1348 CE), at about 3 o'clock in the afternoon,

* Department of History, Wake Forest University, P.O. Box 7806, Winston-Salem, NC 27109, U.S.A. (zhangq@wfu.edu). I would like to thank Steve Marrone who kindly shared with me his insights and work in progress on parallel issues in European history. I am also indebted to Edith Sylla and the anonymous outside reviewer whose comments and suggestions helped me refine the arguments of this paper.

there appeared in the south four black dragons coming down from the clouds, and taking up water. Shortly afterwards another appeared in the southeast, and lasted a considerable time before it disappeared. This was seen at Jiaxing city."[1] Here Yang Yu, the fourteenth-century Chinese author, was not describing *metaphorically* what we would call "a whirlwind with water sprouts." For him and his contemporary readers, the dragons were not only fully present in nature but had an active role to play in what happens in the atmosphere.

Indeed, the dragon had enjoyed such a phenomenal presence in Chinese culture since the Shang Dynasty. The pictographs designating the dragon were found in oracle bone inscriptions dating to 1300-1050 BCE, some of which documented the use of clay dragons in the ritual prayers for rain.[2] In the subsequent millennia, there evolved a full-blown discourse of the dragon, in which this aquatic animal would assume supernatural dimensions, its powers ranging from raising storms, hailstones, and tornados—still called "dragon's whirlwind" (*long juan feng* 龙卷风) today—to siring emperors and transporting the gods and immortals in heaven. The dragon was considered to be spirit-like because of its alleged ability to alter its shape and size. According to the oldest extant comprehensive dictionary of Chinese characters, *Shuowen jiezi* (说文解字 Explaining Words and Explicating Characters) by Xu Shen (许慎 30-122 CE), the dragon is "the longest among the scaly animals. It is capable of appearing and disappearing, contracting and expanding, and shrinking and lengthening. It ascends to heaven at spring equinox and hides in the deep at autumnal equinox."[3] This was the very creature whose medicinal properties entered Li Shizhen's (李时珍 1518-1593) celebrated work *Bencao gangmu* (本草纲目 Compendium of Materia

---

1) Cited with modification from Joseph Needham (with collaboration of Wang Ling), *Science and Civilisation in China*, Vol. 3: *Mathematics and the Sciences of the Heavens and the Earth* (Cambridge, 1959), 479.

2) Qiu Xigui, "On the Burning of Human Victims and the Fashioning of Clay Dragons in Order to Seek Rain as Seen in the Shang Dynasty Oracle-Bone Inscriptions," trans. K. Fowler, *Early China* 9-10 (1983-1985), 290-306.

3) Xu Shen, *Suowen jiezi* (说文解字) (Beijing, 1963), 245.

Medica, 1578).[4] Straddling heaven and earth, the natural and the supernatural, the dragon persisted in the Chinese cultural imagination well into the twentieth century when, at long last, it came to mean the "wondrous animal in *ancient Chinese legends*" that it is typically understood to be today.[5]

The eventual dissolution of the dragon into myth epitomized the wholesale transformation of the structure of natural knowledge and of popular consciousness occurring during the first half of the twentieth century as a result of the establishment of modern science and a series of vigorous campaigns against superstition in the name of science. But this early twentieth-century drive against superstition was not without historical precedents. As will be demonstrated in the course of this paper, an "enlightenment" project of similar tenor and much smaller scope had taken place during the seventeenth century. It was spear-headed by the Jesuit missionaries and some of the Jesuit-inspired Chinese scholars and carried out in the name of universal reason, in its Aristotelian incarnation, and of God. It was in this earlier episode that we see, for the first time in Chinese history, attacks against the traditional dragon lore and its underlying cosmology, launched from the platform of Aristotelian conceptions of nature and meteorological theories, and forming a distinct and coherent discourse, which marked the beginning of a definite decline of the dragon.

The Discourse of the Dragon in Pre-modern China

The belief that the dragon had much to do with the weather originated in the animistic world view of high antiquity which perceived such meteorological phenomena as rain, thunder, hail, rainbows,

---

4) "Dragon Bones," still an essential ingredient in Chinese medicine said to have calming and astringent properties, are defined now as "fossil bones of certain ancient mammals such as elephants and rhinoceros." See *Xiandai hanyu cidian* (现代汉语词典 Dictionary of Modern Chinese) (Beijing, 1984), 730.

5) For example, *Xiandai hanyu cidian* says of *long* (龍 dragon, simplified as 龙): "A wondrous animal in ancient Chinese legends. It has an elongated body and possesses scales and horns. It can walk, fly and swim and is capable of raising clouds and making rain." In *Xiandai hanyu cidian*, 730.

mirages, and earthquakes as the workings of living beings. We have already noted that people of the Shang attributed rain-making powers to the dragons; similarly, some oracle bone inscriptions of the late Shang referred to the rainbow as an animal, one of which described a rainbow as "coming out from the North and drinking from the River."[6] Although by the last centuries before the common era there emerged a rationalistic inquiry into nature which sought to explain natural phenomena in terms of the workings of *qi* (primordial/vital energy), *yin* and *yang*, and the five phases (metal, wood, water, fire, and earth), the animistic world view did not disappear entirely; rather, it persisted in one form or another not only among the uneducated but also within learned circles.

Among the populace, the dragon continued to be revered as a semi-divine being whose major responsibility was to dispense rain. Severe drought or flood was often taken to be an indication that the dragon intended it as a collective punishment for the people in the locale; or perhaps that the dragon simply neglected its duties. Village spirit mediums and sometimes local government administrators led ritual sacrifices to the dragon variously to request, entice or intimidate the dragon to make or stop rain. With the arrival and domestication of Buddhism, the Hindu idea of rain-dispensing Dragon Kings found its way to China and grafted itself onto this native dragon cult. Consequently, an elaborate pantheon of dragon kings appeared, with one dragon king presiding over every considerable body of water throughout the empire. During prolonged drought, Buddhist masters would be summoned to perform rain-making rituals at the local shrine or temple dedicated to the dragon king. In time, the Daoist religion also absorbed the dragon kings into its pantheon and offered its own version of rainmaking rituals.[7]

---

[6] Guo Moruo ed., *Jiaguwen heji* (甲骨文合集 Collection of Oracle Bone Inscriptions), 13 vols. (Beijing, 1978-1983), 1982: 10405b and 10406b. Cited in Aihe Wang, *Cosmology and Political Culture in Early China* (Cambridge, 2000), 31. This animistic understanding of the rainbow is reflected in the construction of the characters designating the rainbow, "hong 虹" and "didong 螮蝀," all bearing an insect radical.

[7] For studies on the rich rain-making ritual culture in traditional China, see A. Cohen, "Coercing the Rain Deities in Ancient China," *History of Religions* 17 (1978), 244-265; Michael Loewe, "The Cult of the Dragon and the Invocation for Rain," in

As Joseph Needham pointed out, the Chinese recognized the water cycle as the cause of rain as early as the fourth century BCE.[8] This physical explanation of rain notwithstanding, few scholars outrightly rejected the popular belief in the dragon's rain-making powers. To the contrary, there seems to have been a continuous effort to rationalize it even among some of the most critically-minded thinkers. One case in point is the Han dynasty philosopher Wang Chong (王充 27-97 CE), a champion of rationalism and relentless critic of the popular beliefs and practices of his time. In his collection of essays entitled *Lun Heng* (论衡 Arguments Weighed), Wang characteristically attacked many aspects of the contemporary dragon lore. He denied that the dragon had any spirit-like quality, arguing that its ability to alter its physical dimensions was only part of its naturally endowed cunning instinct. He also doubted the dragon's ability to sire emperors, suggesting that as a mere aquatic animal, the dragon was not likely to be sexually aroused by human beings. However, he firmly believed that the dragon could cause rain. Speaking the language of correlative cosmology, he argued that:

> clouds and dragons belong to the same category of being [associated with water]; therefore, their respective *qi* responds to each other and causes each other to appear. That is why the *Book of Changes* said: "Clouds follow the dragon, while wind follows the tiger," and again, [the text of *Master Huainan*] said: "When the tiger roars the valley wind stirs, and when the dragon is lively the empyrean clouds descend to meet it." … In midsummer the sun's intense heat is obstructed by clouds and rain. The sun being fire, and clouds and rain being water, their collision causes explosions in the form of thunder. The dragon rises upon hearing the sound of thunder, and when it does so, the clouds descend to meet it. The dragon

---

Chinese Ideas about Nature and Society: Studies in Honour of Derk Bodde, ed. Charles Le Blanc and Susan Blader (Hong Kong, 1987), 195-213; Edward H. Schafer, *The Divine Woman: Dragon Ladies and Rain Maidens in T'ang Literature* (Berkeley, 1973); Mark Elvin, "Who is Responsible for the Weather—Moral Meteorology in Late Imperial China," in *Osiris*, 2nd Series, vol. 13: *Beyond Joseph Needham: Science, Technology, and Medicine in East and Southeast Asia*, (1998), 213-237, and Anning Jing, *The Water God's Temple of the Guangsheng Monastery: Cosmic Function of Art, Ritual, and Theater* (Leiden, 2002), especially 70-73.

[8] See Needham, *Science and Civilisation in China*, 3:467.

> then mounts on the clouds and ascends to heaven. ...when the clouds disperse, the dragon falls down.[9]

Here Wang conceded that the dragon could ascend to heaven and cause rain, but he saw it as nothing more than the automatic result of the sympathetic responses between the *qi* of the dragon and that of the clouds and rain. Thus the alleged rain-making operation of the dragon was not a conscious act: "there were no volitional factors involved in it."[10] Wang further defended the efficacy of the ritual formulas for seeking rain designed by the Han Confucian philosopher Dong Zhongshu 董仲舒 (179-104 BCE), pointing out that although these rituals made use of clay dragons rather than real ones, the clay image of the dragon served the purpose of instilling "absolute sincerity" in the supplicant, and this sincerity of mind would allow the *qi* of the supplicant to reach and stimulate the *qi* of the real dragon, causing the latter to emerge above water, which in turn would attract clouds and rain.[11]

The Neo-Confucian philosophers of the Song Dynasty (960-1279) also displayed a willingness to embrace the dragon lore and other such popular beliefs in spite of their vigorous efforts to establish a rationalistic world view. Zhu Xi (1130-1200), their chief

---

9) Wang Chong 王充, *Lun Heng Zhushi* 论衡注释 (Lun Heng, annotated edition, Beijing,1979), 376. For further discussions of Chinese correlative cosmology, see Michael J. Puett, *To Become a God: Cosmology, Sacrifice, and Self-Divination in Early China* (Cambridge, Mass., 2002), chapter 4; Nathan Sivin, "State, Cosmos, and Body in the Last Three Centuries B.C.," *Harvard Journal of Asiatic Studies*, 55.1 (June 1995): 5-37; John B. Henderson, *The Development and Decline of Chinese Cosmology* (New York, 1984), chapter 1; Li Ling 李零, *Zhongguo fangshu kao* (中国方术考 Inquiries into Chinese Occult Arts) (Beijing, 1993), chapter 2; and Aihe Wang, "Correlative Cosmology: From the Structure of Mind to Embodied Practice," in *Bulletin of the Museum of Far Eastern Antiquities*, 72 (2000), 110-132.

10) Wang Chong, *Lun Heng Zhushi*, 327.

11) Ibid., 910-924. For further discussions on Wang Chong's views on the dragon, see Jocelyn V. Chey, "Wang Ch'ung on Dragons: 'Science' in the Han Dynasty," in *Essays in Chinese Studies presented to Lo Hsiang-Lin* (Shou Luo Xianglin Jiaoshou lunwen ji 寿罗香林教授论文集) (Hong Kong, 1970), 349-359. Dong Zhongshu's rain rituals are contained in his Chunqiu fanlu (春秋繁露 Rich Dew of the Spring and Autumn), *juan* 74. See Zhong Zhaopeng 钟肇鹏, ed., *Chunqiu fanlu jiaoshi* (春秋繁露校释 *Chunqiu fanlu*, a critical edition with annotations) (Wuhan, 2004), 2: 981-1004.

representative whose interpretations of the Confucian classics came to be canonized as the Neo-Confucian orthodoxy, proposed that all phenomena in the world originated from the universal principle, *li*, coupled with the self-propelling primordial energy/stuff, *qi*. Thus all things in the universe acquired their specific principles or natures through embodying *li*. Therefore, for Zhu, to know and cultivate one's own nature as a true Confucian entails not only a life-long conscious effort at moral perfection, but an unending process of investigating the specific principles of all things in order to attain full knowledge of *li* [*gewu zhizhi* 格物致知]. This emphasis on reaching out to the world spawned a new interest in nature, and indeed Zhu himself discussed many natural scientific subjects in his work.

However, a quick reading of, for example, the section on atmospheric phenomena in the recorded conversations he had with his students suffices to show that his efforts to fathom the principles of things often amounted to little more than justifying the "conventional wisdom" in naturalistic terms, that is, by invoking the mechanism of *qi*, *yin* and *yang*. In one of the conversations, a student asked about whether the dragon makes rain as was commonly believed. In response, Zhu said: "The dragon is an aquatic animal; therefore, when it comes out [of water] and encounters the *yang qi* [in sky], rain is formed. But ordinarily rain is formed when *yin qi* is in confrontation with *yang qi*. It does not necessarily involve the dragon."[12] In other words, he saw the dragon's purported rain-making not as an exception to, but a special instance of the normal cause of rain, since aquatic animals were associated with water and thus were considered *yin* in nature. Zhu also conceded to the popular belief that lizards produced hailstones, although he cautioned that "it would be wrong to think that all hailstones came from the lizards … some of them must have been formed in the sky." Zhu cited several eyewitnesses who reported seeing large lizards dwelling on high mountain tops spitting out ice balls during extremely cold early morning hours, which were subsequently carried away

[12] Zhu Xi 朱熹, *Zhuzi yulei* 朱子语类 (Assorted Sayings of Master Zhu), ed. Li Jingde, fascicle 2 (Beijing, 1986), 1:23.

and scattered into low-lying neighborhoods by wind and rain storms. He noted that such occurrences were nothing out of the ordinary, for "lizards resemble the dragons in shape; they, too, belong to the *yin* category, and it is the working of the *yin qi* that brings about the hailstones."[13]

In the following conversations on the causes of thunder, furthermore, we find that Zhu Xi is even attempting to make sense of the "thunder god":

> (1) Question: "Concerning thunder and lightning, Master Cheng said: 'These are the clashes and frictions of [the *yin* and *yang*] *qi*.' Is that true?"
> Answer: "Yes."
> "But some people think that there is a spiritual being controlling it."
> "When *qi* gathers, something must be formed as a result of it, but as soon as the *qi* disperses, the thing will also disappear. The ax of thunder and the like are formed in this manner."
> (2) "Although thunder is merely *qi*, when *qi* is condensed, it must take on a concrete form. For example, a rainbow is nothing but a shadow formed by sunlight falling upon layers of light rain. However, since it does have a shape, it is therefore capable of sucking up water and wine. If it happens to fall in someone's house, it could do good or evil things."[14]

We may notice that in the first conversation, Zhu Xi dismissed the existence of a thunder-god, only to offer a scaled-down version of it by admitting that the gathering of *qi* in the course of thunder and lightning would result in objects such as the "ax," which is what people see the "thunder god" striking at its targets with. In the second passage, he suggested, by means of analogy, that once the *qi* congeals to form a concrete object, it may assume life and consciousness. Indeed, by acknowledging that the rainbow could even "do good or evil things" Zhu has admitted the thunder god through the back door.

By the Ming dynasty (1368-1644), the Neo-Confucian orthodoxy, though endorsed by the State as the standard for the civil service examinations, met serious challenges both as a guide to moral and spiritual cultivation and as a paradigm for understanding the

---

13) Zhu Xi, *Zhuzi yulei,* 1:24-25.

14) Zhu Xi, *Zhuzi yulei*, fascicle 2, 1:24. See also fascicle 3, 1:34.

natural world. Wang Yangming (1472-1529) and his followers regarded the investigation of things that it recommended as an unnecessary and ineffective way to attain genuine understanding of the universal moral principle; they argued instead that one should look inward and reflect on one's own heart and mind, which is in itself a manifestation of the universal principle. On the natural philosophical plane, some scholars such as Wang Tingxiang (1474-1544) were disenchanted by its neat, formulaic explanation of natural phenomena in terms of *yin* and *yang* which, as exemplified in the above examples from Zhu, seemed to be able to explain everything and yet nothing. Concerning the causes of thunder, for example, Wang Tingxiang wrote:

> It is said that thunder is caused by the blocking of *yin qi* by *yang qi*: when the *yin* breaks out of the blockage, it makes a loud clap. While this may explain thunder that comes on suddenly and loudly, it does not seem to apply to thunder that sounds either slow and deep or long-winded and meandering. To call these the sounds of *yin* and *yang* striking at each other is simply nonsense. *Yin* and *yang* are both [tenuous masses of] *qi*; how could their collisions produce such [subtly varied] noises? I suspect that these are the workings of some living beings which come out only when clouds and rain are present. They may entwine themselves and copulate, or engage each other in a fight, while we humans have no way to see such goings-on from far below. In recent years, in Huayin and Wuyang counties, unicorns (*lin* 麟) have been spotted in the fields. They roared like thunder, and their mouths spewed out a fire which resembled lightning. Such is the power of some living beings. Now, if we consider the various sounds of thunder in this light, the rapid ones would seem to come from a beast in rage, the loud ones from wolves engaging in a fight, and the softly chiming ones come after they start to make peace with each other. The ancients said that the divine dragons can expand and contract their bodies; after making rain in the sky, they withdraw their essences, return and hide in the earth. This is why human beings cannot see them well. Could thunder be the work of dragons? It is regrettable that we do not know whether or not dragons are able to make thunderous noises and spew out fire like the unicorns. Or could there be still some other animal that is responsible for it?[15]

[15] Wang Tingxiang 王廷相, *Yashu* (雅述 Graceful Remarks), in *Wang Tingxiang zhexue xuanji* (王廷相哲学选集 Selected Philosophical Works of Wang Tingxiang), ed. Hou Wailu 侯外庐 *et al.* (Beijing, 1965), 136-137.

Wang thus rejected a physical interpretation of thunder in terms of *yin* and *yang* in favor of an organismic one. It is noteworthy that Wang arrived at this conclusion based on the alleged discovery about the unicorns. To what extent Wang tried to verify the discovery is unknown to us, but the idea that unicorns could roar like thunder and spew out a fire resembling lightning certainly was something that suited his vision of the world, one which defies the neat packaging of universal laws and can only be known empirically, discretely and descriptively. As Wang frequently remarked in his works, the universe is vast, and the capacity for the spontaneous transformation of *qi* to bring new things into being is limitless. "What can be too strange to exist?"[16]

Wang was one of the first philosophers whose criticisms of traditional correlative cosmology and enunciations of a mystery-laden vision of the world marked the start of what John Henderson called the "decline of Chinese cosmology."[17] A century after him, another prominent scholar-official Xie Zhaozhe (謝肇淛 1567-1624) also went down the same path. His best known work, *Wuzazu* (The Five Assorted Offerings), a collection of random jottings classified into the five categories of Heaven, Earth, Humanity, Things and Affairs, demonstrated his intense curiosity about cosmic mysteries and his persistent effort at sorting out the vast jungle of remarkable and fantastic phenomena told in folklore, hearsays, and the *zhiguai* (志怪 records of the strange) literature.

Needless to say, the dragon featured prominently in his notes. It seems that Xie not only firmly believed that dragons existed, but also readily subscribed to the core of traditional dragon lore—that the dragon is spirit-like, capable of altering its size and shape, and "has power over clouds, rain, thunder and lightning, winds, and

16) Wang Tingxiang, "Da He Bozhai zhaohua lun," in his *Neitai ji* (内台集.答何柏斋造化论), in *Wang Tingxiang zhexue xuanji*, 226; see also 42.

17) See John Henderson, *Development and Decline of Chinese Cosmology* (New York, 1984); "Chinese Cosmographical Thought: The High Intellectual Tradition," in The *History of Cartography*, vol. 20: *Cartography in Traditional East and Southeast Asian Societies*, eds. J.B. Harley and David Woodward (Chicago, 1994), 203-227.

hail."[18] And there is no wonder why he did so: he had personally seen them in action when he was at sea at the age of ten. In one of the notes, he tells us:

> In the Year Jimao of the Wanli Reign (1577) I accompanied my uncle [Xie] Jie on his diplomatic mission to Ryukyu Island. Halfway to our destination, hurricane winds started to blow. All of a sudden, there was thunder, lightning, rain, and hail. Three dragons appeared, suspending upside down at the front and rear of our ship, their whiskers pulling the sea water up into the clouds. Their heads and horns were clearly visible, but the part below their waist was not. Everybody in the ship was panicked at the sight of them. An elder said: "It seems that the dragons are here to pay court to [the carrier of] the Emperor's message." Then he asked the attendants to help the envoy get back on his feet. Subsequently the envoy wrote two large characters on a piece of paper which said: "Court dismissed!" He showed them to [the dragons], which complied and withdrew from sight.

Reflecting on this childhood experience, Xie commented: "It is often heard that 'the Son of Heaven commands power over all spiritual beings.' There is certainly an indisputable truth in this. If I had not seen it myself, would not I regard this as impossible nonsense?"[19]

Likewise, Xie also investigated the nature of thunderbolt or "thunder god," as it was often called. In a note in which he discussed why thunder strikes certain objects and not others, he said:

> It often happens that when a thunderbolt strikes someone this has been due to a [mischievous] dragon, or because the thunderbolt has risen from the ground and smitten them accidentally, to their misfortune. One opinion is that when a willful dragon is shirking his duty of making rain, he will more often than not escape into the wall of someone's house, or into someone's ear or nose or the horns of an ox. The thunder god is thus obliged to arrest him, and this often causes his hiding places to receive a clap of thunder.

[18] Xie Zhaozhe 谢肇淛, *Wuzazu* (五杂俎 Five Assorted Offerings), fascicle 9 (Taipei, 1618/1977), 212.

[19] Ibid., fascicle 4, 110. In another entry, Xie quoted a description of a similarly close encounter with the dragons from a memoir by Martial Liu Binwen from Zhejiang who led a fleet into the East Sea to prepare for a defense against the pirates in 1595. Ibid., 108.

He gave some eyewitness accounts, including one of his own about how he as a child narrowly escaped a thunder strike when a thunderbolt appeared to have pursued his nanny and struck her, and concluded that in such cases people like himself were unharmed "not because of their sheer good luck but because both dragon and thunder possess consciousness."[20]

But just exactly what is this "thunder god?" In another note Xie reported that in Southeast China "there is a thing that resembles a chicken with very fleshy wings. During fall and winter, it hides in the mountains under the ground. When people dig the place [where it happens to be hiding], it would make a loud explosion and run away. The natives who catch it would kill and eat it. They call it '*leigong*' (thunder god)."[21] Xie doubted that a thing so easily caught and eaten could be the real "thunder god" operating in the sky, but this made him open to the idea that the "thunder god" may be an animal of a similar kind, only much bigger and more powerful. Some of his personal observations seem also to point to that direction, as he noted in yet another entry:

> Years ago I lived at the foot of the Nine Immortals Hills. There was a cypress outside the kitchen, and every year at the beginning of spring the thunderbolts would rise up from the same side of this tree. Its roots and branches were half burnt away, and the color resembled that of charcoal. I stayed there over a period of four years, and thunderbolts rose up [from that tree] four times. Thus it would seem that the thunder lies hibernating in specific places."[22]

For him, there seemed to be enough evidence to conclude that "the Confucian philosophers of the Song were truly ridiculous in simply applying the principles of *yin* and *yang* to explaining thunder and lightning," for, "thunder has a body and produces sound; it rises in spring and hibernates in autumn. It is so obvious that it

20) Ibid., fascicle 1, 16.

21) Ibid., fascicle 1, 17.

22) Ibid. Translation cited with modification from Mark Elvin, "The Man who Saw Dragons: Science and Styles of Thinking in Xie Zhaozhe's Fivefold Miscellany," *The Journal of the Oriental Society of Australia* 25-26 (1993-94), 1-41, 34.

must be a living being which moves in tandem with clouds and rain."[23]

From our perspective today, it is the organismic model of thunder, rain, and other meteorological phenomena upheld by Wang Tingxiang and Xie that appears "truly ridiculous." However, if we examine their arguments and those of the Song Neo-Confucian philosophers such as Zhu Xi, we will find that, in their final analysis, they were not really that different after all. At the methodological level, none of them questioned the empirical basis of the folkloric beliefs about the dragon, lizard, and the thunder god, among other things. It is true that Xie drew on his personal observations as well as hearsay, but he was not aware of the need to evaluate them critically before citing them as evidence.[24] The same was true for Wang Chong as well. At the substantive level, although Zhu Xi, like other Song Neo-Confucian philosophers, tried to apply a strictly "physical" approach to explaining rain, thunder, hail, and other occurrences in the atmosphere, the flexibility afforded by the conceptions of *yin* and *yang*, which presumably existed both in their pure form in the space as tenuous masses of *qi* and in their mixed form in concrete objects, allowed Zhu to accommodate the alleged ability of the dragons and lizards to cause rain and hailstones. Furthermore, Zhu's suggestion that once a tenuous body of *qi* congeals to form a concrete object, it can acquire life and consciousness indicates that his "physical" model and Xie's chicken-model of thunder were not so far apart from each other, nor were either of them far from the theistic view of thunder gods.

The crux of the matter is that all of the thinkers we have examined so far worked within a common cosmological framework that centered around *qi*: everything there is in the universe—from tangible objects to human consciousness to the numinous world of spirits, ghosts and gods—was said to derive its existence from *qi*. This "*qi*" is not to be identified with the prime matter in Aristotelian

[23] Xie Zhaozhe, *Wuzazu*, fascicle 1, 17.

[24] Mark Elvin, in "The Man who Saw Dragons," uses Xie Zhaozhe as a representative example to argue that China failed to develop a set of rigorous scientific methods comparable to that of Baconian science in the early modern era.

natural philosophy, which is inert and represents pure potentiality. Rather, it is both primordial matter and psychic energy; it both originates and constitutes all things; it is spontaneous and yet exhibits certain patterns; and it is self-propelling and engages in eternal motion. In other words, it is the four Aristotelian causes combined into one. Within this cosmological framework, then, there is no rigid boundary between the physical and biological and between nature and the supernatural. Indeed, the same *qi* connects all realms of being—inanimate objects, plants, animals, humans, and the spiritual beings, allowing for the possibility of fluid interchange and boundary crossing among all of them. It was the longevity of this *qi*-centered cosmology that enabled the traditional dragon lore to weather the many storms occurring in Chinese intellectual history, from Wang Chong's rationalist critique to the reformist movement of the Song Neo-Confucian philosophers, to the rebellions against the Neo-Confucian orthodoxy during the Ming.

The Aristotelian Intervention and its Impact

From the very beginning of their mission, the Jesuits in China were keenly aware that the *qi*-centered cosmology undercut their program of conversion from two opposite directions. On the one hand, the idea that all things derive their existence from the cosmic *qi* and that *qi* can generate not only physical but also spiritual or psychic properties implied that all things in nature had the potential to become gods or spirits, hence the animistic beliefs of high antiquity were able to continue to thrive at the popular level in a new guise. The worship of the dragon was only one of a host of popular cults directed at some natural objects, real or imaginary, which presumably acquired spirit-like powers either due to their extraordinary endowment of *qi* (like the dragon) or through their assimilation of the fine essences of the cosmic *qi* over a long period of time (such as the alleged one-thousand-year-old fox or monkey or even a very old tree).

On the other hand, among the Confucian scholar-officials who managed the imperial bureaucracy, it was the opposite sentiment that prevailed. The Neo-Confucian orthodoxy on which the civil service examinations were based taught that the human spirit or soul is an emergent property of *qi* within the body, and thus it perishes along with the body, a position that was supported by the basic tenets of Chinese medicine. In their commentaries on Confucian classics, the Neo-Confucian philosophers also downplayed the theistic element in the conceptions of *shangdi* (上帝 Lord-on-High) and *tian* (天 Heaven) in classical Confucian texts, traces of a transcendental, personal god worshipped in Chinese high antiquity, and interpreted them in terms of *qi*, the primordial cosmic energy, and *li*, the universal principle inherent in the movement of *qi*. Such views spawned a strong atheistic tendency and a disinterest in matters about afterlife among the scholar gentry. Those who were spiritually oriented but disenchanted by Daoism or Buddhism developed a form of Confucian spirituality characterized by the ideal of "sagehood," a state of moral perfection in which the individual becomes an embodiment of the eternal moral principles emanating from the ultimate cosmic source. This offered a viable path to personal salvation, but it is one that could be achieved through one's own effort right here in this life.

Caught between those who worshipped the wrong gods and others who worshipped none at all, Matteo Ricci, who formulated the Jesuit strategy of intellectual conversion in China, decided early on that the future of the mission rested on a foundational project, one aimed at systematically reforming Chinese culture by introducing a European system of learning, the centerpiece of which was Aristotelian-scholastic natural philosophy. Ricci himself attempted this in his major Chinese work, *Tianzhu shiyi* (天主实义 The True Meaning of the Lord of Heaven), a catechism set in eight chapters. The goal of this work, as Ricci himself stated in his journals, was to demonstrate by "universal reason" the basic truths of Catholicism regarding the existence of God, the immortality of the soul, the existence of heaven and hell, and reward and punishment; and to destroy, "at the very roots, with irrefragable arguments, the opinions of the Chinese which contradict those truths."[25] Naturally, Ricci's

[25] *Fonti ricciane: Documenti originali concernenti Matteo Ricci e la storia della prime relazioni tra l'Europa e la Cina* (1579-1615), ed. Pasquale M. D'Elia (Rome, 1942-9), no. 709.

measure of universal reason was Aristotelian-scholastic categories of thought and views of the world, which constituted the bulk of both his demonstrations and his refutations. The book thus provided the Chinese the first significant exposure to Aristotelian logic, metaphysics, and the basic teachings on the four causes, the four elements, the concentric-geocentric model of the cosmos, and the hierarchical model of soul.

The full-scale introduction of Aristotelian-Scholastic natural philosophy in China was carried out by the second generation of Jesuits in a slightly different context. Soon after Ricci's death in 1610, a debate arose among the members of the Jesuit mission in China and Japan over the legitimacy of Ricci's decision to adopt certain Chinese terms to express Christian concepts. The focus of the debate was such Chinese terms as *shangdi*, *tianshen*, and *linghun*, which Ricci and his followers had used to render "God," "angels" and "the rational soul," respectively. Ricci's opponents argued that the Chinese *qi*-centered cosmology made it impossible for these terms to communicate the idea of a pure "spiritual substance." Thus using such terms would necessarily introduce idolatrous elements into their Christian teachings.[26] Although the Jesuits closed their debate by retaining Ricci's policy, the debate made them more acutely aware of the need to address the problems of Chinese religion at their cultural roots. Thus they began an immense project of systematically transmitting Aristotelian learning into China, beginning with two expositions of scholastic psychology: *Xingxue cushu* (性学牔述 A Brief Outline of the Study of Human Nature, 1623) by Gulio Aleni, adapted from the Conimbricenses on Aristotle's *De anima* (Coimbra, 1598) and *Parva naturalia* (Lisbon, 1593 and Lyon, 1594), and *Lingyan lishao* (灵言蠡勺 Some Cursory Remarks on the Soul, 1624) by Francesco Sambiasi, based on the *Conimbricenses* on *De anima* (Cologne, 1617 edition). These were followed by Francisco Furtado's *Huan you quan* (寰有诠 On Heaven and Earth, 1628), adapted from the commentary on Aristotle's *De coelo et mundo* (Lisbon, 1593; Lyon, 1594), and Alfonso Vagnone's *Kongji gezhi*

[26] For further discussion on the history of this debate, see Liam Brockey, *Journey to the East: The Jesuit Mission to China, 1579-1724* (Cambridge, Mass., 2007), 85-89.

(空际格致 Investigation into the Phenomena in the Atmosphere, 1633), adapted from the commentary on Aristotle's *Meteorologica* (Lisbon, 1593; Lyon, 1594), as well as two other works on Aristotelian logic and ethics.

Taken together, these works presented a highly stratified vision of the world, marked by sharp dichotomies between God and Nature, spiritual and corporeal substances, and soul and body, and a neatly structured cosmos with rigid boundaries between the celestial and sublunar spheres and among the various elemental spheres and realms of being on earth—minerals, plants, animals, and humans. Meanwhile, these authors also targeted the core of Chinese cosmology by attempting to demystify the concept of *qi*.[27] Especially worth mentioning in this regard are Aleni's *Xingxue cushu*, which refuted the idea that the human soul is an emergent property of *qi*, and Vagnone's *Kongji gezhi*, which defined *qi* in terms of air, and through describing the respective properties of the four elements and the three layers of the atmosphere, ushered in an entirely new and strictly physicalist way of talking about *qi*: its warmth or coolness, wetness or dryness, density or rarity, and cleanness or putrefaction, and so on.

This body of Aristotelian learning was introduced to the Chinese as the perfect embodiment of universal truth, a point made amply clear by Aleni in his *Xixue fan* (西学凡 An Outline of Western Learning, 1623), a short treatise on the educational system and curriculum in contemporary Europe which he published at the beginning of this wave of translation, where he stated:

> Aristotle's scope of knowledge is extraordinary, his erudition unfathomable, and his talent limitless.... He collected books from far and wide, consulting and carefully weighing the ideas of others. Whenever he heard of any curious objects, he would procure them regardless of cost or personally travel to them regardless of distance, so that he could closely study them. He let no peculiar things in nature escape his attention. Having observed how things are, he would pursue further

[27] I discussed this issue in my article, "Demystifying Qi: The Politics of Cultural Translation and Interpretation in the Early Jesuit Mission to China," in *Tokens of Exchange: The Problem of Translation in Global Circulation*, ed. Lydia Liu (Durham, 1999), 74-106.

> why they are the way they are; having treated what is known, he would investigate further what is not yet known. Hence, from the apparent to the obscure, from the large to the minute, he categorized the myriad things of every order and ascertained their respective nature on solid grounds. *His theories were so much in accord with the true principle that they were indisputable and infallible.* The books he wrote were extensive, reaching the utmost depth and subtlety of every field of philosophical inquiry.... *For more than two thousand years up till now, nobody has been able to raise a single doubt about them.* [28]

In that same work, Aleni envisaged a massive translation project that would take the missionaries and their Chinese collaborators up to ten years to complete. "By then," he suggested:

> we may select young and talented students, and, while their minds are still fresh and impressionable, have them study these works year by year following the order [according to the Western curriculum]. Equipped with proven theories and trained through methodical and diligent mental exercises, they will be able to penetrate the ultimate origin of things and elucidate a broad realm of phenomena, so that the teachings of the sages of the Western and Eastern Seas may converge into one harmonious flow.[29]

It would seem that while continuing to follow Ricci's example and trying to win over their Confucian-educated readers by rational arguments, the second generation of the Jesuits also initiated a new line of action directed at the younger and fresher minds of China, a tactic that reminds us of the famous remark by the quantum physicist Max Planck—that "a new scientific truth does not triumph by convincing its opponents and making them see the light, but rather because its opponents eventually die, and a new generation grows up that is familiar with it."[30]

The talk of the dragon was, of course, not of major concern to the Jesuits at this point, as they were preoccupied with the far more urgent tasks of combating the organized religions of Buddhism and

---

28) Giulio Aleni, *Xixue fan* (西学凡 An Outline of Western Learning), in *Tianxue chuhan* (天学初函 First Collection of Writings on Heavenly Learning), ed. Li Zhizao (Taipei, 1965), 1: 42-3. Emphasis mine.

29) Aleni, *Xixue fan, 59.*

30) Max Planck, *Scientific Autobiography and Other Papers.* Cited in Thomas Kuhn, *The Structure of Scientific Revolutions*, 2nd ed. (Chicago, 1970), 151.

Daoism and coping with the problems of Neo-Confucian orthodoxy. But Vagnone did address part of the Chinese dragon lore in the second volume of his *Kongji gezhi* in a section entitled "Flying Dragons," where he explained how tornadoes were formed:

> The air [current] released from inside the earth is very hot and dense. When this air rushes up into the sky, if it suddenly encounters cold clouds, it will be forced downward in a spiral motion, in the course of which it is ignited, giving the appearance of a flying dragon. Because the head of the ascending current is clear and clean, when it turns downward and is ignited, it resembles a dragon spewing fire. The tail end of the current, on the other hand, due to the pressure from the cold clouds above, is stretched thin and curvy, like a dragon's tail. People think it is a real dragon, but they are mistaken.

Vagnone discussed the origin and prediction of rain in two separate sections, but he did not refute the Chinese belief in the dragon in that context. That topic was taken up by Aleni at the prodding of one of the Chinese converts, Li Jiubiao. In Li's *Kou duo richao* (口鐸日抄 Diary of Oral Admonitions), he recorded a brief conversation that he had with Aleni on August 3, 1639, right before a rain. It went as follows:

> Qixiang (Li) asked: "In my country rain is held to be caused by a dragon. I do not know whether the same [belief] also exists in your country."
> The Master (Aleni) said: "Do you ever get to see a dragon in China? Or have you only got your information from traditional stories that are recorded in your texts?"
> Qixiang said: "There are many stories about dragons in written texts, but I never have seen one myself."
> The master said: "People do not even dare to believe in what they see with their own eyes—how can one decide about the existence or non-existence of something nobody has ever seen? Moreover, in China dragons are said to be capable of transforming themselves by contraction and expansion, and they are supposed to be spirit-like beings, but in my country we never had such a theory."

Li further inquired how rain came about. Aleni referred him to another text where this topic was discussed in detail—in all likelihood Vagnone's *Kongji gezhi*, and then told him what he himself had seen during a sudden rain brought on by a whirlwind when he was sailing to the East. He described:

> At that time the sky was clear, but suddenly I saw how a dark cloud was formed: from the sky it went down till it reached the surface of the sea. Suddenly it soared up again, and a moment later it was dispersed into rain. At the time a friend of mine concluded that the cloud had been blown down by a whirlwind, and then, after having absorbed the [moist] air on the surface of the water, was dispersed and became rain. I have observed this very clearly with my own eyes, but I absolutely did not see any so-called dragon.

Aleni did not categorically deny that dragons existed, as he said that "among the hairy, feathered and scaly animals between heaven and earth any kind of [strange creature] can be found." [31] However, he emphasized that even if they do exist, they are only animals endowed with senses but not with intelligence, much less with supernatural powers.

As Li's diary suggests, topics such as dragon worship must have come up often in the daily conversations between the Jesuits and their converts, but it did not seem to occupy much attention in the Jesuit Chinese publications.[32] The major campaigns against such popular cults and "superstitions" were spearheaded by the Chinese Catholics and other scholars who were positively disposed to the ideals and learning presented by the Jesuits. One of the earliest such

---

31) Translation cited with modification from *口鐸日抄 Kouduo richao*: *Li Jiubiao's Diary of Oral Admonitions. A Late Ming Christian Journal*, trans. with introd. and notes by Erik Zürcher (Sankt Augustin, 2007), 591. Indeed, the Jesuit writings on the world outside China by Matteo Ricci, Giulio Aleni and Ferdinand Verbiest, among others, were replete with such strange creatures.

32) One of the few Jesuit works that dwell extensively on Chinese popular cults and "superstitions" was Joseph de Mailla's *Shengshi churao* (盛世芻荛 Modest Offer for a Flourishing Age, 1732), where, in a section entitled "Praying for Shine and Rain," he attacked the cult of the Dragon King. Taking the dragon to be a snake-like animal, he said that "the dragon is just the longest among the scaly animals; it rises in the spring and hibernates in the winter. What power does it have to make rain?" He further pointed out that when the so-called Dragon King did seem to respond to people's prayers, it was always because God took pity on the people by sending down rain after a prolonged drought and stopping it when it went on for too long; in that regard He was much like loving parents who, even when their children committed offences again them, continued to clothe and feed them in the hopes that they would repent and correct themselves. Joseph de Mailla, *Shengshi churao*, in *Tianzhujiao dongchuan wenxian xubian* (Documents Concerning the Spread of Catholicism to the East), ed. Wu Xiangxiang, 1648-1651. The quote is from p. 1649.

attempts was made by Yang Tingyun (1562-1627), known as one of the three pillars of the Chinese Catholic Church during the late Ming. In his treatise *Tian shi mingbian* (Clarifying the Differences between Heavenly Teaching [i.e. Catholicism] and Buddhism), in the context of refuting the Buddhist doctrine of transmigration, Yang applied the Catholic-scholastic conception of nature and the Aristotelian hierarchical model of soul to discrediting the belief in animal intelligence and morality which not only underlay the cult of the dragon and various forms of animal divination but also had strong resonance within the Neo-Confucian tradition.[33] The paragons of animal intelligence were the dragon, unicorn, phoenix, and tortoise, which together were called "the Four Intelligent Animals." But some other animals, too, were regarded as possessing some degree of intelligence: the roosters announcing the arrival of the day, the dogs keeping watch at night, the spiders spinning webs, and the silkworms crafting cocoons. As for animal morality, bees and ants were often cited as being loyal to their kings and queens, tigers and wolves obedient to their parents, geese and swans loving to their siblings, and so on.

Yang took issue with the basic assumption underlying such beliefs—that the difference between human beings and animals is one in degree rather than in kind, arguing that:

> according to Heavenly Teaching, plants only have vegetative soul; therefore they sprout from their respective seeds and grow according to the season, but have no ability to turn to what they like and avoid what they dislike. Animals have ears, eyes, mouth, nose, flesh and blood; therefore, they have vegetative and sentient souls and know how to get what they like and avoid what they dislike, but they do not have the capacity for reasoning and making moral judgments. Human beings alone are able to learn about the past and the present, observe the Heavens and the earth, and investigate the principles of all things. It is due to this ability that human beings have the five moral constants [governing the relationships between the ruler and the subjects, father and son, husband and wife, elder and younger brothers, and friend and friend] and manage various kinds of social

[33] Zhu Xi, for example, said that "both man and things [i.e., animals] are capable of consciousness and movement, but though things possess humanity (*ren*), rightness, propriety, and wisdom, they cannot have them completely." Cited from Wing-tsit Chan, *Source Book in Chinese Philosophy* (Princeton, 1963), 622.

> affairs. This ability comes from the rational soul. Thus human beings are entirely different from both animals and plants. [How can we] speak of which one of them is the more intelligent or stupid? It is a matter of either having intellect or not.... The difference [between human beings and the rest] is like [that] between bright and dark or hot and cold. There is no mistaking the two.[34]

He further drew the distinction between instinctual behaviors and voluntary actions. He said that what people take to be signs of animal intelligence came from their instincts which were decreed by God. The animals had to follow their instincts and could not do otherwise, "just like fire which naturally ascends and water naturally descends. It is all out of their nature, and neither fire nor water has awareness of its doing so. In the same way, none of the animals do what they do out of conscious intent." "Besides," he continued, "within the universe, all animals of the same kind have the same nature. None of them has discernment or comprehension, which is why they always stay in the same way and never change. If they had intellect and were able to think, like the humans do, how much more unpredictable would the animal behaviors have been?"[35] Turning to the so-called "Four Intelligent Animals," Yang came close to dismissing them as a myth, saying that "today, not even a single unicorn, phoenix, and dragon can be found. As for the tortoises, if they really are intelligent and able to divine, why don't they escape their own fate of being trapped in the net and undergoing the pain of having seventy-two holes bored into their shells?"[36]

Meanwhile, the dragon's alleged role in shaping the weather was also called into question by those who studied and embraced Aristotelian meteorology expounded in Vagnone's text. Xiong Mingyu 熊明遇 (1579-1649), one of the late Ming scholar-officials who took a serious interest in Jesuit works on natural philosophy and the sciences, wrote a series of essays explaining the causes and movements of winds, clouds, rain, dew, snow, hail, thunder and

---

[34] Yang Tingyun, "Tian shi mingbian," in *Tianzhujiao dongchuan wenxian xubian*, ed. Wu Xiangxiang, I:289-290.

[35] Ibid., 292.

[36] Ibid., "Tian shi mingbian," 291.

lightning, comets, and meteors, and other meteorological phenomena, drawing on the ideas presented by Vagnone. In his essay on hailstones, he refuted the idea that they were dropped from the sky by lizards or dragons, calling such ideas "the talk of women and small children." With an authoritative tone, he explained that *qi,* which he now equates with the atmosphere, consists of three layers, the bottom layer, due to its proximity to the earth, is warm, and so is the top layer which is close to the sphere of elemental fire. The central layer is the coldest. It is there that hailstones are formed. He said that they are more common in the summer than in the winter, because in the winter, since the temperature on the surface of the earth is colder, there is less evaporation from the surface of the earth, and the little there is goes up slowly and barely reaches the central layer of the atmosphere, where it meets the cold air and is gradually chilled into fine icy particles or rain drops. But in the summer, the hot and moist steam from the surface of the earth rises up with so much force and in such a great quantity that it is able to penetrate deeper into the coldest layer of the atmosphere, where it crystallizes into large clumps of ice. In sum, he said, "the thicker the layer of moist vapor and the greater the speed with which it travels upward, the bigger hailstones there will be."[37]

By the end of the seventeenth century, we see yet another assault on the traditional dragon lore on the basis of Aristotelian science. It came from the Catholic physician Wang Honghan. Born into a family with a sizable library of Jesuit publications and strong personal ties to the prominent Chinese Catholic Xu Guangqi, Wang

---

[37] Xiong Mingyu 熊明遇, *Luxue lou ji* (绿雪楼集, Collection of Essays from the Mansion of Green Snow), reprinted edition by Siku jinhui shu congkan (Beijing, 2000), vol. 185:83-520, at 109-110. For further studies on Xiong's assimilation of Western learning, see Zhang Yongtang 张永堂, "Xiong Mingyu de gezhi zhi xue (熊明遇的格致之学 Xiong Mingyu's Natural Studies)," in his *Mingmo qingchu lixue yu kexue guanxi zailun* (明末清初理学与科学关系再论 Rethinking the Relationship between Neo-Confucianism and Science during the Late Ming and Early Qing) (Taipei, 1995), 5-48; Feng Jinrong 冯锦荣, "Mingmo Xiong Mingyu *Gezhi cao* neirong tanxi (明末熊明遇格致草内容探析 An Analysis of the Content of *Gezhi cao* by Xiong Mingyu of the Late Ming)," in *Ziran kexue shi yanjiu* (自然科学史研究 Studies in the History of the Natural Sciences), 16.4 (1997), 304-328.

was one of those younger Chinese envisaged by Aleni who would grow up with a mind fully saturated with Western learning. As he reminisced later in his life, he avidly collected and pored over Jesuit Chinese writings on mathematics, world geography, cosmography, meteorology, and other such subjects and found them to be "unparalleled in the teachings of [our] ancients and supplementing our Confucian knowledge regarding the principles of things."[38] Nevertheless, this intense and extensive exposure to Western learning did not make Wang complete the intellectual "gestalt shift" that Aleni had anticipated; instead, he tried to find ways to integrate this body of Western learning with parts of traditional Chinese knowledge. In his major medical work *Yixue yuanshi* (医学原始 Origins of Medicine, 1692), he gave a detailed summary of Aristotelian-Galenic-Catholic teachings regarding the heavens, earth, the four elements, and the human body and soul and attempted to reconstruct Chinese medicine on the foundation of these theories.[39] At the same time, he also sought to reshape the medical profession in the image of science, Western style, by severing the strong historical ties with Daoism, shamanism, the cult of immortals, and a host of divination practices. For this purpose, Wang wrote a treatise entitled *Gujin yishi* (古今医史 History of medicine from antiquity to the present), in which he tried to rectify errors found in traditional "hagiographical" accounts of famous doctors and other historical personages who were believed to have had significant influences on the development of the medical arts, thereby to cleanse the jungle of medical literature of what Wang deemed to be fantastic, harmful, superstitious, and fallacious tales.

---

38) Xu Haisong 徐海松, Qingchu shiren yu xixue (清初士人与西学 Scholars of Early Qing and Western Learning) (Beijing, 2000), 155.

39) For further studies on Wang Honghan's work in this connection, see Zhu Pingyi 祝平一, "Tongguan tianxue, yixue yu ruxue: Wang Honghan yu Ming Qing zhi ji yixue de jiaohui" (通贯天学, 医学与儒学: 王宏瀚 与明清之际中西医学的交会 Medicine East and West: Wang Honghan's Synthesis of Medicine, Christianity and Confucianism), *Zhongyang yanjiuyuan Lishi yuyan yanjiusuo jikan* 中央研究院历史语言研究所集刊, 70.1 (1999), 165-201; Xu Haisong 徐海松, *Qingchu shiren yu xixue*, 146-164; and my article, "Hybridizing Scholastic Psychology with Chinese Medicine: A Seventeenth-Century Chinese Catholic's Conceptions of *Xin* (Mind and Heart)," in *Early Science and Medicine* 13 (2008), 313-360.

It is in this book that Wang launched his attack on the dragon lore. It appears in the biographical entry on the legendary Yellow Emperor, to whom the major canon of Chinese medicine, *Huangdi Neijing* (Inner Canon of the Yellow Emperor) was traditionally attributed. Wang cited the established accounts about the Yellow Emperor, told and retold by generations of earlier medical biographers, including the one that had first appeared in *Huangdi jiuding shendan jing* (黄帝九鼎神丹经 The Classic of the Yellow Emperor's Nine-Crucible Divine Pill), one of the earliest alchemical texts dated around the first century CE. The story recounts that after receiving divine assistance from a female Daoist deity, the Yellow Emperor succeeded in refining the elixirs of life and as a consequence became immortal. Soon afterwards, a yellow dragon descended before him and transported him up to heaven. Wang wrote a 772-character commentary on this story, prefaced with "zhengwu" (正误 correcting errors), in which he said:

> Such a story is entirely fallacious and deceptive. In the past four decades of my life, I have applied myself day and night to the philosophical study of nature and humanity and acquired a little knowledge about the principles of science and the truths about Heaven, Earth, and the Four Elements .... Let me use these principles of science to examine this story about the yellow dragon allegedly taking the Yellow Emperor up to Heaven so as to expose the bizarre and misleading ideas spread by the Classics of the Divine Pills and the followers of Laozi.

First, Wang refuted the idea that the dragon has the ability to alter its size. Wang argued that "the dragon is a corporeal being from the family of scaly animals," and "anything with a corporeal form cannot alter its size. If it appears to be now smaller now bigger, it must be because the eyes of the viewer are temporarily deluded by effects of evil magic, while in reality the object's body remains the same." As for those who claimed to have seen the dragons transform into human-like "dragon kings" and "dragon ladies," "what they saw," he said, "was the work of the devil which attached itself to the dragon and created those illusions." Wang's idea that corporeal beings cannot alter their physical dimensions would have sounded very new to those who were used to thinking in the traditional language of *qi,* an infinite reserve of energy which, though invisible, could be called upon through mental concentration and other forms of manipulation to induce visible changes within one's body. It is also interesting that Wang did not question the empirical reality of sightings about the "dragon kings" and "dragon ladies"; he merely explained them away as the work of the devil.

Turning to the idea that there are dragons in heaven which tend to the needs of the gods, he pointed out that:

> The fact is that dragons are aquatic animals living on earth. They cannot leave the water and rise up in the air any higher than 10-20 *li*.[40] It is not in their nature to enjoy leaping into space. But after spring and summer the air [trapped] in the earth starts to ascend, and the water is not calm enough for dragons to submerge themselves in the deep. So they are forced to rise up, riding on the ascending air from the earth in search for a calmer body of water. In the midst of relocation sometimes they may encounter strong winds in the air, and, failing to maintain balance, they may bump against mountain tops, injuring and killing themselves. Therefore, people often find dead dragons in deep mountain valleys. But when autumn and winter come and the air in the earth descends and stays low, what power would a dragon have to leap in the air all by itself? The heavens are tens of thousands of *li* above, while the dragons can leap just as high as 20 *li*, how can they possibly reach the height of heaven?

Here, Wang reaffirmed the conception of animal nature already expressed by Yang Tingyun, emphasizing that by their very nature as aquatic animals, dragons have to stay in water and have no business to do in the air other than during transit to another body of water. Wang did not challenge the established view about the dragon's annual cycle, but he gave it a purely mechanistic explanation: it was the air trapped in the earth which, when released by heat during spring and summer, rose up (so far a very Aristotelian idea) and brought the dragons along. Without this external momentum, the dragons could not help but stay in water. Thus their rising in spring and summer had nothing to do with their purported task to make rain. We also see here that Wang is applying the Aristotelian model of the world, with its rigid spatial boundaries among the various sublunar spheres and between them and heaven, to circumscribe the dragons, whose proper location is now strictly in water.

---

40) 1 *li* is equal to 0.5 kilometer.

Finally, Wang set out to demystify the core of the dragon's power: its role in rain making. He said:

> People are mistaken in believing that there are "real dragons 真龙"—the flying dragons in the clouds which take up water [to make rain]. Now, in the deep mountains there are large snakes hatched from the eggs of *shen* which eventually turn into *jiao* dragons. Between spring and summer they emerge from below the mountain rocks. [Due to their speed, size and their watery habitat], their movements cause wind storms and heavy rain which may inundate people's houses. Such things do happen from time to time. Once the *jiao* dragons enter lakes or the sea, they would stay quietly there. Or, it could be that between the end of spring and beginning of summer the air [trapped] in the earth starts to ascend, and the water lacks calm, making the dragons uncomfortable. So they may leap up to find another body of water. That's why people reported seeing large formations of migrating dragons [龙阵] in the sky. This may very well be true. But to say that there are "real dragons" which take water up in the air and then sprinkle it back onto the earth and that they have been seen in the act for prolonged periods of time by many people, that is wrong. What people actually see is that the dry air up in the sky, being forced down by cold clouds from above, moves towards the moist air on the surface of the earth. The moist air from the top of rivers and lakes, mountain rocks, or flat land latches onto this dry air stream and together with it ascends to the sky. The dry air from above, after receiving the moisture this way, turns into thick clouds. This then is either followed by rain, or if not, the clouds simply scatter, and the sky becomes clear again. If it were true that there are real dragons up in the clouds taking up water, how is it that when the clouds scatter and the sky clears up again, the bodies of the dragons are nowhere to be seen? Evidently this is the working of dry air and not of dragons.

Wang's rebuttal of dragons' alleged rain-making power, though coated largely in Aristotelian language, is not entirely clear of mythologies, such as the *jiao* dragons causing floods or the dragons migrating in the sky in large formations, both of which were widely reported in late Ming and early Qing popular literature and were taken as a given by Wang. However, he regarded these phenomena as transitory and accidental and explained both in terms of the dragon's nature as an aquatic animal. Comparing Wang's dismissal of the "real dragons" in the above passage and the eye-witness account of the water-sucking dragons by Yang Yu, we see that the era of the dragon, the mighty, spirit-like, and rain-making aquatic animal, was beginning to end.

In the brief account given above, we have shown that during the period bracketed by the active writing careers of Xie Zhaozhe and Wang Honghan, the discourse of the dragon in China underwent a qualitative change, or rather, a definitive step towards its decline, for by Wang's time almost every core thesis of the established dragon lore that Xie had subscribed to was discredited by a small but distinct group of scholars: that the dragon is spirit-like, capable of transforming its own shape and size, appearing and disappearing; that it has "consciousness" or intelligence; and that it has the power of generating rain, hail, tornadoes, and a host of other weather conditions. These critical voices were speaking in a Christianized version of Aristotelian language, in which the all-inclusive, elusive and ineffable *qi* that was so central to Xie Zhaozhe's world became air or atmosphere whose properties can be measured and described in sheer physical terms. Still, for most of these authors, the dragon remained on the natural scene, but it no longer was the boundary-crossing creature that could swim, walk and fly. The cosmic order so familiar to Xie and his contemporaries, one characterized by tremendous fluidity, connections and interchange, spontaneity and creativity, was pitched against a teleological view of a created order in which each category of being possesses its distinct nature and fulfills a distinct cosmic role.

While modernization began in Europe with the jettisoning of Aristotelianism, the shift from "dragonology" to meteorology, though occurring in the minds of only a select few, demonstrated the potential of the Aristotelian intervention in China as a "modernizing event," in the limited sense that it ushered in a de-animated view of nature and the "obsession" with dichotomies—those between mind and body, spirit and matter, God and nature, religion and science, and so on. Aristotelian natural philosophy arrived in China during an opportune time, as the sixteenth and seventeenth centuries marked the beginning of a critical downturn of traditional Chinese cosmology and the Neo-Confucian orthodoxy. While some of the Chinese scholars, like Wang Tingxiang and Xie Zhaozhe, were turning away from any totalizing cosmology altogether, there were others such as Wang Honghan for whom the introduction of Aristotelian learning offered a fresh window of opportunity. However, the historical circumstances resulting from the "Rites Controversy" and the subsequent expulsion of the missionaries from China in the eighteenth century militated against the spread of the new cosmology in China. For the majority of Chinese scholars, the inherent limitations of the Aristotelian teleological world view and its inevitable association with the nascent Catholic culture—which replaced the whole range of explanatory and ritual functions of the native cults, such as dragon worship, with a foreign culture centering on a new realm of spiritual beings (the Angels, the Devil, and an omnipotent God)—inevitably compromised its intellectual appeal as a tool against the popular cults and "superstitions."[41] Thus the "modernizing" potential of Aristotelianism would remain just that, a potential visible only in hindsight.

---

[41] Concerning the scope and limitations of the transmission of Western learning in the context of the Jesuit mission and the progression from traditional "natural studies" to modern science in China, see the extensive and thorough study by Benjamin Elman, *On their Own Terms: Science in China, 1550-1900* (Cambridge, Mass, 2005).

Baroque Fire (A Note on Early-Modern Angelology)

Anne A. Davenport*
Boston College

Abstract
Between Galileo's discovery of the moons of Jupiter (1610) and the publication of Newton's *Principia* (1687), uncertainty regarding the structure of the heavens combined with a lyrical fascination for extraterrestrial life inspired a distinctly Baroque outpouring of speculation in which angels played a key part. English Catholic "recusants," haunted by a feeling of lost unity, vividly illustrate the imaginative character of Baroque speculation.

Keywords
angels, Thomas White, planetary atmosphere, meteorology, Franciscus à Sancta Clara, Kenelm Digby, Antoine Legrand, Robert Boyle

How did the scholastic doctrine of angelic appearances, laboriously crafted by medieval masters to fit Aristotelian parameters, fare in a post-Copernican universe?[1] Did early-modern philosophers who tentatively embraced some form of heliocentrism but who also remained committed to the scholastic project of rational theology continue to defend angelic appearances philosophically? This essay will consider three closely-connected authors who flourished in an anglophone environment prone to vandalizing angelic images and largely antagonistic to scholasticism.[2] My first author is the Roman Catholic

* 96 Bedford Street, Concord, MA 01742, U.S.A. (adavenport@cfa.harvard.edu).
I thank Bill Newman and Edith Sylla for their helpful comments and suggestions.

1) See Tiziana Suarez-Nani, *Les anges et la philosophie* (Paris, 2002), especially Part II, 92-182.

2) See Alexandra Walsham's account, "Angels and Idols in England's Long Reformation," in *Angels in the Early Modern World*, eds. Peter Marshall and Alexandra Walsham

priest Thomas "Blacklow" White, Kenelm Digby's friend and philosophical associate, who in 1652 published a theology, *Institutiones Sacrae*, based on Digby's natural philosophy. Thomas White dedicated his *Institutiones Sacrae* to my second author, the equally idiosyncratic English Franciscan convert Franciscus à Sancta Clara, who, also in 1652, while serving as Provincial of his Order for England, published a philosophical treatise, *Paralipomena philosophica de mundo peripatetico*, aimed at harmonizing Scotism with new astronomical and chemical discoveries. My third author is Sancta Clara's younger Franciscan *confrère*, Antoine Legrand, whose devoutly Cartesian *Institutio Philosophiae* of 1675 includes a detailed section on angels—surprisingly, since Descartes explicitly shunned the topic.[3] How were angelic appearances viewed by these three Catholic "recusant" authors, based for the most part in London, unmoored from the very catholicity for which they yearned?[4]

Let me first briefly characterize the new conditions faced by angels after Galileo's discovery of the Jovian moons (1610). With the demise of Aristotle's geocentric cosmos, crystalline spheres slowly receded

(Cambridge, UK, 2006) 134-168. See also William Godwin's account in *History of the Commonwealth of England* (London, 1824) 81; and Albert Loomie, "The Destruction of Rubens's Crucifixion," *The Burlington Magazine*, 140/1147 (October 1998) 680-682. See also Increase Mather, *Angelographia* (Boston, 1696) "To the Reader," A3: "As for *Philosophical Questions* about the Nature of Angels and the way of their Understanding Things, whether it be *Per Species Intelligibiles Insitas vel Acquisitas* and many the like, which *Thomists* and *Scotists* have such wrangling Contests about, I have wholly declined them."

3) Cf. Descartes's Interview with Burman, in René Descartes, *Oeuvres choisies*, ed. Bridoux (Paris, 1996) 1370: "La connaissance des anges nous échappe presque entièrement, parce que [...] nous ignorons aussi tout ce qu'on demande d'ordinaire à leur sujet, s'ils peuvent s'unir à un corps, quels ont été ces corps que dans l'Ancien testament ils revêtaient souvent, et choses semblables."

4) See the following three articles in the *Oxford Dictionary of National Biography*, Sept 2004; online edn, Oct 2006: Beverly Southgate, "White [Blacklo], Thomas (1592/3–1676)" [http://www.oxforddnb.com/view/article/29274, accessed 4 Aug 2008]; P. Chambers, "Davenport, Christopher (c. 1595-1680)" [http://www.oxforddnb.com/view/article/7199, accessed 4 Aug 2008]; and Richard Acworth, "Le Grand, Antoine (1627/8-1699)" [http://www.oxforddnb.com/view/article/16364, accessed 4 Aug 2008].

into thin air.[5] Planets, suns and angels found themselves thrown together promiscuously in a boundless space bereft of resting place, the same in every direction, without mooring or rim, isotropic at every point.[6] What gave planets their spherical shape and kept them in their orbits? Copernicus, as we know, promoted a new principle of pluri-centrism, speculating that bodies cohere symmetrically around their own center thanks to an inherent inclination infused in their parts.[7] As for the clouds that were raised by Ptolemy to argue against the earth's motion, Copernicus responded that the earth carries "a good part of the air" along with it, either because the air "is mixed up with earthy and watery matter and therefore obeys the same nature as the earth," or because "the movement of the air is an acquired one, in which it participates without resistance on account of the continuity and perpetual rotation of the earth."[8] The intoxicating idea of a planetary atmosphere, whirling swiftly through space while sheathing its own planet, was born. And after Galileo concluded in *Sidereus Nuncius* that the earth is not unique in

---

[5] See A. Koyré's *From the Closed World to the Infinite Universe* (Baltimore, 1968) and, more recently, Jean Seidengart, *Dieu, l'univers et la sphère infinie. Penser l'infinité cosmique à l'aube de la science classique* (Paris, 2006), esp. 88-101. For the possible impact of the new cosmology on iconography, see Tintoretto's *The Creation of the Milky Way*, painted c. 1575 (National Gallery, London), which vividly captures the excitement and turbulence of the demise of Aristotelian cosmology under the multiform assaults of Renaissance Humanism.

[6] See e.g. Mästlin's defense of Copernicus, cited by André Goddu in the present volume, "Copernicus's Mereological Vision of the Universe," [pp. 316-39]: "Quod si is locus alicubi simul sit Mundi centrum, id non nisi per accidens contingit." Goddu translates: "If any such place is the [unique] center of the World, then it is by accident."

[7] *Nicolai Copernici Torinensis De revolutionibus orbium coelestium* (Nurenberg, 1543), liber I, cap. IX, 9: "Equidem existimo, gravitatem non aliud esse, quam appetentiam quandam naturalem partibus inditam a divina providentia opificis universorum, ut in unitatem integritatemque suam sese conferant in formam globi coeuntes."

[8] Ibid., liber I, cap. VIII, 6: "Quid ergo diceremus de nubibus, caeterisque quomodolibet in aere pendentibus, vel subsidentibus, ac rursum tendentibus in sublimia? nisi quod non solum terra cum aqueo elemento sibi coniuncto sic moveatur, sed non modica quoque pars aeris et quaecumque eodem modo terrae cognationem habent? Sive propinquus aer terrea aqueave materia permixtus, eandem sequatur naturam quam terra, sive quod acquisititius sit motus aeris, quem a terra per contiguitatem perpetua revolutione ac absque resistentia participat."

possessing a vaporous envelope, vanguard philosophers made it a point of Copernican honor to defend the possibility of extraterrestrial life.[9] What exactly are planetary atmospheres and how do they support life? One elegant and economical approach, adopted by Sir Kenelm Digby and Thomas White, was to argue that elements convert into one another.[10] All that is needed is the presence on planets of "crass" air, produced by rarefaction out of water and earth.[11] But do elements in fact convert?

Thomas White's younger friend and former student at the English College of Douay, Franciscus à Sancta Clara, much influenced by Jan Baptista Van Helmont whom he visited in his Brussells laboratory, argued that they do not.[12] While Sancta Clara fully agreed that animal life on the moon requires lunar water and clouds, he denied that this implies that elements convert.[13] Instead, the same

---

9) For the ubiquity of planetary atmospheres, see Galileo Galilei, *Sidereus nuncius*, reproduction of the Venice 1610 edition (Alburgh, 1987) 31.

10) Thomas White, *Institutionum peripateticum ad mentem summi viri clarissimique philosophi Kenelmi Equitis Digbaei, pars theorica: item appendix theologica de origine mundi* (London, 1647), liber II, lectio V, 37, article 12: "Rursus quoniam elementa solis rari et densi differentiis sibi invicem opponuntur, clarum est transmutationem unius in aliud non esse nisi rarefactionem et condensationem." Cf. *Peripateticall Institutions. In the way of that eminent person and excellent philosopher Sr. Kenelm Digby* (London 1656), Book II, lesson V, 54, art. 12: "Since the Elements are oppos'd to one another only by the differences of rare and dense: 'tis evident, their transmutation into one another is nothing else but *rarefaction* and *condensation*." Art. 13: "'Tis plain, too, that dense things, being forced against rare, do compresse them, and, if there be no way to escape, do necessarily condense them: which condensation, if it be enough both in time and degree, will, of necessity, change that which is condens'd, into that Element to which such density is proper."

11) See Sancta Clara's summary of White's position in *Paralipomena philosophica de mundo peripatetico* (Douay, 1652), cap. XIV, 188: "Thomas quidem Anglus, respondendo cuidam objectioni, probanti animalia in Luna vivere non posse, quia ibi non est aquarum et terrae varietas, neque pluviae, neque nubes etc. Respondet, si sit aer crassus qui notatus est, ex necessitate futuram aquam; haec enim ex terra fiunt, et prius aquae naturam habent, etsi fortasse nubes non sint tam magnae, ut possint notari etc. Vult igitur unum elementum posse in aliud converti."

12) For Sancta Clara's visits to Van Helmont, see Ibid., 11: "aer communis resolvitur per ignem, ut semel mihi declaravit D. Helmontius celebris medicus Bruxellensis, in alembico tunc operans ad eius resolutionem."

13) Ibid.: "Ego quidem non possum putare nubes esse aerem ex aqua factum."

variety of elements must be assumed, in a homogenous universe, to exist everywhere, *vel hic vel ibi*.[14] As an alternative to White's theory that water transmutes into air to become clouds and then back again into water, Sancta Clara argued that clouds are vapors made up of "simpler parts" of both earth and water.[15] Such vapors are drawn upwards from telluric bodies by celestial influence, then return in condensed form, as rain. Sancta Clara pointed out that a similar condensation is achieved in chemical experiments, without the conversion of one element into another.[16] Inspired by what he understood of Van Helmont's experiments, Sancta Clara thus chose to promote the theoretic stability of elements as a tentative substitute for the lost stability of cosmic emplacement (natural place). Elements that are immutable by divine *fiat* might suffice, in a universe bereft of absolute center, to explain both the stable order and variety of cosmic phenomena.[17]

I. Thomas White (1593–1676)

Doubtful that material atoms could serve as immutable cosmic principles, Thomas White developed a different solution. White reinterpreted the four elements to be degrees of rarity/density that either emit, transmit, absorb, or refract light rays.[18] He then championed the conversion of elements but put atomic (i.e. immutable/indivisible)

---

14) Ibid.: "Non vero ex uno elemento aliud fieri. Nec natura indiget tali varietate vel hic vel ibi."

15) Ibid.: "Sed ex terra et aqua vapores seu simpliciores eorum partes, operatione coelorum superius advehi, et inde coagulatas recidere etc., ut dico de Pluviis."

16) Ibid.: "Consimilis est reliquis naturae alembici operationibus, sine tamen conversione unius elementi in aliud."

17) Ibid., cap. XIV, 194: "Constans vero est ordo universi, nec mutabilis: non igitur partes ejus integrantes absumi vel diminui possunt; nam certus est rerum status in natura; multo minus diminui vel augeri possunt *vera Elementa* (sic), in quibus penitior est effectus et intimior mundi ab illis dependentia."

18) White, *Institutionum peripateticarum*, liber II, lectio V, 36, art. 4: "Sic itaque rarum quod se et alia videri faciat, *ignem* apud nos nominamus; quod vero neque hanc vim habeat, neque aliis tamen per se vim movendi visum prohibeat, *aerem* appellamus; densum quod lumen penitus rejicit, *terram*; quod vero partim admittit quadam tenus remittit, *aquam* vocitamus."

angels in charge of it.[19] In both the *Peripateticall Institutions* of 1647 and the *Institutiones Sacrae* of 1652, White speculates that the activity of angels underlies cosmic phenomena, producing everything from galaxies to frogs. According to White, the universe "rip'ned by degrees to this excellent beauty" and "did not by instantaneous Creation immediately start into perfection."[20] God in the beginning created only three elements, earth, water and air. This made "the matter fit for Angelicall operations" and angels produced fire out of air:

> The Matter of the World being created, it remains that we see what follow'd by the additional operations of Creatures. And because the operation of *Angels* is no other than *rarefaction*: and nature wanted its natural instrument, *viz. Fire*; for This we see principally made use of for almost all natural effects, especially, the generation of substances; and This is not rais'd out of Water and Earth *immediately* without first becoming AIRE; it must be, that the Angels or Angel whose task this was, by *rarefying* the *Aire* rais'd a vast *Fire.*[21]

Fire is thus not an element among others but the product of angelic skill. Angelic fervor is infused, so to speak, directly into physical creation in the form of heat, setting the universe on its course. The highest form of angelic fire is *light*, which "must have been made by the Angels through the rarefaction of pure Aire." Light, according to Thomas White, is thus a sort of angelically distilled *quintessence* that visibly exhibits the ubiquity and radiance of angelic industry. Angels, who know "the whole Plot of the World" and who know how best to be useful, are divinely-appointed chemists who faithfully bring about God's cosmic plan out of elementary matter.[22] From the ordered revolutions of distant planets to the perennial

---

19) Convinced philosophically that immaterial souls are absolutely immutable, White went on to deny that prayers for the dead are effective, provoking Sancta Clara's zealous defense of the practice. See my discussion in "Durandus at the Stuart Court," upcoming in *Theology and Philosophy in the Long Middle-Ages: Papers in Honor of Stephen Brown*, eds. Kent Emery Jr., Russell Friedman and Andreas Speer (Leiden, 2008).

20) Citing White from the English *Peripateticall Institutions* (London 1656), Theological Appendix, chap. 1, 3.

21) Ibid., Appendix, chap. 3, 344 and 348.

22) Ibid., bk. V, lesson XI, 296.

rebirth on earth of vegetation, cosmic phenomena result from angelic fire, as it radiates out symmetrically, in all directions, mirroring angelic simplicity and self-diffusing goodness.[23]

Angels produce not only fire and light, but also winds and tides, which, like fire and light, are nature's active instruments.[24] Modelled closely on Aristotle's old celestial movers, White's angels are immaterial intelligences whose properties are known philosophically, but they are no longer confined to a bounded portion of space. Whether on the moon or in intergallactic space, Whitean angels labor incessantly to distill pure air into light, or "crass" air into rain in order to quench the thirst of extraterrestrial flocks, just as they work in the mineral depths of our own earthly mountains and rivers. Whitean angels are, so to speak, angels without borders. They pervade the whole seamless manifold of creation, blowing winds on our seas, lighting our domestic fires, turning iron deposits to gold, "spiritualizing" plant seeds in April to make them sprout.[25] Angels, White emphasizes somewhat sternly against his literal-minded coreligionists, are said to be "in heaven" only metaphorically, to mean that they are in God's immediate presence wherever they execute God's plan, which is anywhere at all in the universe.[26] Moreover their task is no longer to revolve eternally around a cosmic center but, appropriately enough for a newly expanded and isotropic space, to maintain a homogenous universe in dynamic equilibrium through rarefaction, keeping it from collapsing into a cold, dense clump.[27]

---

23) Cf. Kenelm Digby, *Discours sur la végétation des plantes*, French transl. P. de Trehan (Paris, 1667), 16 and 18.

24) *Institutionum sacrarum peripateticis inaedificatarum, hoc est Theologiae, super fundamentis in Peripatetica Digbaeana iactis, extructae, Pars Theorica. Authore Thoma Anglo, e generosa Albiorum in Oriente Trinobantum Prosapia oriundo* (London, 1652), vol. 2, bk. II (De Angelorum natura et ministerio), lectio IX, 169.

25) Cf. Digby, *Discours sur la végétation des Plantes*, 19-26.

26) White, *Institutionum sacrarum*, tom. 2, liber II, lectio XI, 192: "Angelis itaque esse in coelo, componitur ex his duobus: Deum esse in coelo et Angelos esse praesentes Deo, seu esse ubi Deus est."

27) Ibid., lectio IX, 162: "ne quid praeter rarefactionem immediate agant; et in eorum potestate cuncta esse quae per rarefactionem fieri possunt; et elementalem conversionem (quae naturalium omnium est fundamentalis) consistere in rarefactione."

Much as he prefers to focus on rarefaction, White concedes somewhat reluncantly that angelic operations include making spiritual beings appear to human senses in bodily form, at God's command. Does this mean that angels *assume* bodies? Not really. While an angel is always the immediate efficient cause of what is represented by God's special command to human sense, what appears is whatever God commands the angel to represent.[28] In other words, Whitean angels, who are immutable and immutably in God's presence, do not assume bodies so much as apply their ordinary alchemical skills to make seraphs, or lions, or God, or archangels, or "angels" appear, as though staging a divine Masque. No angel need actually be *sent* anywhere.[29] Each angel stays firmly stationed at its own divinely appointed cosmic post, to serve God and execute both God's ordinary and special commands. Whitean Angels are thus at once perpetually *active* and constantly *at rest*, as continuous in their *energia* as the celestial intelligences of Aristotle's vanished heavens.

There is, however, a key difference between Aristotle's celestial movers and White's angels. Whitean angels do not trigger physical effects through natural necessity and for the sake of cosmic eternity, but instead cooperate freely with God to shape an unfolding cosmic history. Whitean angels act infallibly, but they act deliberately, purposively and irreversibly. Their nature is Aristotelian, but their fervor and their investment in advancing God's sovereign plan are distinctly *Baroque*. Voluntary service constitutes the very essence of Whitean angels: *Palam est volitionem esse actum principalem Angelis*.[30]

White's universe, in short, is not stabilized by a cosmic center, nor by immutable material elements, but by a corps of immaterial angelic volunteers. Whitean Angels love God and love God more dearly than themselves because they know themselves to exist and

---

28) Ibid., lectio X, 170: "Non semper Angelum apparere quoties tale corpus movet: sed saepe Deum. Ille enim apparet quem exhibet, seu cuius repraesentatio est in tali corpore, non vero qui facit talem repraesentatiom. Quoties itaque repraesentatio est Dei, Deus apparet non Angelus; etsi solus Angelus immediata sit causa repraesentationis."

29) Ibid., lectio XI, 192: "Ex quo dilucet, quomodo non sit necesse, Angelis descendere de coelo, ut agant aliquid circa commendatos sibi pupillos."

30) Ibid., 122.

know that their existence implies God's existence by logical necessity.[31] Whitean angels, moreover, cohere infallibly, but freely, willingly, into a stable hierarchy, not through natural place, but through the innate rational love that each angel feels for its immediate angelic superior and for its inferiors for a higher sake.[32] Higher angels (symbolizing wisdom and, below wisdom, science) help to purge lower angels (symbolizing prudence and, below prudence, industry) of their lust for finite things, thus stoking their gratitude and ever-renewed fidelity. Angels of every rank zealously perform the distinctive duty assigned to them and form a voluntary chain of service as tight as numbers: *sicut numeri*.[33] Even God cannot insert a new angel between two existing angels.[34] Not unlike Copernican bodies driven to cohere around their own center, Whitean angels cohere freely to each other and to the transcendent center of their spiritual freedom: *quasi coeant in unum ad Deum, quod unire appellat Dionysius*.[35]

Out of the ruins of Aristotle's geocentric cosmos, dominated at every fatal instant by its own eternal present, there springs a spatially boundless and radically youthful universe, freely created *ex nihilo* by God and administered by angelic servants acting out of personal devotion.[36] From the burning seraphs who do nothing but adore the Trinity down to the alchemical genii who regulate our weather, angels implement God's providential government with

---

31) Ibid., lectio IV, 121: "Omnis creatura rationalis incorporea naturaliter amat Deum plusquam semetipsam. Cum enim constet ipsi, vel constare natum sit, Deo non existente, sese non posse existere; sic amat Deum, sicut suum esse." Cf. Descartes's second proof *quia* of God's existence in Meditation III, and my commentary in *Descartes's Theory of Action* (Leiden, 2006), 145-171.

32) Ibid., lectio VIII, 157: "Quoniam vero omnis cognitio valet ad affectum in animo bene disposito; affectus autem Angeli est ad inferiora et superiora, et bene disponitur, quatenus non amat inferiora nisi propter superiora."

33) Ibid., lectio VIII, 153.

34) Ibid., lectio VIII, 156: "Non posse autem inter duos Angelos creari medium, non est maius inconveniens, quam inter duos numeros vel figuras immediatas, Deum non posse facere medium."

35) Ibid., lectio VIII, 157-8.

36) See Thomas White, *The Grounds of Obedience and Government*, 2nd ed. (London, 1655), 27-34. (The treatise is dedicated to Kenelm Digby.)

diligence and art.[37] In his political writings, White denounced the infantile indignity of mere obedience and promoted instead a new model of freely pledged fidelity, closely tied philosophically to the new importance of contracts and oaths.[38] No longer defined by natural place (custom), Whitean angels are called at every vital instant to participate actively in universal Redemption. They deliberately serve their divine monarch for the good of the whole commonwealth, like a new breed of civil servants. Last but not least, Whitean angels are fully equipped to implement not only God's ordinary plan for creation but also God's exceptional interventions, such as commanding that angels appear. A fierce defender of the apostolic prerogatives of the locally-fixed secular clergy against the encroachments of regulars and their shadowy missions, White saw no need whatsoever to make theoretical room, in his cosmology, for *missionary* angels.

Thomas White's angelology solved the problem posed by a suddenly amorphous space, where cosmic order seemed threatened. Like medieval theologians before him, White recycled Aristotle's celestial intelligences into angels, but he made them into fiery Baroque servants who are specialists of comets and light, tides and vegetation. Man, it turns out, is not at the physical center of a fettered cosmos, but at the moral center of an inscrutable Providence unfolding freely through the commingled inventiveness of human and angelic agency. In White's angelically-wrought universe, we recognize the Baroque cosmology that is embodied in such churches as

37) Cf. Kenelm Digby's definition of separate souls (angels, intelligences) in *Two Treatises* (Paris, 1644), 426: "An indivisible substance, exempted from place and time, yet present to both: an actuall and present knowledge of all thinges that may be knowne; and a skill or rule, even by what itself is, to all things whatsoever. So that *in fine*, a separate soule is of a nature to have, and to know, and to govern all things."

38) White, *Grounds of Obedience*, 27: "The virtue by which they serve is not that of obedience, but of fidelity and truth." 28: "Servants pay their duties, perform their bargaines; and Fidelity is their proper vertue; to do it carefully, and uprightly; not Obedience." 30-31: "Although Fidelity bee a distinct vertue from Obedience, yet is it not such a stranger as not to demand a place in this Treatise; especially because the kinde of Obedience wee desire to looke into, which is that by which Common-wealths and Communities subsist, is founded generally on Fidelity, or the matter thereof which we call Promise, and in vertue of it bindes to observance."

the Val de Grace in Paris, centered at infinity, aflame with myriad fire-pots and torch-bearing genii. English Catholics like Thomas White who spent long periods in Paris knew the design of the Val de Grâce well and saw its exuberant gold-lit dome being raised from the neighboring windows of the English Benedictine Monastery. Dedicated to the Divine Birthing ("Iesu Nascenti") by a queen who was mysteriously saved from sterility, the church of the Val de Grâce, at once Roman and subversive, attests to the lyricism of a generation lost in space, enamoured of cosmic regeneration and of celestial fire.[39]

II. Franciscus à Sancta Clara (1595–1680)

Let us now turn to Thomas White's younger Franciscan friend, Sancta Clara, whose naive zeal for the Old Religion contributed to the downfall of his Stuart patrons not once but *twice*, and who, in his spare time, encouraged Kenelm Digby to try out his "sympathetic powder" on him, taught natural philosophy and denied that elements convert.[40] Sancta Clara could not agree that clouds had actually been observed on the moon.[41] Yet he accepted the premise that lunar clouds *probably* exist since there is no reason why they should not.[42] He even speculated that, since animal respiration requires air that is mixed with telluric particles, air must become

39) On the construction and meaning of the church, see Jennifer Germann, "The Val de Grâce as Portrait of Anne of Austria," *Architecture and the Politics of Gender in Early Modern Europe*, ed. Helen Hills (Aldershot, 2003), 47-61.

40) For more information on Sancta Clara's life, see John B. Dockery, *Christopher Davenport, Friar and Diplomat* (London, 1960). For his philosophic views, see my "Scotus as Father of Modernity. The Natural Philosophy of the English Franciscan Christopher Davenport in 1652," *Early Science and Medicine*, 12 (2007) 55-99.

41) *Paralipomena philosophica*, ch. XIV, 188: "Addit ipse (sc. Thomas White), Teutonicos advertisse aliquid ingenti nubi simile in Luna."

42) Ibid.: "An ipsi viderint tales nubes nescio, ob nimiam distantiam et nubium transparentiam, adde et inconstantiam, sed nubes esse ibi, et in caeteris stellis, nullatenus dubitare possum." "Sicut circa terram nostram est aer valde vaporosus, nempe ex terra et aqua eductus, sic etiam circa Lunam, Solem, et caeteras stellas habitabiles."

progressively less breathable as it becomes progressively purer with distance from its home planet.[43]

In his *Paralipomena philosophica de mundo peripatetico*, Sancta Clara raises the question of angelic appearances in numerous places, most especially in connection with dreams.[44] The context involving angels and meteorology, however, holds a special interest. A key concern of Sancta Clara's at every turn of his treatise is to establish that the rational soul is absolutely immaterial since the human will is radically free.[45] Sancta Clara is thus keen to defend a created hierarchy of pure spirits that includes the immaterial human soul as its least member, immediately below the lowest angel. This leads him to reject Pietro Pomponazzi's opinion that the human soul is an intermediary entity between angels and beasts in the sense of being part-angelic, part-material.[46] Pomponazzi's view prompts Sancta Clara, moreover, to examine John Trithemius's proposal that an intermediate life-form be inserted in the *scala perfectionis* between angels and men for the sake of obtaining a tighter continuum.[47]

What is the evidence? Sancta Clara rejects Trithemius's personal testimony of having seen, "near the sun and at the price of suffering an eye lesion," a genus of "quasi-corporeal" angels or demons.[48]

---

43) Ibid.: "Nam spissitudo illa aeris videtur necessaria habitationi et respirationi." Thus what we could call the relative vacuum of space is, in his scheme, pure air or "ether."

44) Ibid., ch. VII, 124. Angels activate latent images stored in the memory. Hence the pastoral importance of religious art: "Deus per Angelos excitabat species quas habuit (sc. Iuvenis Hispanus) fortissimas B.Virginis in phantasia."

45) Ibid., 100: "Quod meo judicio demonstrative arguit incorporeitatem animae"; 103: "ergo intellectus non dependent ab ullo organo vel causa materiali"; 125-6: "Si anima rationalis licet immortalis et spiritualis sit materialis, ut contendit Bonaventura et Doctor Subtilis de Anima cum Magistro meo cavello (Hugh MacCaghwell); non video cur non eam attingere possint (sc. virtutes coelestes). Sed ego non putem posse materialem et spiritualem consistere, ob rationem supra tactam."

46) Ibid., 200: "Ut Pomponatius asseruit hominem esse *quid* adeo *medium* inter Angelos et Bruta, quod anima rationalis in se sit partim immaterialis et immortalis, in quo participat naturam Angelicam; partim materialis et mortalis in quo participat naturam brutalem."

47) Ibid., 206: "Trithemius videtur velle, inter Angelos et homines esse aliquid tertium quasi intermedium, ut ordo sit magis solidus."

48) Ibid.: "Dixit se vidisse non longe a Sole, quoddam genus Angelorum seu Daemonum, quasi corporeorum; addens, se nos aspexisse non sine laesione oculorum."

Trithemius, it seems, simply mistook clouds with dramatic shapes and iridescent colors for a new type of living being.[49] As Sancta Clara was well aware, reports of *Aireyall Inhabitants* were not uncommon.[50] His own mentor John Gennings believed that his martyred brother Edmund had been called to the Roman priesthood by "a strange spectacle that appeared to him in the ayre" of "armed men fighting" with "bloud running everywhere about them."[51] The sudden proximity of space, the strain of imagining life on distant planets, the unnerving idea of spatial boundlessness, the palpable rush forward of events, converged to turn the earth's atmosphere into a troubling heterotopia at the threshold of confabulation, full of fleeting ciphers—and into a new testing ground for rational fidelity against superstition. Depictions of angels in Baroque painting had become gradually more naturalistic and more atmospheric, more immediately proximate.[52] They attest to a bold suspension of disbelief but also to a mild existential panic, as though the human heart, suddenly exposed to an unbearable vastness, could no longer distinguish the physically possible from the humanly conceivable.

Sancta Clara cites a second empirical report in favor of Trithemius's theory, namely Theophrastus Paracelsus's claim to have seen rational *quasi homunculi* living on the outskirts of human settlements, not quite human in form.[53] Sancta Clara again rejects the evidence but argues, this time, that Paracelsus mistook *demons*, not

---

49) Ibid.: "Forte vidit nubes aliquas mobiles iridis instar diversicolores, quas judicaret esse novum genus viventium, bene quidem cum laesione oculorum, quia difficulter, si omnino conspiciuntur ibi; puto etiam cum deceptione."

50) See John Heydon, *The Harmony of the World* (London, 1662), Preface: "These Aireyall Inhabitants stayd at least ten hours disputing and arguing of sundry things..."

51) John Gennings, *The Life and Death of Mr. Edmund Gennings priest, crowned with martyrdom at London the 10 day of November in the year MDXCI* (St. Omers, 1614).

52) See e.g. Tintoretto's *Brazen Serpent* (Scuola Grande di San Rocco, Venice), painted 1575-6; and Guercino's (Giovanni Francesco Barbieri) *Burial of Santa Petronilla* (Pinacoteca Capitolina, Rome), painted 1621-2; and, perhaps known to Sancta Clara, Rubens's ceiling paintings, installed in 1636, Banqueting House, Whitehall, London.

53) *Paralipomena philosophica,* cap. V, 207: "Ad alterum Paracelsi, de quibusdam quasi homunculis a se visis, et ratione utentibus, extra consortium hominum habitantibus, et forma differentibus."

clouds, for some new species of living being.[54] So eager is Sancta Clara to put the hypothesis of *quasi-homunculi* to rest that he springs into a detailed defense of angelic appearances.

Sancta Clara justifies discussing angelic appearances in a strictly philosophical work on the grounds that the existence of angels is known to natural reason and that angelic appearances are rationally conceivable. The rational non-impossibility of angelic appearances, moreover, is closely tied to the thesis that planetary atmospheres are composed of suspended, but *per se* stable telluric elements, which in turn is closely connected to the rational non-impossibility of extraterrestrial life. For indeed, Sancta Clara argues, since "many suns" (parhelia) have been seen to appear in the air when the air is sufficiently dense for a reflected image of the sun to form, why should we have any difficulty admitting that God can make men appear through the same thickening process?[55] Angelic appearances are thus construed by Sancta Clara to be a special class of *meteors* (in the early-modern sense of the term), analogous physically to parhelia. The key principle is that condensed vapors, like all opaque bodies, scatter light. Sancta Clara had already established this key principle earlier in his treatise while explaining the luminous tails of comets, citing a wide consensus (Thomas White, Tommaso Campanella and Libertus Fromondus).[56]

---

54) Ibid.: "Putem eos esse Demones, quorum consortium non adeo abhorruit, ut in operibus suis ipse fatetur." On early-modern demonology, see Moshe Sluhovsky, *Believe Not Every Spirit: Possession, Mysticism and Discernment in Early Modern Catholicism* (Chicago, 2007); and Nathan Johnston, *The Devil And Demonism In Early Modern England* (Cambridge, 2006).

55) *Paralipomena philosophica*, cap. XV, 207: "Quod vero Angeli vel Daemones (quos non solum credimus ut vult Cremoninus; sed rationabiliter ex natura deducimus, ut patebit postea) possunt, et ex divina dispositione solent in forma humana comparere, ultra Authoritatem Sacrae Scripturae, ratio declarat quoad posse. Nam quod possunt eo usque ingrossare aerem, ut appareant homines, est facile concipi; cum videmus Solem et alias Stellas per reflexionem in aere apparere, et plures Soles componere; Quare igitur non homines?" For the phenomena invoked by Sancta Clara, see David K. Lynch, "Atmospheric Halos," in *Atmospheric Phenomena*, ed. D. Lynch (San Francisco, 1980) 122-130.

56) Ibid., ch. X, 147: "Vapores condensati reflectunt lumen sicut caetera opaca. Sic Thomas Anglus post Authorem Anglicum, De Mundo in Luna, qui etiam citat

Sancta Clara next bolsters his demonological reclassification of Paracelsian *homunculi* by enlisting three medieval authorities: Thomas Aquinas, John Duns Scotus and Durandus of St. Pourcain, who together represent the three major scholastic schools (realism, Scotism and nominalism) and who concur in affirming that angels assume bodies of air. Sancta Clara's chief authority, not surprisingly, is the Franciscan doctor, Scotus, who analysed angelic appearances with characteristic virtuosity. Scotus started by ruling out the possibility that angels assume bodies of ether or of pure elemental air on the grounds that God himself could not give such bodies shape and color.[57] Scotus then argued that, since at least in some cases angels appear and vanish instantaneously, the bodies they assume cannot be perfect *mixta* like our own organic animal bodies, but must be imperfect *mixta* "such as the colored and configured impressions that form in the air suddenly out of clouds only to dissolve again into clouds just as suddenly."[58]

Sancta Clara is chiefly concerned with establishing that clouds are mixtures and are thus suitable to be shaped into bodies by angels, but by referring us to Scotus's much earlier discussion of angelic appearances, he invites us to appreciate the comforting continuity of Catholic *traditio* as well as to explore a wider set of interconnected problems. For example, since enclouded angels have no sensation, how do they move their airy limbs and airy tongues to

---

Camillum Gloriosum. De Com. l. 5. c. 4. Campanella in Apologia pro Galileo. Fromondum Meteror. l. 3. c. 2. art. 6. ubi vero partes sunt rariores, refrangunt radios sicut aqua, et ideo diversimode apparent."

57) Ibid.: "Doctor Quidem Subtilis 2 d. 8 vult corpora quae faciunt, non esse purum aliquod elementum, sed esse mixta, quod puto verum. Nam secundum meam sententiam c.2. Elementa pura non subjiciuntur sensibus, et consequenter optime arguit Doctor, quod corpora quae assumunt, cum sint vere colorata debent esse mixta, sicut sunt elementa hec vulgaria et putatitia. Et hic aer ingrossari potest (ut dixi) et ideo assumi, ut tenent Durand. 2. d. 8. qu. I num. 2 et D. Thom. I. p. q. 51. a. 1. ad 3."

58) *Reportata parisiensia*, lib. II, distinctio 8, quaestio unica, ad quaestionem, 3; in Ioannes Duns Scotus *Opera Omnia*, editio minor, II/2, Opera Theologica, ed. Giovanni Lauriola (Alberobello, 1999), 648: "Igitur sunt mixta imperfecta similia illis, quae sunt perfecta sicut impressiones in aere, antequam generentur mixta, et sic in eis est figura, et color, et sicut statim possunt formari ex nube, ita statim possunt dissolvi in nubem."

shape sounds? Scotus answered that angels, unlike us, know singular things without sensors, by direct intellection.[59] Sancta Clara's references to Aquinas and Durandus, in turn, connect the problem of early-modern "areyall men" to medieval semiotics, since the two Dominican masters speculated that the shape and colors of the aery bodies formed by angels constitute a symbolic language that is meant to convey spiritual properties.[60] Implicitly, a good angel appearing by God's command successfully conveys God's message by appearing in a meaningful form, in contrast to demons, who confuse and deceive (e.g. deceive Paracelsus into the hypothesis of *homunculi*).

Sancta Clara's next task is to meet an objection raised by Campanella's disciples, namely that clouds which are sufficiently compact to be anthropomorphically shaped and colored must, in principle, convert into rain.[61] Sancta Clara answers that the conversion of elements has already been ruled out, on independent grounds, not only in the preceding chapter, devoted to defending the stability of elements, but also in an earlier chapter devoted to parhelia and haloes: *illas conversiones Elementorum negavi; deinde, illud quod dixi de Pareliis et Halo convincit oppositum.*[62] In the same earlier passage, Sancta Clara also established that parhelia and haloes cannot form

---

59) Ibid.: "Quomodo igitur salvabuntur apparentiae? Dico quod quando angelus movet tale corpus sicut videtur per organa sensuum, potest cognoscere singularia, quae homines cognoscunt per sensum, et sic facit actus sequentes, ac si haberet cognitiones per sensus, cum tamen sic non cognoscat singularia, sed per intellectum suum sub ratione singulari, quia nihil potest ibi poni, quin in repraesentando abstrahat ab existentia rei. ideo cum per intellectum suum cognoscit singularia movet linguam in isto corpore, et genas, et alias instrumenta, attrahendoo aerem et verberando, et causat voces, et per imperium voluntatis movet membra organice, ac si haberet verum corpus vivum."

60) See *D. Durandi a Sancto Porciano, in Petri Lombardi Sententias Theologicas Commentariorum Libri IIII* (Venice, 1571), lib. II, dist. VII, q. I, 7: "Cum igitur in his corporibus appareant aliquae proprietates visibiles proprietatibus invisibilibus angeli congruentes tunc dicuntur corpus assumere, ut quando apparet in figura leonis, vel bovis, et huiusmodi per quarum proprietates intelliguntur spirituales virtutes angelorum."

61) Sancta Clara, *Paralipomena philosophica*, 207: "Clamant aliqui, ut Campanella *De sensu rerum* et alii, aerem sufficienter condensatum ad induendam formam hominis, in aquam converti debere."

62) Ibid., and cap. V, 148-9.

above the earth's atmosphere since they require the presence of sufficiently dense bodies, namely clouds made up of particles of earth and water, which never reach the cold uppermost region at the threshold of purer air.[63] Implicitly, parhelia and haloes define the region near planets where angelic appearances are possible.[64] Since Sancta Clara also believes that animal respiration requires the same sort of sufficiently dense air that is required for parhelia to form and for angels to appear, a further implication of his theory is that angelic appearances and planetary life go hand-in-hand, requiring the same atmospheric conditions. Angels, appropriately enough, will not appear in regions where there is no living being to see them.

A new objection is raised: is thickened air really solid enough for angelic actions—such as blessing, kneeling, eating, etc.?[65] Sancta Clara's answer draws on an Aristotelian axiom framed by Thomas Aquinas on the authority of Augustine: angels/demons cannot immediately alter a body but only move it locally.[66] This means that enclouded angels or demons, Sancta Clara explains, can move their cloudy limbs and shape air into sounds, but cannot digest or procreate.[67] Sancta Clara's atmospheric angels thus indirectly validate scholastic tradition anew, since they are sufficiently solid to perform what is licit for scholastic angels to perform, but not solid enough to do anything that they are not supposed to do. A smoother transition from angels robed in Aristotelian clouds to angels materializing out of a post-Copernican planetary atmosphere is hard to

---

63) Ibid., 148: "equitur etiam quod non generantur in superiori regione, quia ibi non est aliquod corpus densum." And 149: "(Interpositio nubium) non potest esse in superiori regione, nam de necessitate, nubes tales vel dissiparentur, vel retunderentur in pluvias a media regione. Et certe Parelii semper portendunt multas pluvias."

64) For a "halo" depiction of angels, see Francisco Zurbaran's *Immaculate Conception* of 1634 (Cerralbo, Madrid).

65) Ibid., 207: "Sed tunc difficultas est, quod aer sic ingrossatus non erit satis solidus ad actiones et passiones consuetas."

66) Ibid., 68: "Nam in propositione octava, ex Aug. 3. de Trin. et ex communi doctorum post Aristotelem, recte ponit Angelos vel Deamones non posse immediate alterare, sed solum motu locali movere corpora."

67) Ibid., 208: "Re quidem vera nihil hic efficiunt nisi per motum localem, et ideo totum potest ab ipsis praestari, hoc non videtur negari pose."

imagine. Cosmologies may be in flux, but a core docrine of tradition, the doctrine of angelic visitations, sails ahead intact.

Appearing through the condensation of the same telluric air that we breathe, Sancta Clara's angels, nonetheless, have a more prosaic, earthy quality, than Scotus's angels—or rather, they bring out the presciently "proto-atmospheric" character of Scotus's doctrine. While Aristotelian air had the abstract dignity of its own cosmic sphere below the moon, the Baroque air invoked by Sancta Clara's theory is tied genetically and dynamically to a telluric mass. It is air that is not merely polluted by earth and moisture as a matter of accident but air that is by its very nature swift with the earth's motion, turbulent and planetary. The old Gothic angels of translucent skin and delicately embroidered garments (think of Jan and Hubert Van Eck's Ghent Altarpiece, 1432) have been preserved but they have also been transformed into the more athletic angels of Tintoretto and Guercino. Sancta Clara's atmospheric angels point to a distinctly Baroque fascination with flesh in motion—vividly combustive in vermillion pigment and linseed oil, to evoke energy, fire, faith, turbulence, and charity, especially in the paintings of Peter Paul Rubens, one of whose Crucifixion scenes hung in the Queen's Chapel in Somerset House, where Sancta Clara knew it well until enraged Puritans came in 1643 to hack it to pieces and feed it to the Thames.[68]

Sancta Clara's approach to angelology is deliberately probabilistic, deliberately inconclusive. It consists in testing various positions and disparate frameworks in order to reach, not firm ground, but modest footings here and there—probable opinions reached by bundles of reasons, not one of them conclusive in its own right. In an exchange with Thomas White, who denied that prayers for the dead have any efficacy since pure souls, like angels, are immutable and, therefore, cannot be improved before the final resurrection, Sancta Clara protested that the ancient opinion in favor of prayers for the dead was sufficiently authoritative ("probable") to be respected.[69]

---

[68] See Loomie, "Destruction of Rubens's Crucifixion," 680-681.

[69] See *The Result of a Dialogue concerning the Middle-State of Souls; wherein is asserted the Ancient Doctrine of their Relief, obtainable by Prayers, Alms, etc., before the day*

Sancta Clara's refutation of Trithemius and Paraclesus shows a similar reluctance to toss out whole bodies of testimonies and doctrines in the name of the latest rational deduction or critical experiment. Sancta Clara appeals to a multiform *nexus* of interlaced arguments, in which ancient and recent theories of material mixture, observations, chemical experiments, saintly authority, scholastic metaphysics and theology are all brought to bear collectively on a new philosophical riddle and spiritual danger.

Less elegant than White's cosmology of fiery angels, Sancta Clara's piecemeal universe is a work in progress, wearing its imperfection on its sleeve, guided by conservatism but also by a new imaginative delight in the sheer disjointedness of phenomena. Folktales and legends, Sancta Clara believes, often contain hidden kernels of truth.[70] All hypotheses deserve a hearing. For example, Kepler's theory that winds are caused by planetary breathing leaves Sancta Clara speechless, he admits,[71] but who knows? Knowledge advances through guessing general laws, but also through a strange gift of fidelity and through taking note of unique events, exceptions, sports of nature and miracles. By the same token, positions that have endured the test of time and help human beings live should not be rashly discarded but instead cautiously harmonized with new discoveries and theories, lest we be swept away by new superstitions and lose our immediate proximity to angels. Until better evidence is offered and stronger arguments are framed, the venerable idea that man is *the lowest servant of the missionary realm*, must be cherished and defended. An overly zealous naturalism that would systematically replace the

---

of Judgment; by F.D. professor of Divinity (Paris, 1660), in response to Thomas White, *Middle-State of souls*, (London, 1659), Chapter XI, 127-139.

70) Sancta Clara, *Paralipomena philosophica*, cap. 15, 211: "Nam praeter poetarum fabulas, quas tamen ut plurimum non puto puras fabulas, sed veritates, nunc Astronomicas, nunc Physicas, sub textura poetica involutas."

71) Ibid., ch. V, 97: "Et eodem modo argumentatur Keplerus in sua Harmonica l. 4. c. 7. de terra, nempe quod sit animata fortiter contendit; et ulterius vult ventum esse halitum ejus, quo pulmones contrahuntur et dilitantur, ut in nobis. De hac posteriori assertione, nescio quid asserem."

Figure 1. Richard Blome's "sculpture" of stormy angelic meteors towering above a small, fragile earth, vividly conveys the Baroque combination of fascination and angst triggered by the novel idea of planetary atmosphere. From the 1694 edition of Antoine Legrand's *Institution of Philosophy*, Houghton Library. By Courtesy of Harvard University.

wondrous (angelic appearances) with the merely exotic (homunculi) must be resisted. *Bleibt, ihr Engel, bleibt bei mir!*[72]

---

[72] "Stay, you angels, stay by my side!" Johann Sebastian Bach (1685-1750), Cantata BWV 19, Aria 5 (tenor, with instrumental Choral), from Martin Schalling (1571).

III. Antoine Legrand (1629–1699)

We turn, finally, to a very special Franciscan missionary to England, Antonius à Sancto Francisco, better known as Antoine Legrand. Born and reared under the stormy skies of Ruysdael, Legrand finished his novitiate at the Douay Friary sometime before 1656 and was sent to London to teach philosophy in the often dangerous and always despised Roman Catholic underground.[73] When Legrand embraced Cartesian philosophy nearly twenty years later, did he forsake Sancta Clara's probabilistic, piecemeal approach to natural philosophy in favor of a Cartesian "single thread" deductive epistemology? In particular, did he expel angels from natural philosophy?[74]

While neither Legrand's 1671 Compendium nor his 1672 *Institutio philosophiae secundum principia Domini Renati Descartes* considers the topic of angels, his revised 1675 *Institutio Philosophiae* includes a detailed angelology, on two grounds: first, taxonomical completeness and, secondly, Legrand's belief that "a number of truths regarding the essence, existence and operations of angels are within the grasp of natural reason" and therefore belong to philosophy.[75] Such truths prominently include defining angelic nature as an immaterial *res cogitans*, endowed with a distinctly Cartesian intellect and a distinctly Cartesian free will. Angels, moreover, if and when they voluntarily assume provisional bodies of air, are in these provisional bodies precisely as "pilots in a ship," in compliance with the Thomist–Cartesian distinction between human beings and pure

[73] For the five Franciscan martyrs of the English mission (Bel, Heath, Colman, Bullaker and Woodcock,) see Angelus Mason, *Certamen Seraphicum* (Douay, 1649).

[74] Sancta Clara was Director of Studies at St. Bonaventure's in 1647 and was elected Provincial for the second time in 1650, during Legrand's novitiate. For Legrand's first philosophical enthusiasm for Scotus, whom he compared favorably to Ramus and to Descartes in 1650, see Thomas Mautner, "From Virtue to Morality. Antoine Le Grand (1629–1699) and the New Moral Philosophy," *Jahrbuch für Recht und Ethik*, 8 (2000), 209-232. Legrand's *Philosophia veterum, e mente Renati Descartes more scholastico breviter digesta* was published in 1671.

[75] Legrand, *Institutio philosophiae*, pars tertia, *Daemonologia: seu de intelligentis, et spiritibus creatis, geniis et daemonibus, quos angelos vocare solemus*, 121-3.

intelligences.[76] Legrand, however, departs from Descartes by including Scripture in his discussion, for example Genesis 6, which he interprets like Thomas Aquinas to rule out materiality and to help establish that it is "contrary to reason to suppose that angels can procreate."[77] With regard to affirming the existence of angels, Legrand first argues, like Descartes, that natural reason proves only the non-impossibility of angels, but then, more like Sancta Clara, lists five independent arguments *quod sic*, none conclusive by itself, but together weighing on the side of angelic existence.[78] Legrand's fifth argument, for example, is that both ancient authors and Scripture report angelic appearances.

Chapter IX asks whether angels assume bodies and what actions they exert when embodied. Legrand accepts as his premise, from Scripture, that angels take up bodies to make themselves visible and denies that such apparitions are imaginary or morbid. He goes on to endorse the scholastic consensus that angels chiefly assume bodies of air, which they form through their own industry (like Durandus's angels) and shape semiotically (like Durandus and Thomas' angels) out of compacted vapors and exhalations (like Scotus's and Sancta Clara's angels).[79] To emphasize that immaterial beings are especially skilled in the science of optics, Legrand cites the twelfth-century master William of Paris, who illustrated demonic deception by citing the case of a soldier who mistook the corpse of a beast for a beautiful young girl.[80]

---

76) Ibid., part III, art. 1: *Quid sit Angelus in sua Natura*; and Arts. III, IV and IX. For Descartes, see *Meditation*, VI.

77) Ibid., art. 2, 128-130: *An Existentia Angelorum sit lumine naturae demonstrabilis et quomodo evinci possit.*

78) Ibid. 129-130: (1) argument by God's majesty (it is fitting to God's majesty to govern through delegated magistrates); (2) unexplained phenomena (uneducated persons speaking in Latin or people carried through the air); (3) Aristotelian argument that no body can move itself; (4) argument by God's creative plenitude (great chain of being without gaps); and, finally, (5) angelic apparitions are recorded both by ancient authors and by Scripture.

79) Ibid., part III, art. 9, 151: "Videtur quod Angeli potissimum aerea sibi corpora assumant, quae propria industria effingunt, et è vaporibus et exhalationibus compacta, sibi pro instituti ratione, adaptant."

80) Ibid.: "Ea ratione Daemon referente Guillelmo Parisiensi, parte ultima de Universo, Militem assumpto bestiae cadavere delusit, qui eum, se formosa virgine...." A simi-

Legrand's theory of angelic appearances turns vividly more Cartesian as he explains the colors that angels display in their atmospheric bodies:

> Thus depending on the order of the various parts, a diversity of colors arises, and the shape of a man, or of a beast, is offered to the eyes of the beholder. For indeed as these bodies have true extension and are terminated by various figures, the same alternating shapes, positions and masses are able to appear through the shifting reflection and refraction of the light as if they were genuine.[81]

Angelic appearances, which cannot be denied since they are attested by Scripture, thus independently corroborate the Cartesian theory that sensible qualities stem, not from an organic "form," but from the geometric and mechanistic properties of matter. Whether in a cloud set aflame by the setting sun or in a river mist shaped deliberately by an angel, colors attest extramentally only to quantitative differences in size, position and motion of constituent particles. But then, Legrand reasons, why should angels bother assuming bodies at all? Why not simply modify the light exactly as it would be modified by the surfaces of a human body? Since Legrand's Cartesian angels understand the mechanistic principles of natural philosophy and are supremely skilled in applying these principles for the purpose of producing effects, they have the power visibly to show themselves without taking up bodies. All they have to do is scatter or focus or refract ambiant light to produce (real extra-mental) special effects. Legrand's angelic appearances are not mental hallucinations, any more than the image that is seen through a *camera obscura*, or in a mirror, is a hallucination. In all three cases, a real optical array is present and mapped onto the human retina.

Legrand makes his case by first enlisting the authority of his contemporary the Minim friar and physician Emanuel Maignan to argue,

---

lar theme is found in Baudelaire, "Voyage a Cythère" and "La Charogne" in *Les Fleurs du Mal*.

81) Ibid.: "Adeo ut ex diverso partium ordine, colorum diversitas oriatur, et hominis brutive animalis figura, intuentium oculis offeratur. Nam cum illa corpora veram habeant extensionem, et variis figuris terminentur, pro varia luminis reflectione, aut refractione, alterius figurae, situs, ac et iam molis apparere possunt, quam revera sint."

not that angels act only by means of local motion (as established by Aquinas and cited by Sancta Clara) but that *bodies* affect one another solely by means of local motion, since bodies possess no other active power.[82] Legrand then applies the Thomas–Maignan theorem to angelic activity:

> Wherefore since we understand that a Spirit can exert all those Motions on our senses which any Bodily Substance can, and likewise resist other Motion as well as any Body whatsoever; we may likewise conceive how a Spirit may be perceived by our senses under the corporeal form of that Body, whose manner and measure of Motion, or resistance of Motion, he shall purpose to imitate. Wherefore an Angel in order to his appearing will not need to make himself a Body of such a Colour, Figure, and other Qualities, as he intends to appear with; but it will be sufficient, if he only refract and reflect the light in such a manner and Measure, as that Body would, in whose shape he desires to appear.[83]

The result is a multi-centered argument made up of disparate claims that converge to reframe angelic appearances as a special case of optics rather than a special case of meteorology. Legrand's discussion of angelic appearances may violate Descartes's *method* of doing philosophy but it beautifully validates Descartes's mechanistic optics. Why leave angels out of philosophy and risk, in Puritan London where John Milton's epic *Paradise Lost* was published in 1668 and Robert Boyle's defense of "the excellency" of theology was published in 1674, being accused of atheism? Legrand, in effect, recognized that he could nicely combine into a coherent doctrine Maignan's theory, scholastic angelology and Descartes's optics, especially Descartes's mechanistic explanation of the sky's azure in *La Métérologie*.[84]

---

82) Ibid.: "Nam cum res corporeae, ut ait Magnanus Philosophiae Naturae, capite 33, nullam aliam agendi vim habeant, quam ut moveant localiter, vel resistant motui locali, consequenter non alia vi se praebant sensibus, quam ea, qua proprios motus, tanquam proprios characteres, seu species imprimunt sensibus, vel qua resistunt alienis motibus." Emmanuel Maigan (1601–1676) joined the Order of Minims and published his four-volume *Cursus philosophicus* in 1652.

83) Citing from the 1694 English translation, *An Entire Body of Philosophy, According to the Principles of the Famous Renate Des Cartes*, part III, ch. 9, 87.

84) See Legrand's long and polished Cartesian Natural History published in 1673 and dedicated to Robert Boyle: *Antonii le Grand Historia naturae variis experimentis et rati-*

Far from discarding angelic appearances from mechanistic philosophy, Legrand invoked angelic appearances to validate mechanistic principles and invoked mechanistic principles to frame an elegant new angelology that dispenses angels from having to work directly with "crass" vapors. By a simple act of will, Legrand's Cartesian angels bend or focus light rays in order to appear, as ethereal in their predilection for "subtle matter" as the old Aristotelian intelligences. Their sphere of appearing activity is thus no longer restricted, as it is for Sancta Clara's atmospheric angels, to where parhelia form: light alone is needed, instantaneous and radiant.

Legrand concludes his Angelology by affirming that Angels take care "of Mankind and the things here below":

> It is manifest that God makes use of the Ministry of Angels, in the Ordering of things below; and there seems to be a twofold care of them; the one General, whereby they are concerned with the whole world; the other particular, whereby they watch over particular men.[85]

The theocentric character of the universe as well as the importance of angelic activity for cosmological stability are also endorsed in Legrand's Cartesian paradigm:

> In the Angelic superintendency, the Providence of God most clearly appears, forasmuch as he governs and rules the things here below by Superior Powers and Administers changeable things by such as are Spiritual and unchangeable.[86]

---

ociniis elucidata secundum principia stabilita in Institutione philosophiae edita ab eodem authore (London, 1673), 104. The Jansenist painter Philippe de Champaigne dematerialized angelic appearances, turning them into translucent effects of pure color, as in his *Annunciation* of c. 1660.

85) Legrand, *Institutio philosophiae*, part III, art. 11 (*De cura angelorum circa genus humanum et haec inferiora*): "Cum manifestum sit Deum Angelorum opera uti et rebus creatis per illos prospicere. Duplex in Angelis cura esse videtur; Generalis una, qua Universo orbi invigilant; particularis altera, qua singulis Hominibus adsunt et praeponuntur."

86) Ibid., part III, art. 11, 158: "In hujusmodi ordinatione maxime Dei Providentia elucet, quod inferiora per superiores potestates regat, et res mutationi obnoxias gubernet per eas quae incorporeae sunt."

Finally, like Thomas White and like Sancta Clara, but armed this time with Descartes's clear and distinct theory of free agency, Legrand interprets angelic activity on the model of voluntary service, not blind obedience.[87] He cites Descartes's Meditation IV to argue that angels love God infallibly but freely, the essence of freedom consisting not in indifference, but in self-determination.[88]

What is new in Legrand's Cartesian angelology is the idea that angels, while useful to all human beings, are especially useful to philosophers. In Legrand's mechanistic cosmos, material phenomena result deterministically from laws of motion, and thus no longer need the incessant work of rarefaction supplied by Whitean angels. Philosophers, on the other hand, need angelic guidance more than ever! With their (angelic) *mentes* substantially united to bodies and affected by sensation, philosophers risk erring and being deceived at every turn. Both as a Cartesian and as a Franciscan, Antoine Legrand glories in the infirmity of unaided human reason. The most important service rendered by Legrand's angels is not to light torches physically in the night sky or to stabilize the heavens, but metaphorically to illuminate man's soul and watch vigilantly over human judgment—"greatly to assist our Minds, by enlightening our Understanding to perceive Truth, and by exciting our Will to embrace the Good that is offered to us."[89] Less cosmological than Whitean angels, less atmospheric than Sancta Clara's angels, Legrand's Cartesian angels need only a single candle flame to keep philosophers from stumbling as they walk "alone and in darkness."[90]

87) For Descartes's theory of free agency, see my *Descartes's Theory of Action*, 173-195.

88) Legrand, *Institutio philosophiae*, part III, ch. 4, 58: "Unde concludit Cartesius in quarta sua meditatione, divinam gratiam aut naturalem inclinationem nunquam libertatem imminuere, sed augere potius et corroborare."

89) Ibid., part III, art. 11: "Mentibus enim favent, dum intellectum vi quadam invisibili illuminant, ad veritatem percipiendam; et voluntatem excitant ad virtutem exercendam, et ad ea quae occurrunt bona amplectenda."

90) Citing Descartes, *Discours de la méthode*, Part VI. No painter depicts Baroque interiority more explicitly than George de la Tour, but see also Francisco Zurbaran's *Christ and the Virgin in the House in Nazareth*, c. 1640 (Cleveland Museum of Art) in which angels appear indistinctly in a shaft of light.

Epilogue

Robert Boyle, to whom Antoine Legrand dedicated his 1673 *Historia Naturae*, had a special predilection for angels, God's "Celestial Courtiers."[91] In Boyle's view, a chief vindication of "the excellency of Theology compar'd with natural philosophy," is that the Bible reveals what reason and observation alone cannot prove, namely that angels exist and "assume Bodies shap'd like ours."[92] Although angels are "wont to appear in dazling Splendor and astonishing Majesty," they modestly "stile themselves fellow servants" in order to make sure that we do not worship them (Boyle's angels want us to be Anglican.) The revelation through Holy Writ that angels exist cures us, Boyle argues, from a double presumption. First, since angels are immaterial substances with intellect and free will, knowing that they exist saves us from the mistake of deriving our idea of God's power and wisdom from the unworthy evidence of inanimate things, which are ruled deterministically by laws.[93] What is "astonishing and ravishing" about God's power is not that God keeps the solar system in its tracks—a natural law will do the trick just as well—but that God has the power "to Guide and Over-rule many thousand Milions of Engins endow'd with Wills."[94] God's rule over an infinite infinity of free agents—over angels and over *us*—attests to God's power infinitely better than God's control of celestial bodies.

---

[91] Robert Boyle, *On the High Veneration Man's Intellect Owes to God* (London 1685), 67.

[92] See *The Excellency of Theology, compar'd with natural philosophy,* (*as both are Objects of Men's Study.*) *Discours'd of In a Letter to a Friend. By T.H.R.B.E, Fellow of the Royal Society. To which are annex'd Some Occasional Thoughts about the Excellency and Grounds of the Mechanical Hypothesis. By the same Author* (London, 1674), 15 and 16: "they can assume Bodies shap'd like ours, and yet disappear in a wince."

[93] Robert Boyle, *Of the High Veneration Man's Intellect Owes to god, peculiarly for His Wisedom and Power* (London, 1685), 46 and 48-9: "... inanimate bodies, such as Stones and Metals: whose Nature determines to a strict conformity to those primordial laws of Motion which were once settled by the great Creator, and from which they have no Wills of their own to make them swerve."

[94] Ibid., 55 and 67-68.

Secondly, the existence of angels reminds us that God's creative inventiveness exceeds us absolutely:

> If the Conjectures formerly propos'd about Worlds differing from ours may pass for probable, then it will be so too, that God in these other Systemes may have fram'd a Multitude of creatures, whose Fabrick and Motions, and consequently whose Properties and Operations, must be very differing from what is usually met with in our World.[95]

For all we know, God's "Government of Spirits" extends to "regulating an innumerable company of differing creatures," including the "intelligent, though not visible, inhabitants" of the Air and Aether.[96] Who knows what creatures lie out there, in God's boundless creation? Implicitly, human scientists need not worry about ever reducing the distance between themselves and God. Once again, angels stand as Baroque sentinels against a vast night and a new dawn and a boundless space: as far as our knowledge may reach, an infinity of worlds and of intelligences lie beyond.

Less perturbed in his religious beliefs than his shadowy "recusant" compatriots, Boyle had the powerful idea of turning angels into signs—into symbols serving to mark unknown realities that lay beyond human knowledge and whose unknown properties were entirely independent of human opinion. In contrast, the angelology invoked by Thomas White, Sancta Clara and Antoine Legrand made every effort to conserve traditional beliefs by involving angels in strategic cosmological functions. The angelology of our "recusant" authors illustrates the sort of metaphisico-musical inquiry that Charles Sanders Peirce places far above the methods of tenacity and authority but below the method of induction.[97] The systems that White, Sancta Clara and Legrand entertained served above all to respond creatively and subjectively to a shared cultural dislocation. As Peirce points out, "such systems have not usually rested upon

---

95) Ibid., 72.

96) Ibid., 72-73.

97) See Charles S. Peirce, "The Fixation Of Belief," *Popular Science Monthly*, 12 (November 1877), 1-15; reprinted in Peirce, *Selected Writings*. ed. Philip P. Wiener (New York, 1958), 91-112.

observed facts, at least not in any great degree."[98] They combine human reason, instinct and taste to frame beliefs that please the mind. They shield us from a wilderness of "rough facts."[99] They veil barren spaces and make the universe habitable. Like angels, they watch over our solitude, bring us solace, warmth, splendor, courage, trust—the desire, yes, but not the power, to proceed to the unknown. *Bleibt, ihr Engel!*

---

98) Peirce, "Fixation of Belief," *Selected Writings*, 105.
99) Ibid., 109-10.

Words, Lines, Diagrams, Images: Towards a History of Scientific Imagery

Christoph Lüthy and Alexis Smets*
Radboud University Nijmegen

To John E. Murdoch, eminent iconographer

Abstract

This essay examines the problems encountered in contemporary attempts to establish a typology of medieval and early modern scientific images, and to associate apparent types with certain standard meanings. Five particular issues are addressed here: (i) the unclear boundary between words and images; (ii) the problem of morphologically similar images possessing incompatible meanings; (iii) the converse problem of comparable objects or processes being expressed by extremely dissimilar visual means; (iv) the impossibility of matching modern with historical iconographical terminologies; and (v) the fact that the meaning of a given image can only be grasped in the context of the epistemological, metaphysical and social assumptions within which it is embedded. The essay ends by concluding that no scientific image can ever be understood apart from its philosophical preconditions, and that these preconditions are often explained during disputes between the protagonists of different iconographical types.

Keywords

scientific imagery, epistemic images, word and image, taxonomy of images, iconography, chymistry, Marsilio Ficino, Giordano Bruno, René Descartes

* Faculty of Philosophy, Radboud University Nijmegen, P.O. Box 9103, NL-6500 HD Nijmegen, The Netherlands (luethy@phil.ru.nl; a.smets@phil.ru.nl). We would like to thank the editors of this volume, William R. Newman and Edith Sylla, for their precious comments and observations.

Introduction

Contemporary science is pictorial and often even picturesque. Hardly any laboratory press release comes without colorful graphics, no lecture seems to be able to do without gaudy slides, no journal article without glossy pictures. The multiple reasons behind this proliferation of images need not concern the historian of medieval and early modern science. But if Martin Kemp is correct in claiming that "the modes of representation in twentieth-century science are very much the heirs of the Renaissance revolution" that led to "the rise of illustration as a major tool of science," then it falls to the historian to elucidate both this alleged revolution and to retrace its wider implications, also for our own times.[1]

In fact, much has in the past fifteen years been written about the visual aspects of science past and present. Historians of art have begun to study non-artistic visual manifestations, including so-called 'epistemic images' appearing in the context of theory formation.[2] Historians of science, in turn, have paid increasing attention to the visual manifestations of scientific theory and practice across the centuries, although they have shown a tendency "to treat scientific pictures only as after-images of verbal ideas."[3] Philosophers of science, finally, are busy analyzing the function of models and diagrammatic representations in the logic of scientific discovery and explanation as well as in the dissemination of knowledge.[4]

---

1) Martin Kemp, "Seeing and Picturing. Visual Representation in Twentieth-Century Science," in *Science in the Twentieth Century*, eds. John Krige and Dominique Pestre (Amsterdam, 1997), 361-90, at 363.

2) In this paper, we use the term 'epistemic image' to refer to any image that was made with the intention of expressing, demonstrating or illustrating a theory. The more frequent term, 'scientific image', which we also invoke because of its greater accessibility, has however an awkwardly anachronistic ring to it when applied to centuries in which *scientia* meant something quite different than 'science' does today.

3) The quote is from David Topper, "Towards an Epistemology of Scientific Illustration," in *Picturing Knowledge. Historical and Philosophical Problems Concerning the Use of Art in Science*, ed. Brian S. Baigrie (Toronto, 1996), 215-49, at 215.

4) Davis Baird, *Thing Knowledge. A Philosophy of Scientific Instruments* (Berkeley, 2004) views as instruments both two- and three-dimensional 'models' (see notably his

Underlying much of this prolific literature, however, one finds the assumption that it is legitimate to speak *systematically* about images, rather than merely *historically* and *contingently*. This view presumes (1) that there are timeless criteria for distinguishing images from non-images; (2) that images possess a fairly stable ontological and epistemic status across the centuries; and (3) that it is possible to develop a stable classification or taxonomy of images.

All three assumptions seem to us not only doubtful, but also open to refutation. Indeed, it is the purpose of this essay to document a number of complications that arise from any supra-historical, essentialist approach to epistemic images and to suggest instead an approach that takes into consideration the epistemological, ontological and pedagogical assumptions that surrounded their production. It is our conviction that the investigation of epistemic images requires a patient study of that endless range of stable or unstable, but always temporally and geographically bounded iconographical traditions, on the one hand, and on the other hand of the way in which each of these traditions was in time extended, subverted, redefined or simply replaced by another.

The reason why this subject matter deserves to be discussed in the present context is that this essay originated in a lecture held during an academic celebration of John Murdoch's 80th birthday. With his monographic contribution to I.B. Cohen's *Album of Science* series, Murdoch has, after all, provided one of the earliest and most trenchant analyses of the logic of historical epistemic images. Unlike the other authors in Cohen's series, who in keeping with the series title word did produce 'albums', Murdoch spent little energy on the display and discussion of images that merely *illustrated* scientific theories and practices (e.g., images of star-gazers, surgical instruments, or plant species), but chose instead to focus on the question of how images attempted to *express*, *prove*, *organize* or *interpret* the arguments found in the texts to which they were

ch. 2: "Models: Representing Things"). In a similar direction go recent attempts to place images in the category of 'paper tools'; see e.g. Ursula Klein, *Experiments, Models, Paper Tools: Cultures of Organic Chemistry in the Nineteenth Century* (Stanford, 2003).

attached (by means of diagrams, geometrical proof, *sciagraphiæ*, etc.).[5]

Problem I: The Unclear Boundary between Words and Images

Particularly praiseworthy seems to us Murdoch's attention to the way in which words can pattern themselves out over a page and in so doing take the first step towards image formation—a sensitivity that is absent from all those contemporary discussions that either bluntly treat images as texts, or alternatively presume with equal bluntness an essential distinction between word and image. A study of the illustrations included in part II of Murdoch's book, "Standard Schemata and Techniques for the Visual Facilitation of Learning," will inevitably remind us of the fact that the Greek word γραφεῖν (grafein) means any gesture that (literally) 'engraves' something on a tablet—irrespective of whether the result is a word (which might, or might not, violate the rules of ortho*graphy*), or a dia*gram*, or indeed a *graphic* design. When the γραφίς (grafis), the slate pencil, is made to draw its lines, the result can be a 'drawing', a 'letter', or some other type of 'text'—but in each case it will be a γραφή (grafê) or a γράμμα (gramma)—for these two all-embracing words mean all of these things.[6] In the particular case of hieroglyphics, the 'drawing' and the 'letter' may even fall together, and only the context will tell you whether the drawings should be read as text or as image.[7]

---

5) John Murdoch, *Album of Science.* Vol. 1: *Antiquity and Middle Ages* (New York, 1984). Under the general editorship of I. Bernard Cohen, three more volumes were published in this series.

6) Noëlle Batt, "L'expérience diagrammatique: Un nouveau régime de pensée," in *Penser par le diagramme: De Gilles Deleuze à Gilles Châtelet*, ed. Batt (Saint-Denis, 2004), 5-28, suggests that the verb γραφεῖν hails from the Indo-European root 'grbh-mn', in which 'grbh' means 'to scratch' and 'mn' 'image, letter, text'. According to this etymological reconstruction, the manual gesture and its outcome would early on have merged into a single verbal unity. This etymology does not differ very significantly from that given by James Elkins in *The Domain of Images* (Ithaca, NY, 1999), 82.

7) The image-text division of hieroglyphics is further complicated by the fact that hieroglyphic texts can choose between phonetic and ideogrammatic scripture. On this

In his ground-breaking *The Domain of Images*, James Elkins challenges Nelson Goodman's division of 'image' into writing, notation and picture.[8] He demonstrates that, upon close inspection, almost any image seems to rely on a blend of all three categories, though examples can be found to document the gradual progression from 'pure scripture' to 'pure picture'.[9] Interestingly, Murdoch's collection of images appears to corroborate Elkins' research, although it documents a different type of progression from word to image. It allows us to witness how arguments occurring in a running text are ordered into columns or what we would now call 'text boxes' by an assiduous copyist (see fig. 1); how lines are drawn between such singled-out words to show logical dependencies, as in the case of dichotomies (see fig. 2); how dichotomies can be organized into the shape of trees (see fig. 3); and how these trees end up being decorated with leaves and apples and sometimes inserted into landscapes or cosmological maps (see fig. 4). This succession of examples, which documents the transformation from a word pattern over a simple dichotomy to a 'tree of Porphyry' (which can in turn be planted within a fully-fledged cosmological *Weltbild*), proves that there is no definite borderline between text and image.[10]

The ease with which words can become patterns and patterns can become images is fascinating—and at the same highly problematic:

---

issue, and for a criticism of the image-text distinction, see W.J.T. Mitchell, *Iconology: Image, Text, Ideology* (Chicago, 1986), esp. 24-30.

8) Elkins' concept of notation differs from Goodman's. For the former, a notation is "an image employing organizational principles other than the formats associated with pictures or writing systems," whereas for the latter, notational systems are syntactically and semantically unambiguous systems that in turn "ensure a one-to-one correspondence between the notation … and real-world denotata" (*Domain of Images*, 257 and 69, respectively).

9) Elkins proposes instead a classification into seven categories that triangulate, as it were, between Goodman's three categories. For his suggested progression from 'pure writing' through 'allographs', 'subgraphemics', and 'hypographics' to potentially 'pure pictures', and the notion of potentially 'pure notations', see his *Domain of Images*, 82-91 and part II.

10) On the relation between world-view, visual view of the world, and *Weltbild*, see *Die Welt als Bild. Interdisziplinäre Beiträge zur Visualität von Weltbildern*, eds. Christoph Markschies and Johannes Zachhuber (Berlin, 2008).

1

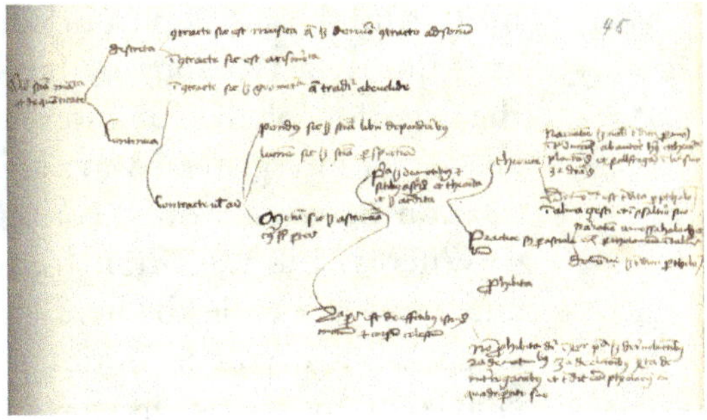

2

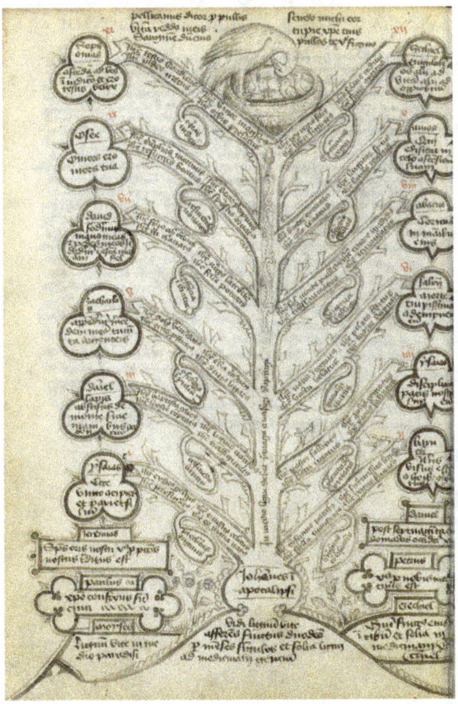

3

4

Figures 1-4: The gradual evolution of a tabulated text into a landscape.
Fig. 1: To facilitate comprehension, an intelligent fourteenth-century scribe has tabulated Cassiodorus' description of the nine valid *modi* of the first *formula* of categorical syllogisms (MS Bibliothèque nationale, Paris, fonds latin, 8500, fol. 34v. Reproduced with kind permission of the BNF). Fig. 2: A fifteenth-century student sets out the then available works on mathematics and physics in a dichotomizing form, which begins, on the left, with the division of mathematics into 'discrete' and 'continuous' quantities. Conceptual relations are indicated by lines (MS Öffentliche Bibliothek der Universität Basel, F.II.8, fol. 45r. Reproduced with kind permission of the

→

for this fluid transition from writing to imaging massively complicates attempts at a typology or taxonomy of epistemic imagery; renders the question of the function and role of such images impossible to answer in any general way; and undermines attempts to define the relation between words, concepts, theories, and images.

Whereas the transition from the flowing prose text to the table and from the table to the dichotomy (figs. 1-2) can be handled by a general theory of semiotics, the introduction of figurative elements eludes attempts at a timeless interpretation of signs. The tree in figure 3, for example, requires an understanding of the reference to the *lignum vitæ* of Apocalypse 22,2, which is mentioned in the root of the tree, that is to say, to "the tree of life, bearing twelve fruits, yielding its fruits every month: the leaves of the tree for the healing of the nations." The twelve leaves in this image are each associated with a prophet, and each branch with four characterizations of Jesus. The pelican doesn't nest for decorative reasons at the top of this tree, but in reference to Christ's self-sacrifice at the Cross. In fact, an inscription on the trunk states that "there should be an image of a crucifix depicted in the center of the tree." Now, given that the Christological pelican is often depicted at the top of a crucifix, this tree can also be read as a twelve-armed cross—a duplication of both morphology and meaning that undermines attempts to fit such an image into any unambiguous category.[11]

[11] Cf. Murdoch, *Album*, 49.

UB Basel). Fig. 3: When maps and dichotomies grow upwards, rather than from left to right, they become trees, as in this fifteenth-century "tree of prophets" (*arbor prophetarum*; MS Bibliothèque nationale, Paris, fonds latin, 3473, fol. 80v. Reproduced with kind permission of the BNF). Fig. 4: The tree has become a pictorial component in a miniature landscape in this fourteenth-century initial 'C', in which a teacher of logic teaches two students the very fundamentals of this art that are shown in the dichotomizing tree in whose shadow he sits (MS British Library, Burney 275, fol. 166r. Reproduced with kind permission of the British Library). (On the four images, see Murdoch, *Album*, 37, 42, 49 and 51.)

Problem II: Similar Shape, Different Meaning

This fluid transformation from a dichotomy of words into a fully formed tree can be documented throughout history. Take that notebook entry by Charles Darwin, that famous impulse-turned-graph, the "I think" followed by a drawing, subsequently annotated and explained (fig. 5). It is a particularly impressive demonstration of how an idea can first announce itself visually before becoming fully verbal—a phenomenon trenchantly analyzed in Rudolf Arnheim's

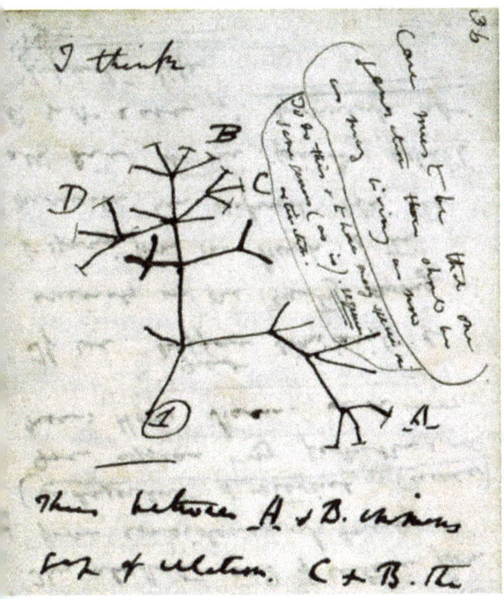

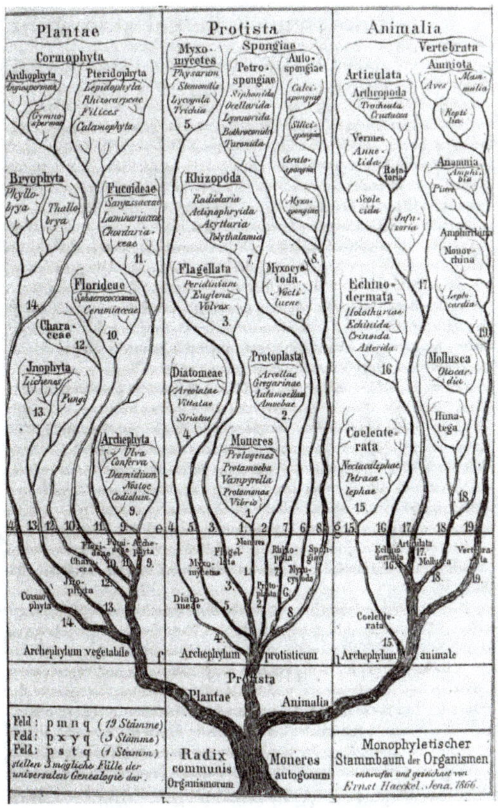

Figures 5 and 6: Two 'trees of evolution'? Fig. 5: Charles Darwin's 1837 sketch of speciation in Notebook B (Cambridge University Library, Dar. Ms 121, fol. 36; reproduced by kind permission of the Syndics of Cambridge University Library). Fig. 6: Ernst Haeckel's *Stammbaum der Organismen*, from his *Generelle Morphologie der Organismen* (Berlin, 1866).

study on *Visual Thinking.*[12] But let us compare these ink lines with Ernst Haeckel's tree of organic life (fig. 6). It seems to us evident that Darwin and Haeckel would have understood each other's images correctly, because figures 5 and 6 are both expressions of the same tradition of representing divisions of conceptual or biological *species* in terms of branching lines. And yet, are the two 'images' equivalent? Are they both 'trees'? Assume that we allow this name also for figure 5, then, does it matter that it spreads out in various directions, whereas Haeckel's tree (fig. 6) only grows upward? We should like to argue that it does matter; and that it makes a marked logical difference whether a dichotomy is unidirectional or not; whether its unfolding suggests goal-directedness (as in figs. 3 and 6) or randomness (as in fig. 5); and finally, whether the branching-out follows logical distinctions or bifurcations in time, or both. In sum, then, a tree isn't a tree. Our morphological names often don't seem to capture the precise logic of a given image.[13]

The meaning of the different exemplars of what might appear to constitute the same morphological type is thus dependent on the logic of the conceptual relations that they are each meant to visualize. We have just seen how easily a mapped-out list of names can turn into a biblical tree possessing soteriological overtones, and how an intuitive drawing of the logic of speciation can turn into an historical taxonomy.

But there even are more radical examples of the diversity of meanings of morphologically similar images. The extremely simple case of six circles of equal diameter grouped around a seventh is, in this respect, very striking.[14] Structurally speaking, figures 7, 8, 9, and 10 differ but little. And yet, they were intended to explain extremely different theories, entities and spatial dimensions. Figure 7 hails from a 1495 edition of Thomas Bradwardine's *Geometria speculativa*,

---

[12] Rudolf Arnheim, *Visual Thinking* (Berkeley, 1969).

[13] On the relation of Darwin's drawings to other types of 'trees', see Horst Bredekamp, *Darwins Korallen. Die frühen Evolutionsdiagramme und die Tradition der Naturgeschichte* (Berlin, 2006), notably ch. 1.

[14] This example has been analyzed in greater detail in Christoph Lüthy, "The Invention of Atomist Iconography," in *The Power of Images in Early Modern Science*, eds. Wolfgang Lefèvre, Jürgen Renn and Urs Schoepflin (Basel, 2003), 117-39.

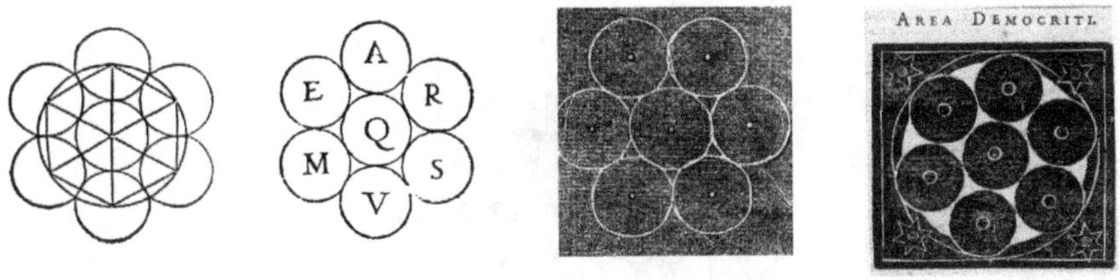

Figures 7-10: How many theories can six circles placed around a seventh circle possibly visualize? Fig. 7 demonstrates the perfection of the number '6' and of the circle (Thomas Bradwardine, *Geometria speculativa* [Paris, 1495], f. 8r). Fig. 8 explains the numerological logic of the six days of creation (Pietro Bongo, *Mysticæ numerorum significationis liber* [Bergamo, 1585]). Fig. 9 shows us 7 worlds touching (Giordano Bruno, *De l'infinito, universo et mondi* [London, 1584], 145). Fig. 10 documents how atoms aggregate to form larger globules (Giordano Bruno, *De triplici minimo et mensura* [Frankfurt, 1591], 50).

where it appears as the symbol of both the number '6' and of "the perfection of the circle" (because the six center points of the peripheral circles coincide with the extremities of the three diameters going through the center of the central circle, and because the six sides formed by the new figure can be inscribed into a new circle). Though crafted like figure 7 in numerological praise of the number '6', figure 8 wishes instead to express the logic of the six days of creation and the seventh of rest: the six days are grouped around the central divinity, whose restfulness, *quies*, is suggested by the inscribed 'Q'. The other letters are also abbreviations: 'E' stands for *Elementale*, 'M' for *Minerale*, 'V' for *Vegetabile*, and so forth. This image, like the previous one, receives its meaning and coherence thus exclusively from the accompanying text, which in this case is Pietro Bongo's *Mysticæ numerorum significationis liber*. Figure 9, by contrast, doesn't purport to reveal any numerological mystery, but intends to visualize cosmological configurations. Personally engraved by Giordano Bruno, this woodcut is meant to illustrate the way in which worlds (*mundi*) touch each other. According to Bruno's teaching, each one of the infinite worlds has a sun at its center (which he rendered visible with an aggrandized compass needle point). Figure 10 is also by Bruno, but this time, what he engraved were not worlds, but atoms. Importantly, the morphological similarity between his two woodcuts was not only intended, but was, in Bruno's eyes,

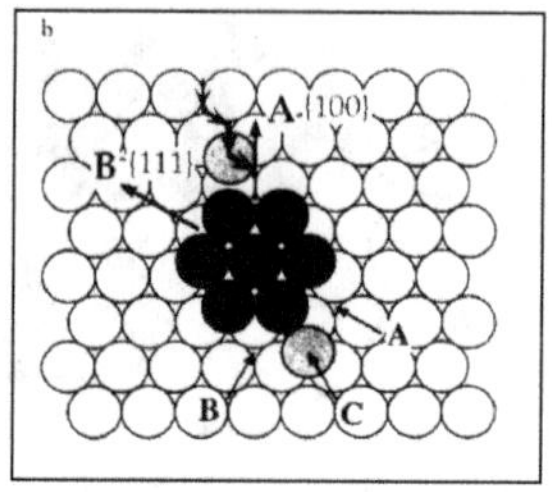

Figure 11: The same seven circles modelling "dendritic growth" of silver atoms on a close-packed surface (J.V. Barth & H. Brune, "Atomare Prozesse," 256).

pregnant with meaning. In fact, the stars placed in the four corners of figure 10 are a mnemonic device to remind the viewer of the similarity of the cosmological and the atomic configurations. Bruno, after all, subscribed to Cusanus' doctrine of the coincidence of the opposites (according to which contraries, including minimal and maximal geometrical forms, coincide) and also appreciated Democritus' atomistic interpretation of the infinite worlds. This is precisely why he felt that he could depict minimal material atoms by means of essentially the same image as maximally large worlds, and why the seven circles represented, in his eyes, an 'archetype'.

Figure 11, by contrast, is taken from a recent article in a physics journal. As its authors explain, the seven black disks represent a "heptamer of 7 silver atoms," which are "growing in the direction preferred by snowflakes" on top of a "hexagonally close-packed surface."[15] Iconographically speaking, the heptamer is the direct historical descendant of figure 10 and thereby adds confirmation to Martin Kemp's claim that contemporary conventions in scientific iconography often have a Renaissance background.[16] Nevertheless, these visually similar atoms refer to entities that belong to incommensurable theories of matter: the (composite and hence divisible)

---

15) Johannes V. Barth and Harald Brune, "Atomare Prozesse an Oberflächen," *Physik in unserer Zeit*, 29 (1998), 251-60, at 256.

16) The iconographical continuity in representing clusters of atoms—from Bruno through Johannes Kepler's crystallographic *Strena* to contemporary imagery produced by scanning tunneling microscopy—is documented in Christoph Lüthy, *De draad van Ariadne: een pleidooi voor de wetenschapsgeschiedenis* (Nijmegen, 2007).

silver atoms represented by the dark disks in figure 11 share no common trait with Bruno's (indivisible and ensouled) minima.[17]

Figures 7 through 11, then, place us before the problem that even where morphologically speaking, one seems to be confronted with a single type of image that survived the centuries quite intact, it is impossible to give it a name, let alone a unitary meaning. The different geometrical, numerological, theological, cosmological, and atomistic meanings are just too diverse to allow for a unitary description, either as a geometrical construction, as a symbol or structural model. In fact, just like the tree in figure 3, which also invoked a cross, figures 8 to 10 quite explicity rejoice in their multiple iconographic meanings. For Bongo, the geometrical and numerological properties enhanced the power of his image of creation (fig. 8), and for Bruno, the ubiquitous applicability of the same growth pattern to the smallest and largest dimensions added force to his general archetype of generation (figs. 9 and 10).

Possibly the most striking evidence for the divergence between shape and meaning is provided by what is defined as a 'diagram' by some and as 'graph' by others, and which consists in a line between two coordinates, as in figure 12.[18]

---

17) Aby Warburg would have been fascinated by this particularly powerful example of pictorial 'Nachleben'—which is obviously quite counterintuitive, given that it seems to transcend the various theoretical revolutions occurring in the domain of matter theory between Bruno and today's physical theories. On Warburg, see Georges Didi-Huberman, *L'image survivante: Histoire de l'art et temps des fantômes selon Aby Warburg* (Paris, 2002), notably 103.

18) See, for example, Jacques Bertin, *Sémiologie graphique: Les diagrammes, les réseaux, les cartes* (Paris, 1973), 50: "quand toutes les correspondances dans le plan peuvent être établies entre toutes les divisions d'une ordonnée et toutes les divisions d'une autre ordonnée, alors la construction est un diagramme." John J. Roche, "The Semantics of Graphics in Mathematical Natural Philosophy," in *Non-verbal Communication in Science Prior to 1900*, ed. Renato G. Mazzolini (Florence, 1993), 197-233, at 212: "The invention of coordinate geometry by René Descartes made possible and stimulated the representation of the quantitative laws of physics by diagrams, today commonly called graphs." The mutual relation between the terms 'diagram', 'graph' and 'schema' remains problematic, as becomes evident in Batt, "L'expérience diagrammatique," and in François Dagognet, *Écriture et iconographie* (Paris, 1973).

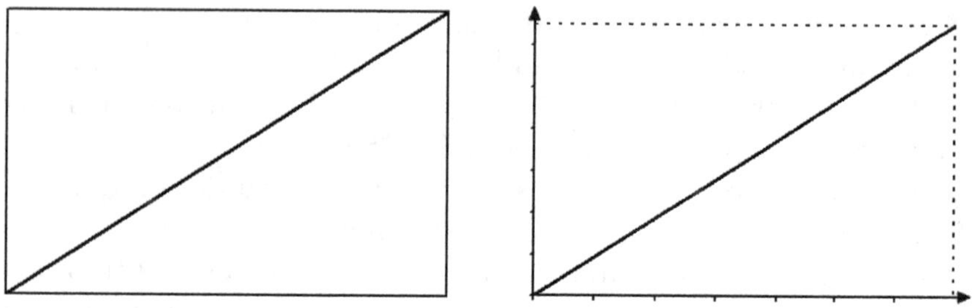

Figures 12a and b: The transformation of a diagram (12a) into a graph (12b). When the diagonal becomes the curve in a graph, the bottom left angle becomes the origin and the left vertical and bottom horizontal sides of the former rectangle become directional axes, which can be indefinitely extended in the direction of the arrows.

Although the 'graph' is, iconographically speaking, an indirect descendant of the mathematical diagram of Greek extraction and may still resemble its ancestor (as in figs. 12a and 12b), any such similarity is deceitful. For there are two abysses that separate the graph from the traditional geometrical proof. The first is constituted by the definition of the horizontal and vertical lines as axes carrying a numerical scale, with the line drawn between the two to express a functional relation to both. Despite all visual resemblances, this numerical functionality (in which the line can either be the visual expression of a given function, or conversely, in which the function can be derived from the line as the combination of empirical measurement points) has a radically different status than the old geometrical drawing, which formed part of a formal proof. The second abyss is constituted by the fact that these lines need not express any spatial extension. Whereas the traditional geometrical figure not only *represents* space, but *gives* us the space of which the mathematical proof speaks, the horizontal axis of the graph can represent any kind of magnitude, such as time, pressure, a probability density, or any other magnitude that can be meaningfully correlated with another magnitude. Having crossed these two abysses, the diagram turns into a graph which functions, in Michael Mahoney's words, "in a mathematical space wholly divorced from the physical space," a mathematical space moreover that is intended for the "mind's eye ... peering into the structural relations among quantities belonging to many different conceptual (rather than perceptual) spaces."[19]

The seven circles in figures 7-11 as much as figure 12 forcefully demand answers to the following questions: (1) Which iconographical similarities are essential (in the sense of pointing to an inherent logical or ontological similarity in a theory), and which ones are accidental? (2) Is the historical ancestry of an image of any importance, or is the only aspect that counts its function within the system in which it is assigned its current role? (3) What does the fact that similar types of representations are used in allegedly incompatible scientific paradigms across the ages tell us about these paradigms, our mental structures, or the conventionality of the visual language being used across numerous scientific generations?

Problem III: Identical Signifiers, Different Types of Representation

We have just seen why visually similar images do not necessarily carry identical meanings. But what about the inverse relation? Do things that are denoted by the same name necessarily require representation by the same type of image? To see that the answer is once again a 'no', it will suffice to take a look at the astonishing range of visual references to mercury in chymical texts across the centuries.[20] In fact, the observed iconographical latitude echoes the definitional latitude: 'mercury' was, after all, not only a specific metallic substance, but for a long while also one of the three philosophical principles of chymistry, and as such less specific and often more spiritual than the quicksilver obtained by reduction from cinnabar. But the observed iconographical latitude is also due to the even more fascinating issue of how to represent chemical properties

---

19) Michael Mahoney, "Diagrams and Dynamics: Mathematical Perspectives on Edgerton's Thesis," in *Science and the Arts in the Renaissance*, eds. J.W. Shirley and F.D. Hoeniger (London, 1985), 198-220, at 209.

20) On the term 'chymistry', see William R. Newman and Lawrence M. Principe, "Alchemy vs. Chemistry: The Etymological Origins of a Historiographic Mistake," *Early Science and Medicine* 3 (1998), 32-65.

as manifested both in different aggregate states and during operational interaction with other substances or principles.

Figure 13 is the oldest of the chemical images presented here.[21] While clearly a representation of mercury, it also stands in an older and long-lasting tradition of religious alchemical imagery. Christ and the Virgin Mary are shown here in mystical, hermaphroditic union.[22] As Barbara Obrist has shown, mercury is omnipresent in this image. Relying on a lexicon of alchemy of 1612, she points to the dragon at the bottom as a conventional manner of denoting "quick-silver or mercurial vapor."[23] But the dragon also denotes Christ (because of Psalm 22,7, in which David, prefiguring Jesus, claims to be a 'worm'). Furthermore, mercury is also represented more kaleidoscopically in the rest of figure 13, notably in the recoiling and raising snakes in the hermaphrodite's hands. Rolled up, the single snake refers to the properties of fixed mercury, while the three snakes emerging from the vessel designate its volatility. But the same metal is concomitantly identified with the philosopher's stone, because of its essential contribution to the generation of gold and silver. Visually, this special power is represented by the entire image (minus the two trees). This leads us to our first general observation: given that chymistry was less a science of individual materials than an operational science, it was the relational properties of mercury and its specific agency during interaction with other substances and principles that had to be visually documented. Nothing could be further from such an understanding than a representation of mercury's inner material structure.

---

21) This illumination is reproduced and analyzed at length in Barbara Obrist, *Les débuts de l'imagerie alchimique* (*XIV*^e^*-XV*^e^ *siècles*) (Paris, 1982), 152 ff.

22) The process of sexual metaphorization and its usefulness for chymistry is studied in Allison B. Kavey, "Mercury Falling: Gender Malleability and Sexual Fluidity in Early Modern Popular Alchemy," in *Chymists and Chymistry: Studies in the History of Alchemy and Early Modern Chemistry*, ed. Lawrence M. Principe (Sagamore Beach, 2007), 125-136.

23) However, the attribution of 'symbols' to specific substances was subject to great variations throughout the history of chymistry. See Obrist, *Les débuts*, 253; Maurice P. Crosland, *Historical Studies in the Language of Chemistry* (London, 1962).

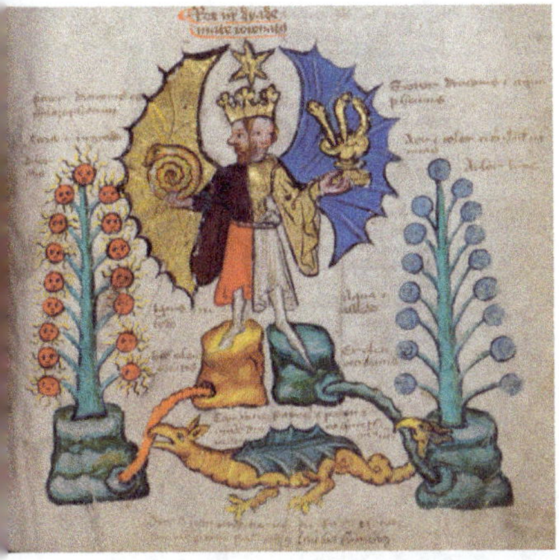

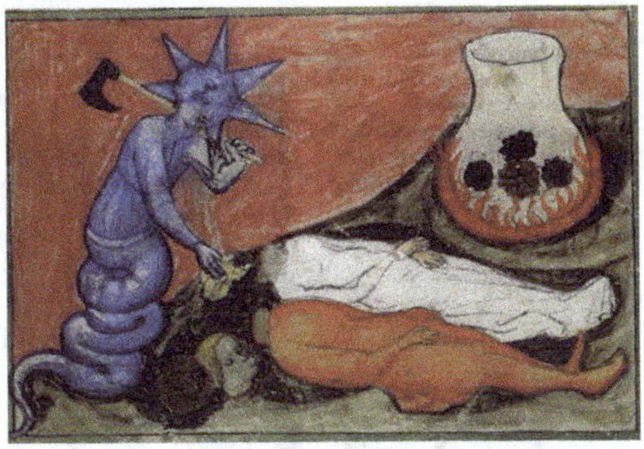

14

16

Figures 13 to 20: Diverse representations of mercury.
Fig. 13: Mercury of the philosophers (*Das Buch der heiligen Dreifaltigkeit*, ca. 1420; Germanisches Nationalmuseum Nürnberg, MS 80061. Reproduced with kind permission by the Nationalmuseum). Fig. 14: Mercury dissolving gold and silver (*Aurora consurgens*, Zentralbibliothek Zürich, MS Rh. 172, fol. 27v. Reproduced with kind permission by the Zentralbibliothek). Fig. 15: Mercury protected by iron against the effect of fire (Michael Maier, *Atalanta fugiens* [Oppenheim, 1618], Emblem 20). Fig. 16: Mercury represented as the messenger god Hermes-Mercury, with caduceus and winged helmet (Johann Joachim Becher, *Parnassi illustrati pars tertia: Mineralogia, Das ist: Deß erläuterten medicinalischen Parnassi dritter Theil, nemlich das Berg-Buch* [Ulm, 1663], 40).

17

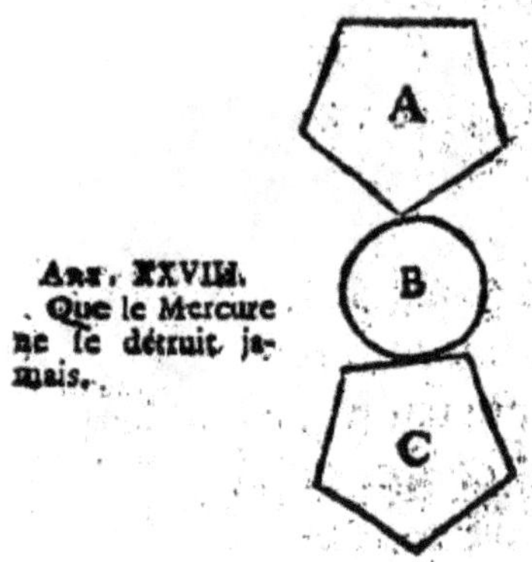

18

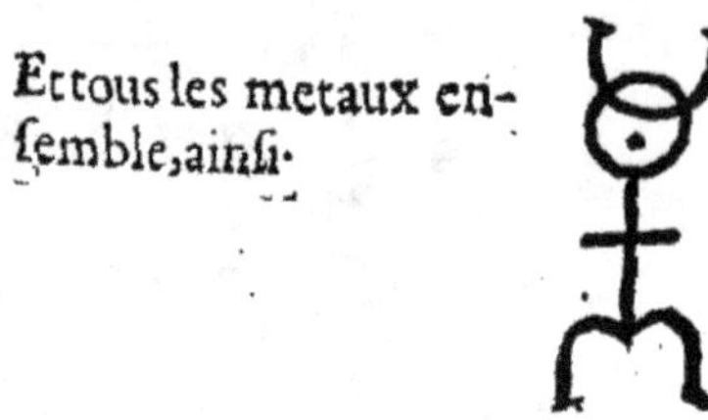

20

19

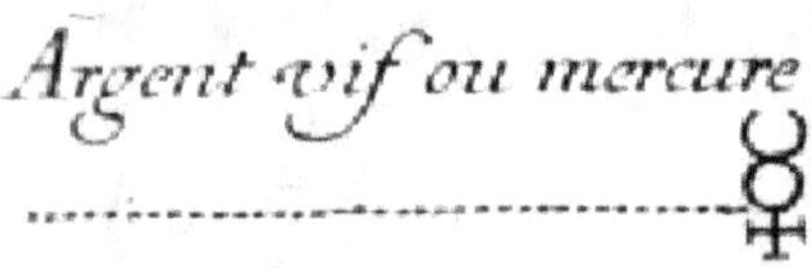

Fig. 17: Mercury immobilized (Giovanni Battista Nazari, *Della transmutatione metallica* [Brescia, 1572], 11). Fig. 18: Mercury dissolving gold (Nicolaas Hartsoeker, *Conjectures physiques* [Paris, 1706], 130). Fig. 19: The symbol of mercury (Nicaise Lefebvre, *Traité de la Chymie* [Paris, 1660], 153). Fig. 20: 'Combinatorics' with alchemical symbols (David de Planis Campy, *Bouquet composé des plus belles fleurs chimiques* [Paris, 1629], 991).

In figure 14, we are once more confronted with a representation of mercury, gold, and silver.[24] This time, however, the viewer cannot observe the birth of gold and silver out of mercury, but rather their demise in the process of calcination. According to Obrist, the pictorial

---

[24] Reproduced and commented in Barbara Obrist, "Visualization in Medieval Alchemy," *Hyle* 9 (2003), 131-70.

elements of the snake and the spiky crown might either refer to Egyptian mythology or to the first decan of the Lion.[25] While figures 13 and 14 seem to share a common 'style' of metaphorization and allegorization of the same three metals, they represent their object differently and according to the canons of different traditions and methods. Figure 13 is Christian by inspiration and attempts to describe the physical properties of mercury. Figure 14, by contrast, looks to Egyptian and astrological sources. Moreover, apart maybe from the ax that 'Mercury' bears on his left shoulder, this personification does not seem to indicate any physical property of the metal, unless one interprets the color and the crown as such indications, in analogy to the snakes and the dragon of figure 13. From a relational or operational viewpoint, however, Mercury decapitating Sun and Moon may be a reference to its putrefaction or calcination in the amalgamation of the three metals; yet there is no visible element announcing its further transformation. As is manifest, the two images also represent gold and silver quite differently. While figure 13 explains chemical change in a biological mode, representing silver and gold as the fruit of trees that are nourished by mercurial rivers, figure 14 personifies the two metals, allowing their identification only because of their specific coloring. The duplications at work in these images are also quite noteworthy: in both cases, the precious metals are indicated by their colors and by sun and moon symbols (a moon-face is recognizable in figure 14).

We have just now spoken, however provisionally, of a common style of 'metaphorization' and 'allegorization'. For the iconographer, the question arises as to whether these two terms can be clearly defined and applied as classificatory terms to epistemic images. On the one hand, our dictionaries would suggest that a metaphor is a figure, and thus a single element, whereas an allegory connotes a longer process, a story, poem or picture in which several elements are at play.[26] On the other hand, such a categorical distinction raises

---

[25] Obrist, *Les débuts*, 236 ff.

[26] This distinction between process and element is already drawn in early modern dictionaries, as in the first edition of the *Dictionnaire de l'Académie française* (Paris, 1694), *s.v.*: "Discours par lequel …" (allégorie); "Figure de discours …" (métaphore).

various problems, as contemporary studies on metaphorical and allegorical styles amply document. Furthermore, the reduction of images such as figures 13 and 14 (together with the texts in which they are embedded) to metaphors and allegories is also historically inaccurate, because the rhetorical vocabulary at the disposal of chymical authors was both larger and more subtle than that. In the fourteenth century, Petrus Bonus conceived of no fewer than a dozen rhetorical means of referring to a chymical substance, of which only the first was by its 'proper name'.[27]

Such 'improper' visual means of representing chymical substances, principles and operations as those that we have encountered in figures 13 and 14—whether they be allegorical, metaphorical, hyperbolical, or enigmatic—were enduring enough to be in use as late as the beginning of the seventeenth century, the period in which our figure 15 was crafted.[28] This figure, however, carried the official title of 'emblem', a classification that refers to a Renaissance conceptualization of image making.[29] We see a girl being protected by a knight against the deleterious effects of fire. According to Helena de Jong, the "vulnerable girl ... is the personification of the volatile Mercury."[30] To those acquainted with this particular mercurial

[27] Petrus Bonus, *Margarita pretiosa novella correctissima*, in *Theatrum chemicum* (Strasburg, 1659-1661), 6 vols., 4:507-713, at 516: "Et est sciendum, quod ars ista quasi sola inter omnes mundi, in sui doctrina utitur nominibus propriis, et extraneis, et inusitatis, et allegoriis, et ænigmatibus, et metaphoris, et æquivocationibus, et transsumptionibus, et involucris, et prosopopoeis, et hyperbolis, et ironiis." Quoted from Obrist, *Les débuts*, 48.

[28] This engraving is reproduced and commented in Helena M.E. de Jong, *Michael Maier's* Atalanta fugiens*: Sources of an Alchemical Book of Emblems* (York Beach, 2002), 162-166.

[29] The concept of 'emblem' originated in Andrea Alciati's *Emblemata* of 1531. The expression 'emblema vermiculatum', meaning 'mosaic work', is indicated as the origin of the concept of 'emblem' by one of Alciati's numerous editors: "Quare et ipse Alciatus, sua hæc epigrammata appellatione convientissima inscribi voluit Emblemata. Sunt eum *Emblemata vermiculata* opera extessellis insititiis apta et composita, interprete Budæo, quod et ipsa vocabuli Greci origo ostendit." (Andrea Alciati, *Emblemata*, eds. Macé Bonhomme and Guillaume Rouillé [Lyon, 1551], 4; see also Mario Praz, *Studies in Seventeenth-Century Imagery* [Rome, 1964], 23).

[30] De Jong, *Maier's* Atalanta fugiens, 166.

personification, figure 15 may indeed have illustrated the limited resistance of this metal to the action of fire. Similarly, the knight would have alluded to the fact that when allied with another substance, mercury could avoid evaporation in the face of fire. But although the textual sources informing this image don't differ much from those used for our previous two figures, this image is strikingly different.[31] One of the differences lies in the fact that not all elements in this copper engraving seem to be part of the allegory. Only the initiated will understand whether the landscape surrounding the woman, the knight and the fire has an emblematic role or is purely decorative. If the latter, figure 15 would possess the double nature of showing an emblem within a landscape or a chymical image within a non-chymical one. This doubleness reminds us, in some strangely significant way, of the transition from text to image in figures 1 to 4, above. In both cases, particular and informed habits of reading and of viewing seem to be required to recognize or 'read' the significant image in the image, or the visual pattern in the text, respectively.[32]

In figure 16, we encounter yet another incarnation of mercury. Instead of the dragon, the snake, and the naked woman, we have here the classical representation of the mythological messenger-god Hermes, or Mercury to the Romans, with wings on helmet and sandals and the caduceus with its two snakes, which was to become the pharmacists' emblem. The text surrounding the image leaves no doubt: this emblematic figure means the metal mercury. However, given that it might with equal right refer to the planet of the same name, the question arises as to whether his portrait tells us anything specific about the metal for which it here made to stand. If there were no chemical 'symbols' inscribed in Mercury's helmet and foot, the answer would be 'no'. But by means of the location of

31) The subsequent emblem XXI of Maier's *Atalanta fugiens* depicts mercury also as a naked woman, but this time in company of a naked man, who expresses sulfur, the principal chemical counterpart (cf. De Jong, *Maier's* Atalanta fugiens, 169).

32) This raises the question of the image in relation to its frame. See on this Johannes Grave, "On the Aesthetics of Scientific Objects. Three Case Studies," in *Wandering Seminar on Scientific Objects*, eds. Sophia Vackimes and Konstanze Weltersbach (Max Planck Institute for the History of Science, preprint 339, 2007), 35-48.

these 'symbols', Becher distinguishes between the volatile and fixed forms of mercury: pure mercury and mercury sublimate are assigned to the air, and mercury precipitate and cinnabar are assigned to the earth.[33] The style of representation has also changed. Figure 16 is heterogeneous, in that it superimposes letter-like signs on a personification, as if on a map.[34] The accompanying text is, as in some emblem books, divided into three parts: there is a title that carries the different names of the substance (*Mercurius*, *Argentum vivum*, and *Quecksilber*); there is an allegorizing poem; and finally, there is concrete, sober and non-allusive prose in which an account of mercury is given.[35]

In figure 17, which is taken from Giovanni Battista Nazari's *Della transmutatione metallica* (1572) and is thus a century older than figure 16, Mercury is recognizable only thanks to his winged helmet, given that his hands and feet are missing. In contradistinction to figure 16, however, figure 17 has strong allegorical overtones, as several of its pictorial elements (the basin with aquatic plants, the

---

33) The circle horizontally cut into two equal parts and surmounted by a cross is given as a denotation of cinnabar in Nicaise Lefebvre, *Traité de la chimie* (Paris, 1660), "l'explication des caracteres chymiques," 152-53. This 'sign' minus the line that cuts the circle horizontally is also presented as cinnabar in Nicolas Lemery, *Cours de chymie* (Paris, 1757), *s.p.* [789].

34) Different types of heterogeneity and duplication are found in the earlier astrological representations of metals; cf. Françoise Cannella, "Alchemical Iconography at the Dawn of the Modern Age: The *Splendor Solis* of Salomon Trismosin," in *The Power of Images*, eds. Lefèvre et al., 107-16; and Guy de Tervarent, *De la méthode iconologique* (Bruxelles, 1961).

35) In the first, playful verses of the poem, the god Mercury is described as being serious with the chymists, but playful with the alchemists: "Es komt Mercurius, der schnell geflüglet Gott / Er brauchet einen Ernst / und lässet seinen Spott; Den er zu treiben sonst mit Alchymisten pflegt / hier aber sich bey ihm / ein andrer Ernst erregt." The textual description of mercury is, by contrast, of empirical sobriety: "Das Queck-Silber Erß wird gewürket in seinen eigenen Bergsteinen / von seiner Natur der Sals-Erden / und behendiger flüchtigen Erden / einer feuchten schwierichten wässerigen Olität / die vermenget wird mit der allersubtilesten rothschwefelichten gekochten Erden / mit der allerschwächesten gemachsamen Verbindung / als eine ohnzeitige angenehme Frucht aller besonder Metallen." Cf. J.J. Becher, *Parnassi illustrati pars tertia, Mineralogia, das ist: Deß erläuterten medicinalischen Parnassi dritter Theil, nemlich das Berg-Buch* (Ulm, 1663), 40 f.

podium, and notably the lack of hands and feet) refer to chemical processes, which are discussed in the accompanying text. In fact, the entire image can be understood as fixed mercury. Curiously enough, figure 17 displays fewer pictorial elements than those indicated in the text (which mentions, for example, a village in the background and an infinity of canes), and is thus the exact counterpart to figure 15 (where the image is pictorially richer than the text).[36]

Figure 18 constitutes a clear break with all the images of mercury we have so far considered. All human-like figures have disappeared in favor of a circle surrounded by two pentagons. But this image, too, attempts to describe some properties of mercury, although it does so in a radically different, geometrical manner. Nicolaas Hartsoeker, who wished to walk in the footsteps of Descartes' reduction of all physical qualities to micro-particles, was convinced that we could induce the invisible corpuscular structure of matter from the empirically recorded properties. Wishing to explain how mercury could dissolve gold, he sought to locate this property in the 'intimate structure' or geometrical make-up of the mercury particle. Being of spherical shape (and, *pace* Descartes, also heavy), such a particle can force its way into the pores of gold molecules and separate these into 'parcels' (which Hartsoeker sometimes calls 'atoms'). What figure 18 purportedly shows is the soft amalgam that results from the intermixture of such gold 'parcels' with mercury particles.[37]

---

36) This translation of the textual 'infinity' of canes into a pictorial handful demonstrates one of the inherent peculiarities of the image-text relation: it shows one of the inherent limitations of imaging, which must rely on the imaginable and can therefore not do visual justice to certain concepts. (See on this issue Elkins, *The Domain of Images*, 40-1.) That this has implications regarding the difference between conceptualization and imagination became clear, for instance, in the controversy between Leibniz and Stahl (cf. Gottfried Wilhelm Leibniz, "G.G. Leibnitii animadversiones circa assertiones aliquas Theoriæ Medicæ veræ Clar. Stahlii. Cum ejusdem Leibnitii ad Stahlianas observationes Responsionibus," in Leibniz, *Opera omnia*, ed. Ludovicus Dutens, 6 vols. [Geneva, 1768], 2-ii:131-61, at 151: "Contendit *Responsio* actualem cujuslibet partis subdivisionem esse *supra* omnem conceptibilitatem; quia scilicet conceptum cum imaginatione confundit.")

37) Nicolaas Hartsoeker, *Conjectures physiques* (Paris, 1706), 130 f.

With figure 19, we have arrived in the domain of what are alternatively called 'symbols', 'signs' or 'characters'. A circle topped by a semi-circle, with a cross underneath: what had originated as a sign referring to the planet Mercury had long become also a symbol of the homonymous metal.[38] But what is the relation between 'sign', 'symbol', and 'image'?[39] The question is unanswerable, for it leads us once more into that swampy frontier area between letters, words, and images that we have entered from a different direction earlier on in this article. What is clear is that figure 19 is more than a simple abbreviation such as the modern 'Hg', which stands for 'hydrargyrum', because it can once again become an iconographical element of a more complicated, combinatorial figure. This is precisely what happened in figure 16, in which the signs for sublimed and precipitated mercury are developed out of the sign for pure mercury.[40] And it is also what happens in figure 20, in which a symbol of the philosophical stone (*monas hieroglyphica*) is composed of the signs for gold (the character's head), silver (its hat, as it were), and mercury. But lo and behold: the abstract symbols have been crafted into a human-like figure, so that, much to our bewilderment, we find ourselves somehow back in the figurative section of our little iconography of mercury.

Problem IV: Typological Names of 'Images'

In problem sections I and II, we have simply spoken of 'images', applying this term indiscriminately to Darwin's hasty sketch in figure

---

38) On the relation between planetary signs and metals, see for instance Diderot and D'Alembert, *Encyclopédie*, *s.v.* "Talisman."

39) The mutual relation between 'sign' and 'symbol' constitutes a battlefield on its own. The respective entries in D'Alembert and Diderot's *Encyclopédie*, for example, strongly overlap. By contrast, Barbara Obrist, *La cosmologie médiévale. Textes et images.* Vol. I. *Les fondements antiques* (Florence, 2004), 301 ff., clearly distinguishes between the two terms in her discussion of an image from Isidore of Seville in which the seasons are linked (*symbolon* = link) by the four qualities and elements.

40) Cf. Marco Beretta's account of the combinatorial hopes of new chemical symbolism in the last decades of eighteenth century, in "The Role of Symbolism from Alchemy to Chemistry," in *Non-Verbal Communication*, ed. Mazzolini, 279-319.

5 and to the artistic composition of figure 4, whereas in section III, we have begun to speak of allegorical, metaphorical and emblematic images as well as of signs and symbols. This leads us inevitably to our next problem, that pertaining to typological names.

Much recent literature has relied on a single, all-inclusive term with respect to all graphic non-textual appearances. Nelson Goodman has sought to include all images, artistic and scientific alike, in his theory of 'symbols', while James Elkins uses the term 'image' for the "entire field of meaningful marks," not unlike Gottfried Boehm, whose term '*Bild*'—whose meaning includes both the 'image' and 'picture'—refers to all static visual manifestations.[41] But this all-encompassing use of 'image' is undermined by more specific definitions of this word in the technical literature. Jacques Bertin, for example, defines 'image' as "a significant visual form that is perceived in one perceptual instant."[42] Whatever requires more than a single *coup d'œil* would thus no longer qualify as an 'image'. This definition, though seemingly dictated by the professional needs of a technical image-maker, happens to coincide with the original signification of '*imago*', which, in contradistinction to 'pictura', referred to the mentally graspable components of a given visual manifestation.[43]

---

41) Nelson Goodman, *Languages of Art. An Approach to a General Theory of Symbols* (Oxford, 1968); Elkins, *Domain of Images*; Gottfried Boehm, ed., *Was ist ein Bild?* (Munich, 1994); id., *Wie Bilder Sinn erzeugen* (Berlin, 2008).

42) Jacques Bertin, *Sémiologie graphique*, 50: "Forme visuelle significative perçue dans un instant de perception (dans un seul coup d'œil). C'est l'unité temporelle de perception significative."

43) On this, see Marion G. Müller, "What is Visual Communication? Past and Future of an Emerging Field of Communication Research," *Journal of the Swiss Association of Communication and Media Research* 7 (2007), 7-34. Müller distinguishes not only between the mental and material in words such as '*imago*' and '*pictura*', but also between the dominant meaning of the English term 'image' and the dual meaning of the German '*Bild*'. She recognizes a relation between the divergent meaning of these terms and the divergent focus of Anglo-Saxon and German research on visuality. The problem of idiomatic diversity in nomenclature can already be recognized in the early modern period. Note the German translation of the Latin title in: *Icones mortis sexaginta imaginibus, totidemque inscriptionibus insignitæ, versibus quoque latinis et novis*

By contrast, historians have been more hesitant in their typology. John Murdoch begins by distinguishing between "pictorial and diagrammatic materials," which he subsequently divides further into 'distinct types' possessing different functional and morphological properties, such a 'dichotomies', 'rotæ', or 'squares of opposition'.[44] Willem Hackmann, in turn, speaks with respect to early modern physics of "four types of textbook illustrations which are distinguished by their function," namely (i) "allegorical titlepages and frontispieces," (ii) "illustrations of the actual experimental configurations described in the text," (iii) "engravings depicting the actual observed phenomena produced by the instruments," and (iv) "diagrams of the supposed underlying structure of these phenomena." A fifth type, "mapping of data in the form of graphs," is briefly mentioned.[45] Barbara Obrist, finally, in her careful enquiries into the evolution of alchemical and cosmological imagery in the Middle Ages, wields an immensely rich vocabulary, which includes, in alphabetical order, the following terms: 'allegories'; 'analogies'; 'functional', 'mnemonic' and 'synoptic diagrams'; 'enigmata'; 'graphs', 'hieroglyphics'; 'icons'; 'ideograms'; 'pictorial metaphors'; 'cosmological' and 'geometrical schemata'; and 'symbolic signs'.[46] Obrist defines a number of these terms, albeit in a partly overlapping manner, while limiting herself in many other cases to mere descriptions or to a simple illustration of the meaning of the term by means of the very image to which it refers.

From this small range of representative examples, it becomes clear that there is neither a standard vocabulary nor an agreed-on typology with respect to 'epistemic images' past and present. The problem is aggravated by the fact that most descriptive terms in circulation don't coincide with historical terms. In other words, however refined or precise our own typology might be, it will never match earlier typologies.

---

germanicis illustratæ. Vorbildungen deß Todtes in sechtzig Figuren durch alle Stände und Geschlechte derselbigen nichtige Sterblichkeit fürzuweisen… (Nuremberg, 1648).

44) Murdoch, *Album*, x.

45) Willem D. Hackmann, "Natural Philosophy Textbook Illustrations 1600-1800," in *Non-verbal Communication*, ed. Mazzolini, 169-96, at 170-72.

46) Obrist, *Les débuts*, passim; and eadem, *La cosmologie médiévale*, passim.

This inevitable mismatch is due to three principal factors. The first is the lack of terminological precision in the historical sources themselves. '*Figura*', for example, frequently possessed the same catch-all quality as our own term 'image'. Indeed, the phrase *sicut hæc figura docet* ("as this figure shows") is found to refer to the entire spectrum of mathematical, diagrammatic or figurative 'images' that may accompany a text. Interestingly, this nondescript use of *figura* lives on in our modern references to visual text-inserts as 'figures' (as is testified by the present essay).

The second factor responsible for the inevitable mismatch between our tentative terminologies and historical ones is the exact contrary of the first factor. One encounters, notably in the Renaissance, authors who draw distinctions between types of images that are very hard to capture by means of our own terminology. The most extreme case is possibly Giordano Bruno, who in his treatise *On the Composition of Images* of 1591 distinguished between "*1. idea, 2. vestigium, 3. umbra, 4. nota, 5. character, 6. signum, 7. sigillum, 8. indicium, 9. figura, 10. similitudo, 11. proportio, 12. imago.*"[47] And as if this were not enough, Bruno—who may have been the only natural philosopher to illustrate his work with woodcuts of his own making—employed even more terms in his other works, including *archetypus*, *forum*, *atrium*, *rota*, *area*, etc.[48] Sure enough, it is possible to translate Bruno's terms into English, or into any other modern language. But apart from the fact that his *idea* is not the same as our 'idea' nor his *vestigium* the same as our 'trace', his (translated) vocabulary would still not constitute a reliable historical typology, but only Bruno's own, and nobody else's. The reasons for this extreme limitation will be explained shortly.

The third complicating factor is a mixture of the previous two. In the scholastic tradition, it was customary to study the opinions of previous authors before arriving at one's own conclusion. A late sixteenth-century author working on vision, perception, imagination

---

47) Giordano Bruno, *De imaginum, signorum et idearum compositione, ad omnia inventionum, dispositionum et memoriæ genera libri III* (Frankfurt, 1591), 97.

48) See the beautiful facsimile reprints of Bruno's woodcuts in Mino Gabriele, *Giordano Bruno: Corpus iconographicum. Le incisioni nelle opere a stampa* (Milan, 2001).

and understanding was, however, confronted with a truly irreconcilable range of views on how the things perceived ended up in our mind. Intromissionist and extramissionist theories, the *multiplicatio specierum* described by the so-called perspectivist authors, Avicenna's and Averroes' theories concerning the *species intelligibiles* and *intentionales*, and *simulacra* theories of atomist extraction, they all presented themselves as possible modes of description, and each with its own technical vocabulary.[49] The consequences of this situation can be beautifully witnessed in the following hair-raising description of what it is that light carries to the eye. "Light takes on nothing else than the *imago* of things, which, whether you call it *forma*, or *simulachrum*, or *idolum*, or *species*, or *spectrum*, does not matter, if you understand but that alone, that it represents the thing."[50] Here we have an author (Fabrizio di Acquapendente in his treatise *De visione, voce, auditu* of 1600) who clearly has lost his way in the forest of possible, but mutually contradictory terms.

Problem V: No Iconography without Epistemology and Metaphysics

It is precisely because of this unmanageable multiplicity and ambivalence of the available terminology that Giordano Bruno attempted to clarify his use of terms by clear definitions. But contrary to most authors, both contemporary to him and to us, Bruno understood the direct dependence of any iconographical terminology on its philosophical framework, and notably on metaphysical and epistemological assumptions. After all, each epistemic image is relational. As

---

49) On the proliferation of models of perception and intellection in the early modern period, see notably Leen Spruit, '*Species intelligibilis*'. *From Perception to Knowledge*, 2 vols. (Leiden, 1994, 1995).

50) Fabrizio di Acquapendente, *De visione, voce, auditu* (Venice, 1600): "[lux] non aliud quam rerum imaginem assumit, quam sive formam, sive simulachrum, sive idolum, sive speciem aut spectrum appelles, nihil interest, si modo id solum, quod rem repræsentat, intelligas." Quoted from Isabelle Pantin, "*Simulachrum, species, forma, imago*: What Was Transported by Light into the Camera Obscura? Divergent Conceptions of Realism Revealed by Lexical Ambiguities at the Beginning of the Seventeenth Century," *Early Science and Medicine* 13 (2008), 245-69, at 259n45.

Aristotle states in *De memoria* (450b23-29), it has a double nature, referring to itself as an image (*eikon*), and also to something else, to that which it represents. But that 'something else' will possess a very different status depending not just on the type of 'image' one is dealing with, but also on the type of philosophy within which it was created.

For someone who assumes, as Aristotle roughly does, that understanding ideally consists in grasping the essence and causal embedding of an object or event by the mind's higher faculties, there won't be much need for an additional visualization of natural phenomena. This explains, as Murdoch puts it, why "the illustrations are notoriously few in the manuscript copies of Aristotle and Galen and of their medieval translations and seemingly endless commentaries."[51] For the typical Aristotelian, the five senses are adequate to the perception of the things that are. This is why he may illustrate rare things or events, simply because not everyone has seen them, but will have no need for abstractions or models of things that are well known. Aristotle, for example, excludes structural models of micro-levels by means of his Lynceus argument (*De generatione et corruptione*, 328a13ff), that is, by the argument that the atomists' philosophy entails the absurd idea that a sharp-eyed Lynceus would see a different reality than we commonly do, namely one made up of atomic particles interrupted by little void spaces. On the other hand, the forceful presence of quaternities in the Aristotelico-Galenic tradition (four primary qualities, four elements, four humors, four seasons, four directions, etc.) allowed for the well-known maps of logical relations and squares of opposites, although these were not inherently necessary to a comprehension of reality.[52]

By contrast, if one assumes, as René Descartes does, that the senses are deceitful and that the world must be explained on the basis of a hypothetical reconstruction from the most clear and evident

---

[51] Murdoch, *Album*, x.

[52] For the Lynceus-argument and its implications for the visualization of matter, see Christoph Lüthy, "Atomism, Lynceus, and the Fate of Seventeenth-Century Microscopy," *Early Science and Medicine* 1 (1997), 1-27; on the evolution of squares of opposites and maps of quaternities, see Murdoch, *Album*, chs. 6 and 7; and Obrist, *La cosmologie médiévale*, passim.

ideas encountered in the mind, then the metaphysical and epistemological framework shifts considerably. Mental representations now come to take center stage. Indeed, there are moments when Descartes appears to be incapable of thinking and explaining without the use of images and objects to which these relate. For example, in one of his conversations with Frans Burman, there is a moment in which it appears that Descartes,

> despite the fact that he has accustomed his mind to imagining, was scarcely able to conceive of [the effect of vortex motion] without the balls. So others will find it much more difficult. For these things depend on mathematics and mechanics, and can be demonstrated better in a visual demonstration than they can in a verbal demonstration.[53]

Famously, for Descartes, these mental representations ought to coincide with the spatio-material particles that are held to be responsible for the secondary qualities provided by our senses. Colors are, for example, only perceived qualities that are provoked by the way in which the pressure of sunrays affects our organs of sight. The pressure of these rays can be represented by lines G, which in the case of figure 21 (which explains the generation of heat) are seen affecting the globular air particles inhabiting the pores of a collection of earth particles. Descartes' own question of "How we may arrive at knowledge of the shapes and motions of particles that cannot be perceived by the senses" raises indeed one of the central epistemological issues of his pictorial physics.[54] The answers this philosopher provided in the course of his life were partially of an inductive, probabilistic nature; they partly relied on a physiological translation mechanism (as in fig. 22); and partly on analogies with imprinting (fig. 23); and in part they yearned for the possibility of deductive proof. His statement, in a letter to Mersenne, that "I imagine or rather find [a particular type of particle] by demonstration,"

---

[53] *Descartes' Conversation with Burman*, transl. with introduction by John Cottingham (Oxford, 1976), 67.

[54] René Descartes, *Principia philosophiæ* (Amsterdam, 1644), part IV, marginal title of §203.

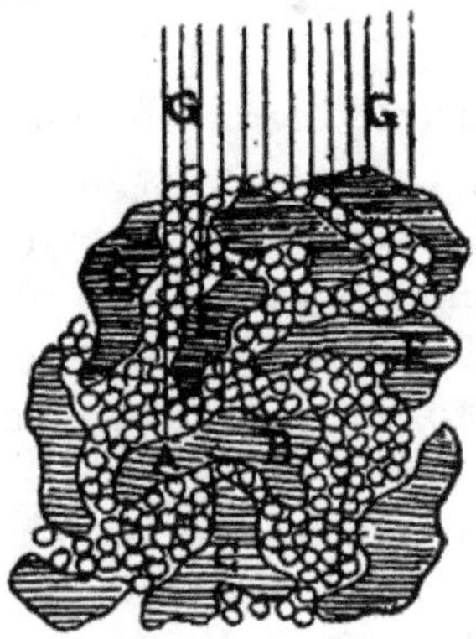

21

22

23

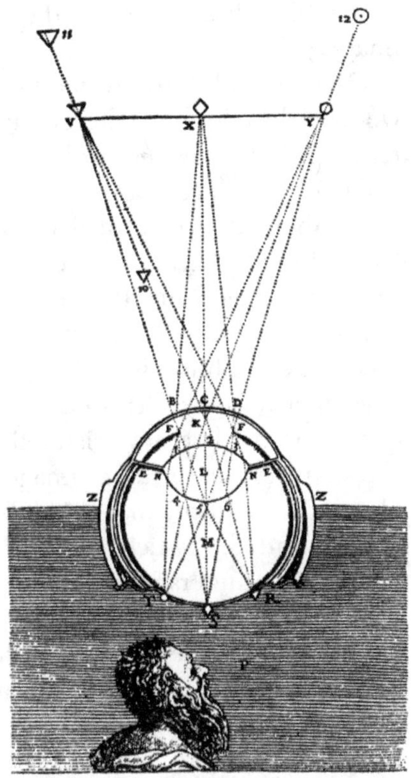

24

Figures 21-24: Cartesian images and their self-justification. Fig. 21: Sunrays *G* heating up a body (composed of the hatched particles) by affecting the air globules within it and bringing AB to shake (*Principia philosophiæ* [Amsterdam, 1644], 204). Fig. 22: Optical fibres being affected by a star, the nerve end of fibre no. 1 being affected most strongly, those of no. 2 less strongly, etc., thereby translating a pressure pattern into visual patterns (*La dioptrique*, in *Discours de la méthode* [Leiden, 1637], 68). Fig. 23: Comparing memorization to the imprinting of a pattern on a linen cloth; the more often you imprint the pattern, the clearer it will be recognizable on the cloth (*De l'homme* [Paris, 1664], 75). Fig. 24: Looking at 'looking' without seeing 'seeing' (*La dioptrique*, Discours 5, 36).

is a telling indication of the multiple origins of his corpuscular imagery.[55]

Descartes' philosophy allows for the perplexing situation in which we can thus look at the eye (fig. 24), understand its function in terms of a *camera obscura*, and yet at the same time know that this *looking* at the eye doesn't tell us anything about *seeing* as such, let alone about the truth of what we see. This truth can only be established by an inner eye whose relation to the outer eye remains to this day a question of controversy.[56]

It is obvious that Descartes' images can only be understood within their unique philosophical context. And the same will hold true for any other maker of 'epistemic images'. It must, for example, be evident that neither an Aristotelian nor a Cartesian framework will allow us to understand images fabricated in that Neoplatonist-Hermetic tradition that developed notably in the wake of Marsilio Ficino, and for which certain 'images' (*imagines*) possessed magic powers. To understand such images, one must first look to chapters 15 to 20 of Ficino's *De vita cœlitus comparanda* (finished by 1489).[57] There, we will encounter a cosmos in which celestial bodies are alive, just as their rays are "living and perceiving," carrying "marvelous gifts from the imaginations and minds of the celestials." There, we will also hear why 'images' engraved on stones under the right astrological circumstances will bundle, as it were, and reinforce

---

55) Descartes to Mersenne, 9 January 1638, in *Œuvres de Descartes*, eds. Charles Adam and Paul Tannery, 12 vols. (Paris, 1897-1957), 2:483.

56) See on the issue of Descartes' images Klaus Zittel, "Abbilden und Überzeugen bei Descartes," in *Cognition and the Book: Typologies of Formal Organisation of Knowledge in the Printed Book of the Early Modern Period*, eds. Karl A.E. Enenkel and Wolfgang Neuber (Leiden, 2005), 535-601; Christoph Lüthy, "Where Logical Necessity Becomes Visual Persuasion: Descartes's Clear and Distinct Illustrations," in *Transmitting Knowledge. Words, Images, and Instruments in Early Modern Europe*, eds. Sachiko Kusukawa and Ian Maclean (Oxford, 2006), 97-133.

57) Ficino, in turn, could rely on medieval traditions of magical practices; see Nicolas Weill-Parot, *Les "images astrologiques" au Moyen Âge et à la Renaissance: Spéculations intellectuelles et pratiques magiques (XIIe-XVe siècle)* (Paris, 2002); Richard A. Kieckhefer, *Magic in the Middle Ages* (Cambridge, 2000); and Michael Camille, "Visual Art in Two Manuscripts of the Ars Notoria," in *Conjuring Spirits. Texts and Traditions of Medieval Ritual Magic*, ed. Claire Fanger (Stroud, 1998), 110-39.

Figure 25: A magically efficacious 'image' of Saturn following Ficino (from Giordano Bruno, *De umbris idearum* [Paris, 1582]).

these gifts—a notion that Ficino took from the Hermetic *Picatrix*.[58] These engraved 'images'—which notably in the guise of amulets can provoke strong curative or deleterious reactions—may carry either geometrical shapes like a circle or a cross, or else allegorical images of planets (like that of Saturn as "an old man sitting on a rather high throne or on a dragon, his head covered with a dark linen cloth, raising his hands above his head, holding in his hand a sickle or some fish, and clothed in a dusky robe," cf. fig. 25).[59] Their efficacy is due to the fact that the "figures and numbers" observable in the skies "have the greatest affinity with the Ideas in Mind, the Queen of the World"—an affinity that allows for a reverberation like that observed between two lutes, of which one resonates as the other plays.[60]

Obviously, this concept of 'image' is incompatible with both Aristotle's or Descartes'. Ficino himself is quite aware of abandoning the prevalent theory concerning images, and he cites with feigned approval Thomas Aquinas' injunction against the belief in their efficacy.[61] This injunction was based on solid Aristotelian reasoning: "Natural matter is not in any way disposed towards [a new form]

58) Marsilio Ficino, *Three Books on Life*. A Critical Edition and Translation with Introduction and Notes by Carol V. Kaske and John R. Clark (Binghamton, NY, 1989), 323ff. (quote from Ficino, *De vita cœlitus comparanda*, ch. 16).

59) Ibid., 335 (quote from Ficino, *De vita*, ch. 18).

60) Ibid., 328-29 (quote from Ficino, *De vita*, ch. 17).

61) Ibid., 341-2 (quote from Ficino, *De vita*, ch. 18).

by figures. [...] So the bodies on which these figures are put have the same readiness to receive the celestial influence as any other bodies of the same species."[62] This argument is worth reflecting on in our present context, because it states that the notion of an efficacious *figura* makes no sense in a philosophy that attributes efficacy to *forma*, and in which *forma* has nothing to do whatsoever with *figura*, because it designates a thing's essential nature, not its shape! Thomas Aquinas' opposition of *forma* to *figura* is important because of that gradual and drawn-out transformation, so exquisitely documented by Norma Emerton, of the concept of *forma* into that of *figura* in the seventeenth century—a transformation that was to culminate precisely in Descartes' corpuscular images of which we have just spoken.[63]

By using one term only, namely *imago*, to denote things as different as the constellations of stars, the natural conformation of stones as well as the geometrical and emblematic shapes and figures that we might engrave on them, Ficino deliberately or otherwise rendered the connection between these various phenomena both close and obscure. With his meticulous list of terms, Giordano Bruno, who stood very much in the Ficinian tradition, attempted to bring clarity into the relations that "images, signs and ideas" held among themselves and vis-à-vis the world of objects. If we return to Bruno once more at the end of this section, it is because he was more aware than anyone at the time that one's iconographic taxonomy depended on one's metaphysics and epistemology. With methodological acumen, he opened his *On the Composition of Images, Signs and Ideas* (1591) with a characterization of his philosophical system on which the relational functions of his various types of images depended. His text begins with a Neoplatonic distinction between the three worlds. There is, first of all, the divine world (*mundus metaphysicus*) of ideas (*ideæ*). It is responsible for the origin of the *mundus physicus*, the created, natural world, which contains the traces (*vestigia*) of the divine ideas. The third world

---

62) Thomas Aquinas, *Summa contra Gentiles*, 105, in *Opera omnia*, Leonine edition (Rome, 1882-1971), 331a.12-17; cited in Ficino, *Three Books*, ed. Kaske and Clark, 445.

63) Norma Emerton, *The Scientific Reinterpretation of Form* (Ithaca, NY, 1984).

(*mundus rationalis*) comes to exist in our soul, which attempts, by conceptual and logical means and by abstracting from the *vestigia*, to fathom the *ideæ*, of which it can, however, only capture the shadows (*umbræ*). With respect to the things in the natural world, "the ideas are the causes of things before these things; the vestiges of the ideas are the things themselves or in the things; and the shadows of the ideas are from these things or after them."[64] Our minds are thus derivative, "like some living mirror, in which there is the image (*imago*) of the natural things and the shadow (*umbra*) of the divine ones."[65] But whereas a mirror might only receive the images and figures (*imagines atque figuræ*) of things, Bruno is convinced that the human mind is able not only to recognize the causally active substantial forms (*formæ*) and the species (*species*) of the things perceived, but is able to process these images in such a way as to render them useful both in theory (e.g., in the intellection of the divine unity) and in practice (e.g., in the *ars inveniendi*). Having received the *imagines* of the natural world, the mind is able to construct more meaningful images, signs and ideas, which, however, need to be carefully distinguished. It is at this point that Bruno introduces his classification into the twelve types that we have cited above (p. 423).

In circumstances where the visible world is itself considered an image, however faint or distorted, of a divine world, of which in turn the mind, based on the visual perception of physical reality, attempts to reconstruct the ideal idea and image, images proliferate, crisscross, and threaten to become second- or third-order mirror images of one another. This is why one must be careful to define each type in terms of its relation to the perceived world, on the one hand, and to the ideal world, on the other. Bruno therefore

64) Bruno, *De imaginum, signorum et idearum compositione*, 1-2: "Ideæ sunt causa rerum ante res, idearum vestigia sunt ipsæ res seu quæ in rebus, idearum umbræ sunt ab ipsis rebus seu post res ..." Because of its oftentimes anachronistic translations, we are not relying here on Charles Doria's translation in Bruno, *On the Composition of Images, Signs and Ideas* (New York, 1991).

65) Bruno, *De imaginum compositione*, 3: "veluti speculum quoddam vivens, in quo est imago rerum naturalium et umbra divinarum."

dedicates an entire little chapter to the "definitions of various [visual] denotations" (*variorum notaminum rationes*). Let us paraphrase his definitions in order to document the strong dependence of the vocabulary on the general philosophical parameters that constitutes the theme of our essay.[66]

> *Nota*: anything that by direct or indirect reasoning manages to demonstrate something.
> *Character*: organized lines or dots signifying something else, e.g., elements.
> *Signum*: a generic term denoting anything capable of signifying, either *qua* idea (*idea*), trace (*vestigium*), or shadow (*umbra*).
> *Sigillum*: a diminutive form of the *signum*, which represents an element or a contracted form of the *signum*, as when we represent 'man' by a hand or a head.
> *Indicium*: something the function of which is not to represent or to signify, but to show, just as the index finger doesn't express in itself that to which it points, but simply invites us to look at the item to which it points.
> *Figura*: whereas the previous types of *notamina* can refer to both the inside and the outside of things, the *figura* only refers to the outside. (*Pace* Ficino, Bruno thus returns to the Aristotelian usage of the term *figura*.) Furthermore, a *figura* must enclose a space, whereas the above-mentioned types of images needn't do that.
> *Similitudo*: in contradistinction to all of the above, each similitude must be of the same type as that which it represents; as a painting (*pictura*), a statue or the species that is captured by the sense of vision and stored in the *phantasia*.
> *Proportio*: designates relations between more than two things (unlike *similitudo*, which is limited to two). Mathematical relations like 2:4 = 4:8 belong to this type.
> *Imago*: possesses "more energy, emphasis, and universality" than the above, because it univocally links two things of the same genus.

This emphatic definition of *imago* would be fairly incomprehensible, were it not for the striking clue that Bruno's discussion of the difference between similitude and image contains: "Such as an artifact is said to be similar to some artificer, it is yet not said that it is to, or in, his image, nor in the proximate genus or in the same species."[67] *Ad eius imaginem*—to any of his readers, that must have evoked Genesis 1,26: "And God said: 'Let us make man in our

---

[66] Ibid., 5-6.

[67] Ibid., 6: "Sicut et artificium simile dicitur quodammodo artifici, non tamen ad eius imaginem vel in eius imagine dicitur, nisi vel in proximo genere vel in eadem sit specie."

image, after our likeness.'" (*et ait faciamus hominem ad imaginem et similitudinem nostram*). But if we humans are ourselves images of the Deity, there is no end to the mutual mirroring between the divine macrocosm and our microcosmic selves.

The Evidence from Clashes

In our five problem-raising sections, and by means of a number of examples, we have attempted to document the lack of precision and the fluidity of the historical vocabulary. We have seen, for example, that neither *figura* nor *imago* mean the same thing for Aquinas, Ficino and Descartes, and that none of these authors uses the large iconographical terminology that Bruno proposes. We have also been at pains to demonstrate that not just the names themselves, but also the very meaning and status of scientific images depend on the philosophical framework within which they are employed.

But if the meaning of images was really as historically contingent as has just been suggested, should one then not be able to find, in the historical sources, disputes about the standing of images? Yes, one should; and in fact, one does. There are numerous such disputes, although they seldom take central stage. In fact, they always accompany clashes over incompatible scientific and philosophical theories. This is why the history of theoretical changes in any given discipline can profitably be approached through a study of its visual conventions and the disputes that erupted around them.[68]

Let us mention just a small number of such disputes. Marsilio Ficino, for example, had to defend himself against those who accused him of an idolatrous usage of images: "Marsilio is a priest, isn't he? ... What does a Christian have to do with magic or images?"[69] More detailed and fascinating is that famous dispute between Robert

[68] One of the co-authors of this article, Smets, is currently working on a doctoral dissertation devoted to the evolution of images expressing chemical matter theories from the Renaissance to the eighteenth century. In this project, the clash between rivaling modes of visualization is a recurrent theme.

[69] This quote from Marsilio Ficino's *Apologia quædam* is taken from Kaske and Clark's translation in Ficino, *Three Books*, 395.

Fludd and Johannes Kepler. Fludd had attempted to demonstrate the truth of his macro-microcosmic philosophy by means of very elaborate and expensive copper plates (notably in his multi-volume *Utriusque cosmi … metaphysica, physica atque technica historia* of 1617-21). As Robert Westman has rightly stressed, these engravings must not be viewed "as *illustrations* but rather as *ways of knowing, demonstrating, and remembering*."[70] In fact, Fludd believed that his expensive prints, though emblematic in nature, possessed demonstrative force. His famous foldout depicting the macrocosm, for example, was meant to be an "Emblematic mirror *demonstrating*" how "the more liberal arts" (*artes liberaliores*), which significantly enough also included the art of painting, led to knowledge of God, demons, and the creation.[71] But in Kepler's eyes, Fludd merely indulged in "pictures forged from air," and he pitted his own *diagrammata*, which he took to provide geometrical proof, against Fludd's airy *picturæ*, *figuræ*, and *hieroglyphica*.[72] His disagreement with Fludd's images was an important, though integral, part of a more comprehensive disagreement about the nature of numbers, the meaning of symmetries, and the structure of the universe.[73]

Every time that images express world-views, as is clearly the case with Kepler and Fludd, the charge is inevitably brought up that a given image is not understood because the theory informing it is not understood, and *vice versa*. In this vein, Thomas Browne, when chiding certain implausible depictions of animals, was in turn accused of not understanding their theoretical underpinnings: "The doctor

---

70) Robert S. Westman: "Nature, Art, and Psyche: Jung, Pauli, and the Kepler-Fludd Polemic," in *Occult and Scientific Mentalities in the Renaissance*, ed. Brian Vickers (Cambridge, 1984), 177-229, at 181, original emphases.

71) Ibid.

72) See, for example, Johannes Kepler's statement: "Tuis picturis mea comparavi diagrammata; fassus librum meum non æque atque tuum ornatum esse, nec futurum ad gustum lectoris cuiuslibet: excusavi hunc defectum a professione, cum ego mathematicam agam." ("Apologia," *Gesammelte Werke*, vol. 6, ed. Max Caspar [Munich, 1940], 396).

73) On the Fludd-Kepler controversy and its implications for iconography, see besides Westman, "Nature, Art, and Psyche," also Frances A. Yates, *Giordano Bruno and the Hermetic Tradition* (Chicago, 1964), ch. 22, and Judith V. Field, "Kepler's Rejection of Numerology," in Vickers, *Occult and Scientific Mentalities*, 273-96.

quarrels with some pictures ... but for this he hath not great reason; for [the author of the animal images] tels him that this and divers other pictures are rather Hieroglyphical and Emblematic then truly historicall."[74] Or take, as a further example, Andreas Libavius' displeasure, in 1615, with various *schemata*, *picturæ*, *tabulæ* and *icones* found in Heinrich Khunrath's lavishly illustrated *Amphitheatrum sapientiæ eternæ*. Libavius took particular exception to one specific but recurrent iconographical element, namely the Latin word "Omnia" descending from the heavens. Like all "Panspermic doctors," so Libavius charged, Khunrath with this image willfully suggested a conflation of Anaxagoras' "all is in all" (ἐν παντί πάντα) with the identical "Omnia in omnibus" of 1 Corinthians 12,6. In Libavius' eyes, no biblical source could legitimately be used for the pagan notion that the seeds of all things are in everything—a notion that for him carried obvious magical and talismanic implications.[75]

A final example: an entire book could be written about the very diverse reactions to Descartes' innovative corpuscular images. We encounter one type of critic that only expressed bewilderment at the logic informing one particular image. This was the case of Henry More, who politely enquired why the *particulæ striatæ* responsible for magnetism managed to maintain their nice screw structure and

74) Alexander Ross, *Arcana Microcosmi* (London, 1652), 156. The quote is from Kevin Killeen, "'The Doctor Quarrels with some Pictures': Exegesis and Animals in Thomas Browne's *Pseudodoxia Epidemica*," *Early Science and Medicine* 12 (2007), 1-27, at 6.

75) Andreas Libavius, *De philosophia vivente seu vitali Paracelsi iuxta Petrum Severinum Danum ex repetitione I. Hartmanni chymiatri Marburgensis* (Frankfurt, 1615), *Examen*, 101: "Henricus Kunrath amphitheatro suo præfixit schema, cuius apex triangularis triplicatum nomen Dei tetragrammaton habet Magica forma scriptum intra nubem ex qua manus porrecta coronam capiti Kunrathi imponit, cum inscriptione: OMNIA. In pictura graduum septem portæ amphitheatri profani, item iubentur procul abesse, subscribiturque: *omnia in omnibus*. In tabula sequente quæ est globus cum decalogo nominibus Dei, & Icone Christi repetitur triangulus Tetragrammati cum subscriptione OMNIA." I owe this quote to Peter J. Forshaw; cf., by Forshaw, "Paradoxes, Absurdities, and Madness: Conflict over Alchemy, Magic and Medicine in the Works of Andreas Libavius and Heinrich Khunrath, *Early Science and Medicine* 13 (2008), 53-81.

did not break up, as did other particles described by Descartes.[76] But there were also those, like Henricus Regius in his *Fundamenta physices* of 1646, who reproduced Descartes' corpuscular physics with all its imagery, while rejecting the very metaphysics that, in Descartes' eyes, provided these images with their legitimacy. There were yet others who did not understand that Descartes' particles should be viewed with the inner eye only, and sought for them with the microscope. Nathaniel Highmore even announced in 1651 that magnetic effluvia "by the help of Glasses" had been observed—those very *particulæ striatæ* that More had found so difficult to imagine![77] And finally, there were those who laughed about the corpuscular imagery, such as the Newtonian John Keill, who ridiculed the allegedly 'mechanical' philosophers' picturesque explanations of natural phenomena by means of "Figures, Ways, Pores and Interstices of Corpuscles, which they never saw."[78]

The question of the legitimacy of certain *types of images* is explicitly raised in all of these debates. What is it that can possibly endow a given epistemic image with any sort of power? And what power would that be? Is it a magical power, a power of proof or one of persuasion? And does the picture derive its alleged power from the text (of which it offers ocular proof, demonstration, or elucidation), or does the text instead derive its force from the visual proof in the way in which in Euclidean geometry, the proof is only completed when the figure is fully drawn? *Quod erat demonstrandum*—but "to be proven" verbally or pictorially, or both?

Only in exceptional, but all the more fascinating cases does nature draw her own images. In the world of diagrammatic representations, one such case is represented by the near-parabolic line generated by an ink-dripping ball that Guidobaldo del Monte let roll over on an inclined sheet. The other case, which belongs to the opposite extreme, namely to the realm of figurative realism, is represented

---

76) Henry More to Descartes, 5 March 1649, *Œuvres de Descartes*, eds. Adam and Tannery, V: 346-47.

77) Nathaniel Highmore, *The History of Generation, Examining the Several Opinions of Divers Authors*, (London, 1651), 117.

78) John Keill, *An Introduction to Natural Philosophy: or, Philosophical Lectures Read in the University of Oxford, Anno Dom. 1700* (London, 1726), iii.

by Johannes Kepler's *camera obscura* drawings, which were simply the retraced outlines of objects projected by a lens onto a screen. Importantly, his experiments with the *camera obscura* forced Kepler to change his iconographical vocabulary. He concluded that the image appearing on the screen and *a fortiori* on the camera-like human retina had to be called *pictura*, whereas what appeared in our perception had to be called *imago*. Although his new nomenclature did not persuade others, this is probably the first case in which 'machine-generated' images led to a marked shift in the conception of what it is that an 'image' does.[79]

Conclusion

Generations of scientists have quarreled over the exact definition of such key concepts as 'force', 'vacuum', 'organism', 'attraction', or 'entropy'. Without an exact definition, no exact science seemed possible. The contrast with the domain of epistemic imagery could not be any more extreme: there, utter chaos reigns. The iconographical and semiotic categories available to us are underdeveloped and ill defined. The term 'diagram', for example, is applied to everything ranging from a construction plan of an airplane to a statistical distribution curve. The word 'model' fares no better, being used for a miniature train as much as for a hypothetical description of quantum mechanical interactions.

With respect to historical modes of visualization, the terminological situation is probably even worse. Admittedly, there are fewer types of epistemic images to be reckoned with. After all, the complicating factor of machine-generated images is absent from the medieval and early modern landscape, if we disregard the exceptions just mentioned. By recompense, the epistemological and metaphysical premises within which a given image functioned, the practices that generated it, and the meanings and allusions that it could carry, are difficult to fathom and, to make things worse, appear to be of little interest to most scholars. The bizarre asymmetry with which

---

79) See on this the collection of essays in *Kepler, Optical Imagery, and the Camera Obscura*, ed. Alan E. Shapiro (= *Early Science and Medicine* 13.3/2008).

editors tend to treat historical texts and images is indicative of this situation. While text variants are obsessively registered in multiple footnote systems, the images accompanying these very same texts are either entirely suppressed or are 'updated' to be 'legible to the modern reader', as it were. Obviously, if an editor applied the same procedure to the text itself, he or she would at once be sacked.[80]

One reason for this recklesness is this behavior of the historical actors themselves. As we have seen, most 'producers' and 'users' left the status of their *figuræ* undefined and vague. This situation must give pause for reflection. For it suggests that, like other paraphernalia of scientific practice, images were routinely regarded as instruments and as subsidiary to scientific theory itself, and as such as unworthy of scientific attention. Yet, like all instruments, images create scientific entities. As a vacuum pump creates the void of which it is at the same time the principal scientific investigator, so a taxonomic tree is the mental instrument that guides research, for example, into missing branches and links. And, incidentally, just as the vacuum pump could become one of the visual emblems of the new science, so the taxonomic tree could become the icon of an evolutionary worldview.[81]

It is important to realize that it is always in the years that a new type of epistemic image is being introduced that the awareness of its status, function and role within a given theory are explicitly discussed. In case the new type is accepted and becomes embedded in a shared scientific paradigm, it will inevitably become—not unlike language—an integral part of the scientific practice, and the awareness of its specific philosophical premises will disappear. This phenomenon can be observed across the centuries, and it is as true of the quaternary patterns of interwoven primary qualities, elements, humors, seasons, etc., which characterized natural philosophy and medicine until the seventeenth century, as it is true of twentieth-century Feynman

80) Cf. Murdoch, *Album*, 113: "variations in diagrams are often every bit as important as variations in the text. They, too, are part of the transmission [of a given work] and as such must be considered fully in the editing of such works, a desideratum that has, until recently, been too often ignored."

81) On the vacuum pump becoming the emblem of the new science, see Hackmann, "Natural Philosophy Textbook Illustrations," 178.

diagrams. In both cases, visual structures can be seen to have constituted the unquestioned starting point of scientific theory formation and practice.[82]

This often entirely unquestioned use of visualizing tools in the sciences puts the historian of epistemic imagery in a complex situation. In those cases where new types of imagery were created to express new theories, and in which both theory and images were explicitly discussed or contested, the historian is in a comfortable situation. But in that large majority of cases where types of images figure as the unquestioned visual backbone of 'normal science', the historian's wish to decode these images will in all likelihood exceed the level of awareness and the intentions of their makers.

Fortunately enough, historians of epistemic imagery find themselves, in this respect, in the same position as historians of science and intellectual historians, who are accustomed to dealing with the phenomenon of the internalization of concepts, ideas, and practices. An understanding of analogous processes is thus present. What is still lacking, however, is a philosophical history of epistemic images, which attempts to draw up, for each style of thought, an analytic taxonomy that reflects the actors'—the authors' *and* the artists'—understanding of the epistemological value and the functionality of the images that they produced, and within which they and their disciples ended up thinking.

82) See Gernot and Hartmut Böhme, *Feuer, Wasser, Erde, Luft. Eine Kulturgeschichte der Elemente* (Munich, 1996); David Kaiser, *Drawing Theories Apart: The Dispersion of Feynman Diagrams in Postwar Physics* (Chicago, 2005).

Index

The following index has been compiled from separate indices submitted by each author. Individuals living before 1400 are alphabetized by their given names; those living after 1400 by their last names. Arabic /Islamic authors mentioned primarily in articles about Latin science appear by their Latin names (e.g. Averroes, Avicenna); those primarily discussed in the section on Arabic/Islamic science appear under their Arabic names. Ibn is counted in alphabetization, al- is not.

Manuscripts Cited

www.ingramcontent.com/pod-product-compliance
Lightning Source LLC
La Vergne TN
LVHW010623300926
849162LV00001B/43
* 9 7 8 9 0 0 4 1 7 8 7 8 6 *